Progress in Computer Science and Applied Logic

Progress in Computer Science and Applied Logic
Volume 22

Igor Shparlinski

Cryptographic Applications of Analytic Number Theory

Complexity Lower Bounds and Pseudorandomness

Springer Basel AG

Editor:

Igor Shparlinski
Department of Computing
Macquarie University
NSW 2109
Australia
e-mail: igor@comp.mq.edu.au

2000 Mathematics Subject Classification 03D15, 11T71, 11Yxx, 68Q15, 68Q17, 68Q25, 94A60

A CIP catalogue record for this book is available from the Library of Congress, Washington D.C., USA

Bibliographic information published by Die Deutsche Bibliothek
Die Deutsche Bibliothek lists this publication in the Deutsche Nationalbibliografie; detailed bibliographic data is available in the Internet at <http://dnb.ddb.de>.

ISBN 978-3-0348-9415-9 ISBN 978-3-0348-8037-4 (eBook)
DOI 10.1007/978-3-0348-8037-4

Originally published by Birkhäuser Verlag, Basel, Switzerland in 2003

Softcover reprint of the hardcover 1st edition 2003

Printed on acid-free paper produced of chlorine-free pulp. TCF ∞

9 8 7 6 5 4 3 2 1 www.birkhäuser-science.com

Contents

Preface vii

Acknowledgments xi

Introduction 1

I Preliminaries 15

1 Basic Notation and Definitions 17

2 Polynomials and Recurrence Sequences 27

3 Exponential Sums 37

4 Distribution and Discrepancy 61

5 Arithmetic Functions 67

6 Lattices and the Hidden Number Problem 83

7 Complexity Theory 103

II Approximation and Complexity of the Discrete Logarithm 107

8 Approximation of the Discrete Logarithm Modulo p 109

9 Approximation of the Discrete Logarithm Modulo $p-1$ 123

10 Approximation of the Discrete Logarithm by Boolean Functions 129

11 Approximation of the Discrete Logarithm by Real Polynomials 143

III Approximation and Complexity of the Diffie–Hellman Secret Key 157

12 Polynomial Approximation and Arithmetic Complexity of the Diffie–Hellman Secret Key 159

13 Boolean Complexity of the Diffie–Hellman Secret Key 179

14 Bit Security of the Diffie–Hellman Secret Key 189

IV Other Cryptographic Constructions **195**

15 Security Against the Cycling Attack on the RSA and Timed-release Crypto 197

16 The Insecurity of the Digital Signature Algorithm with Partially Known Nonces 201

17 Distribution of the ElGamal Signature 207

18 Bit Security of the RSA Encryption and the Shamir Message Passing Scheme 211

19 Bit Security of the XTR and LUC Secret Keys 217

20 Bit Security of NTRU 223

21 Distribution of the RSA and Exponential Pairs 231

22 Exponentiation and Inversion with Precomputation 239

V Pseudorandom Number Generators **247**

23 RSA and Blum–Blum–Shub Generators 249

24 Naor–Reingold Function 271

25 $1/M$ Generator 279

26 Inversive, Polynomial and Quadratic Exponential Generators 283

27 Subset Sum Generators 295

VI Other Applications **301**

28 Square-Freeness Testing and Other Number-Theoretic Problems 303

29 Trade-off Between the Boolean and Arithmetic Depths of Modulo p Functions 309

30 Polynomial Approximation, Permanents and Noisy Exponentiation in Finite Fields 325

31 Special Polynomials and Boolean Functions 333

VII Concluding Remarks and Open Questions **341**

Bibliography **367**

Index **409**

Preface

The book introduces new techniques that imply rigorous lower bounds on the complexity of some number-theoretic and cryptographic problems. It also establishes certain attractive pseudorandom properties of various cryptographic primitives. These methods and techniques are based on bounds of character sums and numbers of solutions of some polynomial equations over finite fields and residue rings. Other number theoretic techniques such as sieve methods and lattice reduction algorithms are used as well. The book also contains a number of open problems and proposals for further research.

The emphasis is on obtaining unconditional rigorously proved statements. The bright side of this approach is that the results do not depend on any assumptions or conjectures. On the downside, the results are much weaker than those which are widely believed to be true.

We obtain several lower bounds, exponential in terms of $\log p$, on the degrees and orders of

- polynomials;
- algebraic functions;
- Boolean functions;
- linear recurrence sequences;

coinciding with values of the discrete logarithm modulo a prime p at sufficiently many points (the number of points can be as small as $p^{1/2+\varepsilon}$). These functions are considered over the residue ring modulo p and over the residue ring modulo an arbitrary divisor d of $p-1$. The case of $d = 2$ is of special interest since it corresponds to the representation of the rightmost bit of the discrete logarithm and defines whether the argument is a quadratic residue. We also obtain non-trivial upper bounds on the degree, sensitivity and Fourier coefficients of Boolean functions on bits of x deciding whether x is a quadratic residue.

These results are used to obtain lower bounds on the parallel arithmetic and Boolean complexity of computing the discrete logarithm. For example, we prove that any unbounded fan-in Boolean circuit of sublogarithmic depth computing the discrete logarithm modulo p must be of superpolynomial size.

The sequential complexity of the discrete logarithm in a certain very powerful computational model is considered as well.

Similar results are also obtained for breaking the Diffie–Hellman key exchange protocol.

Several other applications of our methods are indicated as well. These applications can be put in a much broader context of the theory of complexity of computation.

Although in many cases our results about the structure and algebraic properties of the discrete logarithm and other related functions are quite strong and close to the best possible, the current methods of complexity theory are not able to use their full power. As a result our complexity bounds are much lower than what is really expected. On the other hand, they are about of the same shape and strength as all other similar known bounds for various explicitly given functions. That is, they are essentially as strong as is known for any problem in the complexity class **NP**.

We also show how similar ideas and techniques can be used to establish some attractive properties, such as uniformity of distribution, high linear complexity, bit security and several others, of various cryptographic constructions such as the RSA, XTR, LUC and NTRU schemes, Digital Signature Algorithm and ElGamal signature schemes, Shamir message passing scheme and so on. Although these properties do not have any direct security implications, violation of any of them would be disastrous for the corresponding cryptographic construction.

The area is very dynamic with several important results appearing literally every week. We have tried to present as complete picture as possible, in particular referencing to many preprints and taking the risk that their final versions may have different titles, contain slightly different results and in some cases even have a different list of authors.

We hope both cryptographers and number theorists will find this book useful. The former can learn about new number theoretic techniques which have proved to be invaluable cryptographic tools, the latter can learn about new challenging areas of applications of their skills.

Disclaimer: This book contains no confidential information. Views expressed in this books are not of the author or his employer but, hopefully, rigorously proved statements. If you got this book by mistake, please still try to take a quick glance before destroying it and feel free to pass it to anyone else. Dissemination of the information contained in this book is deeply appreciated.

Acknowledgments

The author wishes to thank his many (too many to name!) colleagues, co-authors and friends, both mathematicians and computer scientists, from whom he has benefited enormously, scientifically and socially. They contributed to this book in the form of joint work as well as fruitful and stimulating discussions, in our offices and in pubs. The author hopes that the chalk and beer have not been completely wasted.

The author would also like to express his deepest gratitude to Tanja Lange, Mats Näslund and Arne Winterhof for their careful reading of the manuscript, making many valuable suggestions and helping to sweep away numerous misprints and errors. There is no need to say that all the remaining weeds are solely the author's responsibility.

The last but not the least, it has been a real pleasure to work with the highly efficient and professional Birkhäuser publishing team, and especially with Thomas Hempfling whose great help and patience have made this book possible.

Acknowledgment

Introduction

It is well known that number theory provides a fertile and powerful background for cryptography, especially for public key cryptography. Through the whole history of cryptography, it has been used for both constructing new cryptosystems and design and analysis of attacks, which often require quite sophisticated number theoretic techniques. The main aim of the book is to extend the area and direction of applications of number theory, and especially analytic number theory, to cryptography. Our goal is to establish some useful (and in some cases harmful) properties of various common cryptosystems, signature schemes, pseudorandom number generators and other related constructions. In many cases this has led to new and unexpected points of view on some number theoretic problems. Thus we believe that our approach can enrich and crossfertilise both areas.

In the literature one can find two different approaches to pseudorandom number generators:

- One of them is more classical and related to Monte Carlo methods. The emphasis is made on the period length, distribution properties, linear complexity, and the simplicity of generation.

- The other approach is more modern and related to cryptographic and complexity theory applications. Such properties as unpredictability of the next term,

  indistinguishability from truly random numbers, and the difficulty of "reverting" (that is, backward generation) after observation of polynomially many elements of this generator, become of primal interest.

For example, the monograph [426] is a typical example of the first approach, while [214, 353] are typical representatives of the second approach. We do not here favour either of the above points of view, in fact our results are somewhat in between. We establish new periodicity, linear complexity and uniformity of distribution properties of generators which are normally considered within the second approach. Although the results we prove do not guarantee the security of the corresponding generator, a failure of any of them would be catastrophic for its security applications.

In the first part of this book we consider various representations and approximations of the discrete logarithm via some other functions over finite fields such as polynomials and their combinations with exponential functions (linear recurrence sequences, essentially) and algebraic functions (that is, via functions $f(X)$ satisfying a polynomial equation $F(X, f(X)) = 0$ over a finite field).

The aforementioned functions form a basic set of "easily computable" functions, at least when they are of small degree or order. For polynomials this is obvious. For algebraic functions it is justified by recent progress in solving polynomial equations over finite fields and finding points on algebraic curves [206, 508]. For linear recurrence sequences one can use a kind of repeated squaring as for computing a single exponential function. Also, studying representations via linear recurrence sequences is essentially equivalent to studying linear complexity, which is a widely accepted characteristic of cryptographic importance [138, 389, 425, 439, 483, 484].

So, the principal motivation of this approach is to show that no such simple representation of the discrete logarithm holds. In some cases we also show that the polynomials involved contain sufficiently many monomials. Moreover, here we deal with partial representations, which hold only for some subsets of the set of all possible values of the argument of the discrete logarithm.

Such results lead to lower bounds on the *parallel and sequential complexity* of computing the discrete logarithm in several different computational models.

We should warn the reader that although in many cases our results about the structure and algebraic properties of the discrete logarithm are quite strong and close to the best possible, the current methods of complexity theory are not able to use their full power. As a result our complexity bounds are much lower than what is really expected. On the other hand, they are about of the same shape and strength as all other similar known bounds for various explicitly given functions. That is, they are essentially as strong as is known for any problem in the complexity class **NP**. Moreover, although similar lower bounds have been established for much simpler functions, it is not clear how to conclude that an "obviously more complicated function", such as the discrete logarithm, has at least the same or higher complexity.

The upshot is: the discrete logarithm is so far the only known function for which they all hold *simultaneously*. This probably can be considered as a good indication that the discrete logarithm is indeed a computationally hard function. We remark that the recent paper [8] provides yet another evidence of such hardness.

We also hope that further progress in complexity theory will allow us to extract essentially better and more realistic complexity bounds from results presented on various algebraic characteristics of the discrete logarithm.

Here we consider the case of prime fields, using the language of congruences modulo a prime p rather than finite fields.

Let us fix a *primitive root* g modulo a prime number $p \geq 3$. For an integer x such that $\gcd(x,p) = 1$ we denote by $\operatorname{ind} x$ its *discrete logarithm*, also known as the *index* of x, that is, the smallest non-negative integer u with $g^u \equiv x \pmod p$.

In some setting it makes sense to define $\operatorname{ind} 0 = p - 1$, but in this book we follow the usual convention and leave $\operatorname{ind} 0$ undefined.

Thus the discrete logarithm defines a bijective mapping from the group of units of the residue ring modulo p, essentially the set $\{1, \ldots, p-1\}$, onto the set $\{0, 1, \ldots, p-2\}$. Hence one can ask about the polynomial representation of this mapping, that is, a polynomial $f(X) \in \mathbb{Z}[X]$ of degree at most $p-1$ such that

$$\operatorname{ind} x \equiv f(x) \pmod p, \qquad x = 1, \ldots, p-1.$$

Indeed, it has been shown in [399] that the polynomial

$$f(x) \equiv -1 + \sum_{k=1}^{p-2} (g^{-k} - 1)^{-1} x^k \pmod p$$

is the unique interpolation polynomial of the discrete logarithm modulo p. We note that this polynomial is of the largest possible degree (any function over $\mathbb{F}_p$ can be approximated at $p-1$ points by a polynomial of degree at most $p-2$) and is dense (that is, it contains $p-1$ monomials).

Here we show that these two properties are preserved even for partial representations and approximations of the discrete logarithm.

More precisely, for many practical purposes it would be enough to have a simple polynomial representation of the discrete logarithm for almost all $x = 1, \ldots, p-1$ rather than for all of them. We show that even such a polynomial must be of high degree and contain many non-zero coefficients. That result is quite simple and completely elementary. Then using more involved arguments we consider

- approximation on small intervals $[N+1, N+H]$;
- approximation on very sparse sets;
- approximation on random sets.

The approximating function can be

- a polynomial;
- a Boolean function;
- an algebraic function;
- a linear recurrence sequence.

Our argument x "lives" in the residue ring modulo p; this is why we consider congruences modulo p. On the other hand, the function $\operatorname{ind} x$ resembles the logarithmic function in the residue ring modulo $p-1$. Thus studying polynomial and other approximations modulo $p-1$ is another natural question which we also address in this book. Bitwise approximations of the discrete logarithm by Boolean functions on bits of the argument are dealt with as well.

The aforementioned approximations are studied respectively

- over the residue ring modulo p (to which the argument of $\operatorname{ind} x$ belongs);
- over the residue ring modulo $p-1$ (where the behaviour of $\operatorname{ind} x$ resembles the behaviour of $\log x$);
- over the r-dimensional Boolean cube where r is the bit length of p (if we consider the argument x and the value $\operatorname{ind} x$ as sequences of bits);
- over the field of real numbers (if we consider the argument x and the value $\operatorname{ind} x$ as real numbers).

We use the results obtained to derive various complexity lower bounds.

In particular, our results provide non-trivial lower bounds for the linear complexity [138, 389, 425, 439, 483, 484] of rather short segments of values of the discrete logarithm modulo any divisor $d > 1$ of $p-1$. These results complement those of [157, 158] where an explicit formula for the linear complexity of the quadratic character modulo p has been given. The question of non-linear complexity is dealt with as well.

In fact for a small divisor d of $p-1$, the residue of $\operatorname{ind} x$ modulo d can be found in $d^{1/2} \log^{O(1)} p$ Boolean operations. Thus the sequential Boolean complexity of this question is known to be polynomial.

The case $d = 2$ corresponds to studying the rightmost bit of $\operatorname{ind} x$. This bit is of special interest of course since its parity determines whether x is a quadratic residue; a good outline of known algorithms is given in [390]. Using estimates of character sums we obtain a lower bound of order $p^{1/4}$ on the number of monomials of a Boolean function on bits of x computing the rightmost bit of $\operatorname{ind} x$. We apply it to obtain the lower bound $\Omega(\log \log p)$ on the depth of Boolean circuits deciding whether x is a quadratic residue or not. Our result supplements some of the results of [200, 204] on arithmetic circuits (over $\mathbb{F}_q$) deciding whether $x \in \mathbb{F}_q$ is a quadratic residue or not. Those papers are based on the observation that this question is equivalent to computation of values of the polynomial $x^{(q-1)/2}$. Here, in the same fashion, we use our bound on the degree of Boolean functions giving the values of the rightmost bit of $\operatorname{ind} x$.

We also estimate from below some other characteristics of such functions which in turn gives a lower bound on their CREW PRAM complexity; see Chapter 1 for a definition.

Then, a non-trivial upper bound is given for the Fourier coefficients of Boolean functions on bits of x deciding whether x is a quadratic residue. This estimate is applied to derive a lower bound on the depth and size of unbounded fan-in Boolean circuits computing such functions. In particular, any such circuit of sublogarithmic depth must be of superpolynomial size.

We note that, for a given depth, unbounded fan-in Boolean circuits are essentially more powerful than the bounded fan-in circuits.

Informally, it is no great surprise that any "reasonable" Boolean function of r variables requires a bounded fan-in Boolean circuit of depth $\Omega(\log r)$. Thus certainly the result of Theorem 10.2 is not trivial; moreover examples of explicitly given functions for which a superlinear lower bound has been proved are still unknown; see Section 2.3 of [68]. However, it is not surprising either. On the other hand, for unbounded fan-in circuits such a "trivially expected" lower bound does not exist. Moreover, there are examples of non-trivial functions which can be computed by unbounded fan-in Boolean circuits of constant depth and polynomial size. Such are the threshold functions $T_k(U_1, \dots, U_r)$ with $k = \log^{O(1)} r$, which take the value of 1 if and only if the input has at least k ones [256]. Also, unbounded fan-in Boolean circuits of constant depth and polynomial size are powerful enough to add integers; see Section 3.3 of [265]. Here, in particular, we show that this is not the case for the rightmost bit of $\operatorname{ind} x$.

Our bound $d \log\log S \geq (1 + o(1)) \log\log p$ on the depth d and size S of unbounded fan-in Boolean circuits computing the rightmost bit of the discrete logarithm modulo a prime p is quite tight. This is because using the minimal *disjunctive normal form* representation one can compute any Boolean function of r variables by a circuit of depth $d = 2$ and size $S \leq 2^r$, thus $d \log\log S \leq 2 \log r$ for this circuit.

We also obtain an exponential lower bound on the decision tree complexity of the rightmost bit of $\operatorname{ind} x$.

For the integer factorisation problem similar results have been obtained in [38, 39, 40, 41, 460]. To be more precise, the above papers study the complexity of testing square-free numbers which is obviously no harder than the integer factorisation problem. For primality testing similar results are given in [9]. Moreover, in [9] this is shown even for more general and powerful circuits.

It is useful to remark that any lower bound on the complexity of deciding whether x is quadratic residue modulo p immediately implies the same lower bound on the parallel complexity of computing the discrete logarithm as well as on the complexity of the irreducibility testing of polynomials over $\mathbb{F}_p$.

We remark that several results on the complexity of bits of the discrete logarithm have been already obtained but all of them are based on some unproven assumptions, in particular on the assumption that finding the discrete logarithm itself is hard. A variety of results of this type can be found in [145, 149, 159, 210, 216, 220, 255, 455, 457, 496].

We also consider various representations of the discrete logarithm over real numbers. For such representations, besides estimates on the degrees and the number of monomials of the corresponding polynomials we also obtain lower bounds on their additive complexity [88, 237, 293, 473, 474, 479, 480] and $M_f(r)$-invariants [366, 380].

The results of this book we have mentioned provide some insight into the parallel complexity. Our results also imply a non-trivial lower bound on the sequential arithmetic complexity of the discrete logarithm in such very powerful computational models as the *computational tree* model and the *random access machine* or *RAM* model over real numbers whose operation set consists of the four standard arithmetic operations and the truncation (or "floor") function; see [365, 366]. We remark that these computational models are surprisingly powerful and many natural problems on r-bit integers can be solved in time $O(\log^{1/2} r)$ [81, 365, 366, 380].

More precisely, to compute the discrete logarithm modulo p over the real numbers we use our method to obtain a lower bound of the form $\Omega(\log^{1/2} \log p)$ for its complexity on the two above-mentioned (incomparable) models.

Then, we show that the same considerations are applicable to some questions related to the *Diffie–Hellman key exchange protocol* discovered in [153] and based on the discrete logarithm. This question is studied over arbitrary finite fields.

First of all let us recall the very nice basic idea of the Diffie–Hellman key exchange protocol.

Assume that two users $\mathcal{A}$ and $\mathcal{B}$ want to exchange some secret information by using an "open" communication channel. To do so they select, by using that channel, a field $\mathbb{F}_q$ and a primitive root g. Then $\mathcal{A}$ selects a secret number x, $0 \le x \le q-2$, and computes $u = g^x$; $\mathcal{B}$ selects a secret number y, $0 \le y \le q-2$, and computes $v = g^y$. Next they exchange these numbers: $\mathcal{A}$ sends u to $\mathcal{B}$, $\mathcal{B}$ sends v to $\mathcal{A}$. Finally, $\mathcal{A}$ knows x and v so he can compute $v^x = g^{xy}$, $\mathcal{B}$ knows y and u so he can compute $u^y = g^{xy}$, thus they get a common *Diffie–Hellman secret key*

$$K(x, y) = g^{xy}$$

which can be used for any other *private key* cryptosystem. So the following information has been exchanged openly: the description of the field $\mathbb{F}_q$, the primitive root g, the values $u = g^x$ and $v = g^y$. To break the cryptosystem one has to find g^{xy} from this information. One of the most obvious ways (and indeed the only known practical way) to attack this cryptosystem is to solve the discrete logarithm problem first, say to find $\operatorname{ind} u = x$ and then to compute v^x.

In [59, 375, 376] it is demonstrated that, under certain conditions, breaking the Diffie–Hellman key exchange protocol is indeed polynomial time equivalent (in the Boolean model of computation) to computing the discrete logarithm; see also [112]. However, there are no unconditional results known in this direction. Relations between their parallel complexities are not known either. Thus obtaining lower bounds on the parallel complexity of breaking the Diffie–Hellman protocol by a general algorithm is of independent interest.

We apply our method to obtain a linear lower bound for the depth of randomized arithmetic circuits over $\mathbb{F}_q$ breaking this cryptosystem. We show that the depth of such circuits is of order $\log q$ at least. Thus it cannot be done in parallel logarithmic time $(\log \log q)^{O(1)}$. This holds for probabilistic circuits giving the correct answers for very sparse sets of values of the argument. Moreover, we show that even probabilistic verification of whether the given $u, v \in \mathbb{F}_q$ satisfy $u = g^x$, $v = g^{x^2}$, for some $x = 0, \ldots, q-2$, cannot be done in parallel logarithmic time.

The aforementioned result concerns the arithmetic model of computation when each element of $\mathbb{F}_q$ is considered as a whole without access to its bits. Then, over $\mathbb{F}_{2^n}$, we also deal with the Boolean model of computation. We assume that each element $u \in \mathbb{F}_{2^n}$ is given by a binary vector $u = (u_1, \ldots, u_n)$ containing the "coordinates" of u in some fixed basis of $\mathbb{F}_{2^n}$ over $\mathbb{F}_2$. Then we give a lower bound on the degree of Boolean functions expressing the coordinates of g^{x^2} via the coordinates of g^x. The bound is rather weak but nevertheless provides some non-trivial results on the parallel Boolean complexity of breaking the Diffie–Hellman key exchange protocol. For example it cannot be done by a Boolean circuit of constant depth. Unfortunately, this method does not work for other finite fields.

Nevertheless, using a new and very general approach developed in Chapter 29 to estimate the complexity of functions over $\mathbb{F}_p$, we obtain the lower bound $(0.25 + o(1)) \log \log p$ on the CREW PRAM complexity of breaking the Diffie–Hellman key exchange protocol modulo a prime p. This lower bound (as well as several others) is doubly logarithmic in terms of the field size, so it does not rule out the possibility that the question belongs to the complexity class **NC**, but at least shows that it cannot be done "superquickly" even with unlimited parallelism. Then we also use another result from Chapter 29 in order to show that any unbounded fan-in Boolean circuit of sublogarithmic depth breaking the Diffie–Hellman protocol modulo p must be of superpolynomial size.

We remark that the methods used in Chapter 29 can also be used to study complexity properties of polynomial and rational mappings over arbitrary finite fields. For example, for a field $\mathbb{F}_{2^r}$ one can identify each element $x \in \mathbb{F}_{2^r}$ with an r-bit string $x = (x_1, \ldots, x_r)$. Thus each function $f(x)$ of elements of $\mathbb{F}_{2^r}$ has an associated system of r Boolean functions $B_1, \ldots, B_r$ with

$$f(x) = (B_1(x), \ldots, B_r(x)).$$

Accordingly, one can study the complexity characteristics of these Boolean functions. The inversion function $f(x) = x^{-1}$ is of special interest because it is an essential ingredient of *Rijndael* [139], and in particular has been a subject of several discussions about possible weaknesses it may create, see [135, 178, 400]. Inversion has also been discussed in [450] as a potentially very promising transformation for so-called S-boxes in private key cryptography.

We remark that several lower bounds are also known on the complexity of deterministic [413] and probabilistic [505] sequential algorithms for computation of discrete logarithms. Several further results in this direction have been presented in [77, 109, 146, 297, 373, 374, 387, 496, 497, 498, 499, 500].

Some interesting aspects of obtaining lower bounds on the complexity of computing the discrete logarithm have also been discussed in [8] and [472]. However, the results and the approaches of those papers are quite different from those of the present work.

It could also be relevant to mention the works [64, 65, 222] where the complexity of finding some small portion of bits of the Diffie–Hellman transformation (over a prime field $\mathbb{F}_p$) is considered and is shown to be expected polynomial time equivalent to the whole problem of breaking the Diffie–Hellman key exchange protocol, see Chapter 14. For the *Shamir message passing scheme*, see Protocol 12.22 from [389], similar results are obtained in [223], see also Chapter 18. In Chapter 18 we also show that this approach can be used to derive a certain new result on the bit security of the RSA cryptosystem. We remark that the papers [222, 223, 523] use exponential sums to close some gaps in the proofs of similar results in [64]. An exhaustive overview of similar results, obtained for other cryptosystems has recently been given in [220], see also the original papers [105, 145, 149, 159, 182, 210, 216, 254, 255, 298, 342, 411, 455, 457, 496, 518]. We recall that the property these result imply is often called the *bit security* of the corresponding cryptographic scheme.

We also obtain an $\Omega(\log p)$ lower bound on the communication complexity of the rightmost bit of the Diffie–Hellman secret key.

The bit size r of p is defined by security considerations (essentially by the progress in computing the discrete logarithm). However then only a small fraction of bits of $K_p(x,y)$ is to be used as a key for some *private key* cryptosystem. A natural question arises whether the communicating parties can exchange fewer bits in order to be able to compute a few most significant bits of $K_p(x,y)$. Our estimate shows that unfortunately this is impossible.

Several interesting results about relations between breaking the Diffie–Hellman key exchange protocol modulo a composite number and the integer factorisation problem have been given in [44, 378].

Some of the aforementioned results have been extended in [342, 518, 521] to the secret key of the *XTR* [332, 333, 334, 335, 551] and to the *LUC* [47, 536] key exchange protocols.

We apply our method to derive quite a general estimate showing that for any non-linear and non-constant function modulo sufficiently large prime p, its arithmetic and Boolean depths cannot be smaller than $0.124 \log\log p$ simultaneously. For linear functions similar results have been obtained in [409]; see also [212, 213, 220, 298, 408, 410].

Although many results are available showing that if one of those depths is small then the other one is not too large [204], estimates of the type which we obtain here seem to be previously unknown.

These results provide the background for the aforementioned lower bound on the CREW PRAM complexity of breaking the Diffie–Hellman key exchange protocol modulo p.

We also show that several other related questions about permutation polynomials, polynomial representations of non-linear pseudorandom number generators, powers, Zech's logarithm, primitive root testing and some special Boolean functions can be dealt with along the same lines.

Furthermore, we consider a certain generalisation of the message concealing problem of [45] for the RSA cryptosystem and also message concealing properties of the subset sum mapping, which appears in the iterative subset sum generator of pseudorandom numbers of [280].

Many of our results can be interpreted as certain pseudorandom behaviour of the corresponding cryptographic construction. In particular, in Chapter 23 we present some results of [191, 194, 192, 236, 511] providing essentially first rigorous evidences of good pseudorandom properties of the power generator which includes, as important special cases, the RSA generator and the Blum–Blum–Shub generator, see [389, 541]. In Chapter 24 we present several similar results from [22, 235, 512, 513, 517, 528] for the Naor–Reingold pseudorandom function, see [405]. Generally the above results can be described as upper bounds on the statistical distance between the output of these generators and the uniform distribution in the appropriate space.

In Chapter 28 we give some results from [9, 38, 39, 40, 41, 460] about the complexity of such natural number-theoretic problems as co-primality testing, testing for square-freeness and primality testing.

We note that although we obtain several lower complexity bounds for

- bounded fan-in arithmetic circuits $\mathrm{AC}_p(d)$, $\mathrm{PAC}_q(d)$;
- bounded fan-in Boolean circuits $\mathrm{BC}(d)$;
- unbounded fan-in Boolean circuits $\mathrm{UBC}(d, S)$;
- combinatorial complexity;
- decision trees;
- computational trees;
- communication complexity;

we do not try to extract all possible complexity bounds for various computational models which would easily follow from our results. For example, combining our results about approximation of the discrete logarithm via real polynomials with results of [31], one can obtain lower bounds on the complexity of the discrete logarithm on a quantum computer, see also [83]. Several other results can be obtained as well, however we do not do it.

Instead we prefer to concentrate on estimating some intrinsic characteristics of the functions of interest such as

- polynomial degree (over various algebraic domains);
- sparsity;
- sensitivity;
- Fourier coefficients;
- additive complexity;
- $M_f(r)$-invariant;
- formula size.

Numerous useful relations among these parameters, certain complexity bounds depending on them, as well as many relevant references, can be found in [79, 80, 151, 152, 179, 226, 447].

Moreover, our results concerning the above-mentioned characteristics are quite strong and in several cases are close to the best possible. However, when we derive our complexity lower bounds we use standard complexity theory approaches which unfortunately are not able to use the complete power of our results. This, unfortunately, is quite a common situation implied by the current very unsatisfactory state of affairs with complexity lower bounds. Indeed, although the corresponding methods of complexity theory make a very ingenious use of very deep results from other areas of mathematics and computer science, they nevertheless produce surprisingly weak statements.

Finally, in Chapters 30 and 31 we study several other properties of polynomials and other functions over finite fields which are related to some complexity theory problems. The same method also applies to some linear algebra problems. In particular, Theorem 30.5 implies certain new "non-approximability" results for permanents of structured matrices over $\mathbb{F}_p$.

Our method is based on such classical tools of the theory of finite fields as:

- bounds for the number of solutions of equations and congruences;
- bounds for various exponential sums;

see [306, 309, 341, 343, 508, 557].

In order to use the first approach we find a certain polynomial equation or a system of polynomial equations (generally different in different statements) which is satisfied by the discrete logarithm or other functions of interest, and show that the approximating function satisfies the same equation or system of equations for sufficiently many points, but does not satisfy it identically. We can now apply the Lagrange and Bézout theorems to bound the number of zeros of the polynomials, the system of polynomial equations, respectively. Theorem 8.1 is a fairly simple but quite instructive example of the use of this idea.

In order to use the second approach we find some property of the function of interest which has a good number-theoretic interpretation and using bounds of exponential sums, show that our function behaves pseudorandomly with respect to this property. For example, we use the fact that the rightmost bit of $\operatorname{ind} x$ determines the value of the quadratic character at x, while the most significant bits determine the order of magnitude of $\operatorname{ind} x$ and thus have to do with uniformity of distribution of the discrete logarithm. All other bits can be interpreted in a somewhat similar way and thus apparently can be studied within the same approach. This approach has been used in [133]; the results of that paper are presented in Chapters 8 and 12 of this book.

Several more results about pseudorandomness of the rightmost bit of $\operatorname{ind} x$, that is, the quadratic character, or the Legendre symbol, of x, have been obtained in [59, 145, 155, 369, 371, 458, 475] using the Weil bound. Similar results can also be obtained for quadratic characters with $f(x)$ for polynomials $f(X) \in \mathbb{F}_p[X]$, satisfying certain additional conditions. See also [11, 266] for several more possible generalizations which make use of similar constructions in function fields. It has been shown in [510] that one can possibly use similar constructions modulo extremely large (doubly exponential) but sparse numbers. The algorithm of [510] shows how to make this computation efficient.

Estimates of exponential sums are also used in [205] in a similar way to obtain lower bounds on the complexity of modular inversion. Some of the results of [133] are also based on bounds of exponential sums and are included in Chapter 8 of this book.

The works [94, 95], providing some results towards the so-called *Diffie–Hellman indistinguishability assumption*, see [51, 93, 159, 405], are based on new estimates of exponential sums. The assumption claims that for any subgroup $G_l \subseteq \mathbb{F}_q^*$ of a sufficiently large prime order $l|q-1$ and any generator ϑ of this group, the Diffie–Hellman triples $(\vartheta^x, \vartheta^y, \vartheta^{xy})$, for x, y selected randomly and uniformly from the set $\{0, \ldots, l-2\}$, are polynomial time indistinguishable from the uniformly distributed triples $(u, v, w) \in G_l^3$. Here we present some of the results of those papers which imply that the distribution of the above triples is close to uniform; see Lemma 3.21 and Theorems 13.6 and 13.8 below. In particular, these results imply that the statistical distance between some small portions of bits of g^{xy} (which are usually used as a key for some private key cryptosystem) and uniformly distributed binary strings of the same length is exponentially small. The results of [195, 196] show that the same property holds even if the exponents x and y are selected from arbitrary (but sufficiently large) sets $\mathcal{X}$ and $\mathcal{Y}$. For example, $\mathcal{X}$ and $\mathcal{Y}$ can be selected in a way which helps to speed up exponentiation, see Theorem 13.7. In [189] a similar result is obtained for the Diffie–Hellman triples modulo an arbitrary composite number.

For just the Diffie–Hellman key ϑ^{xy} some interesting statistical results have been given in [554].

Besides the above mentioned result on the distribution of Diffie–Hellman triples $(\vartheta^x, \vartheta^y, \vartheta^{xy})$, we also present a number of other similar results on the uniformity of distribution of several other cryptographic primitives such as

- RSA pairs (x, x^e) and pairs (x, g^x) generated by exponentiation in finite fields, see Chapter 21;
- Power generator of pseudorandom numbers, which includes, as important special cases, the RSA generator and the Blum–Blum–Shub generator, see Chapter 23;
- Naor–Reingold pseudorandom function in finite fields, see Chapter 24;
- ElGamal signature, see Chapter 17.

Several other results of this type, including their elliptic curve analogues, can be found in [168, 169, 166, 261, 417, 418, 512, 528].

Although such uniformity of distribution results do not have any immediate implications for the security of the corresponding cryptographic constructions, they still provide some intuitive motivation for such conclusions. In particular in each of the above cases the inverse statement (about non-uniformity of distribution) would be disastrous for the corresponding construction. Several such results have been known before, for example, see [145, 391, 430, 457], but this area has never been studied systematically.

On the other hand, surprisingly enough, such statements could be a tool for an attack as well. For example, a statement about the uniform distribution of the ratios $r(k)s(k,\mu)^{-1}$, corresponding to the pairs $(r(k), s(k,\mu))$ produced by the *Digital Signature Algorithm*, has been used in [417] to give a rigorous proof of a certain attack on this scheme. We recall that the Digital Signature Algorithm is a modification of the ElGamal signature with $p-1$ replaced by a prime divisor l of $p-1$, see [389, 541]. Brief descriptions of the Digital Signature Algorithm and of the ElGamal signature scheme are also given in Chapter 16 and Chapter 17, respectively. Some generalisations of the results of [417] to other similar signature schemes, including the *Nyberg–Rueppel* scheme and the elliptic curve based analogues of the Digital Signature Algorithm, can be found in [166, 418].

Links of a different kind between complexity characteristics of Boolean functions and exponential sums have been used in [89, 231, 232, 233]. Although it is not clear how to apply the Weil bound or other similar bounds to exponential sums of these papers, this direction of research looks very promising and definitely is worth further study.

Various bounds of exponential sums have been used in [306] to establish unpredictability properties of some pseudorandom number generators, including the $1/M$ generator of [50], see also Chapter 25, and the truncated linear congruential generator, which complement some results of [197]. In particular, some predictability results of [197] almost match one of the unpredictability results of [306]. We also use exponential sums study several other important pseudorandom number generators in Chapters 26 and 27. In particular, in Chapter 27 we consider the iterative *subset sum* generator of [280] which is based on iterating of a very different function without any explicit algebraic structure which makes its study extremely difficult. On the other hand, it has some important cryptographic properties and certainly deserves to be better known and studied. A certain variant of this generator which is based on using linear recurrence sequences for subset sums rather than on iterations is a little easier to study and we obtain a new result on the distribution of this generator.

We also remark that exponential sums have been used in [515] to obtain some new results about the distribution and the output rate of the *shrinking generator* of pseudorandom number. Those results are analogues of those of [132] but obtained for individual sequences rather than "on average" as in [132].

We also draw attention to the paper [15], where the Weil bound is used in a very elegant way to produce a lower bound on the communication complexity of computation of the quadratic character of $x_1+\ldots+x_k$ where $x_1,\ldots,x_k$ are known to k collaborating parties. Several more results in this direction have recently been obtained in [14].

Besides the aforementioned number-theoretic tools we also use some standard number-theoretic and cryptographic techniques. In particular we widely use methods, facts, and notions of analytic number theory [250, 468], of the theory of finite fields which one can easily find in [343], of the theory of uniformly distributed sequences [161]. The lattice basis reduction algorithm of Lenstra, Lenstra and Lovász [329] is another important ingredient. In particular, it has turned out that a combination of the lattice basis reduction with bounds of various exponential sums provides a very important tool to establish rigorous results about both security and insecurity of several cryptographic constructions. The examples include but not limited to a proof of the bit security of the Diffie–Hellman secret key and a rigorous attack on the Digital Signature Algorithm.

We also present, in Chapter 20 several rather strong bit security results for the NTRU cryptosystem which have recently been obtained in [411].

We collect all necessary results and definitions in Part I. Many of them are very well known or, if not known, would be just easy exercises for an expert in the corresponding area. However, some of these auxiliary results are rather new and quite non-trivial. They introduce some new techniques and are probably of independent interest.

In obtaining the lower bounds on the depth of circuits in Theorem 10.2 our arguments are quite close to those of [200, 201, 204] (see also the references given therein). However, it seems that in the proof of Theorem 12.7 some new arguments appear. We also use several other notions and results of complexity theory [453, 556].

We note that questions of the sequential complexity of several of the problems considered here are the most important open questions of computational number theory; see Problems 18, 21, 26, 35 of [4]. Unfortunately no non-trivial lower bounds are known on their sequential complexity.

Finally, we mention that surveys of many modern efficient algorithms (including heuristic ones) for various number-theoretic and cryptographic problems, such as computing the discrete logarithm, primality testing and integer factorisation, can be found in [1, 4, 16, 74, 82, 127, 136, 171, 203, 326, 328, 352, 379, 389, 395, 451, 462, 463, 464, 465, 494, 508, 541]. A spectacular progress has recently been achieved in [5], where an unconditional deterministic polynomial time primality testing algorithm is designed. Its simplicity and elegance are the most striking features.

Part I

Preliminaries

Chapter 1

Basic Notation and Definitions

As usual $\mathbb{F}_q$ denotes a finite field of q elements, $\mathbb{Z}$ denotes the ring of integer numbers, $\mathbb{Q}$ denotes the field of rational numbers, $\mathbb{R}$ denotes the field of real numbers and $\mathbb{C}$ denotes the field of complex numbers.

For a real γ, as usual, we define $\lfloor \gamma \rfloor$ and $\lceil \gamma \rceil$ as two integer numbers with

$$\lfloor \gamma \rfloor \le \gamma < \lfloor \gamma \rfloor + 1 \qquad \text{and} \qquad \lceil \gamma \rceil - 1 < \gamma \le \lceil \gamma \rceil .$$

We also use $\{\gamma\}$ to denote the fractional part $\{\gamma\} = \gamma - \lfloor \gamma \rfloor$.

For an integer $m \ge 2$ we denote by $\mathbb{Z}_m$ the residue ring modulo m which we identify with the set of integers $k \in \{0, \dots, m-1\}$; $\mathbb{Z}_m^*$ is the subset of $\mathbb{Z}_m$ which consists of integers $k \in \mathbb{Z}_m$ with $\gcd(k, m) = 1$.

We recall that the function

$$\operatorname{Tr}_{\mathbb{F}_{q^m}/\mathbb{F}_q}(z) = \sum_{i=0}^{m-1} z^{q^i}$$

is called the *trace* and the function

$$\operatorname{Nm}_{\mathbb{F}_{q^m}/\mathbb{F}_q}(z) = \prod_{i=0}^{m-1} z^{q^i}$$

is called the *norm* of $z \in \mathbb{F}_{q^m}$ in $\mathbb{F}_q$. These functions play a very important role in the theory of finite fields, see Chapter 2 of [343].

For integers s and $m \ge 1$ we denote by $\lfloor s \rfloor_m$ the remainder of s on division by m. We also apply this notation to rational numbers. Namely, for a rational number z and $m \ge 1$ we denote by $\lfloor z \rfloor_m$ the unique integer a, $0 \le a \le m-1$ such that $a \equiv z \pmod{m}$ (provided that the denominator of z is relatively prime to m).

For an integer m and a real $k > 0$ we denote by $\mathrm{MSB}_{k,m}(t)$ any integer which satisfies the inequalities

$$\frac{m}{2^k}\mathrm{MSB}_{k,m}(t) \le \lfloor t \rfloor_m < \frac{m}{2^k}\left(\mathrm{MSB}_{k,m}(t)+1\right). \tag{1.1}$$

Thus, roughly speaking, $\mathrm{MSB}_{k,m}(t)$ is the integer defined by the k most significant bits of $\lfloor t \rfloor_m$. However, this definition is more flexible and better suited to our purposes. In particular we remark that k in the inequality (1.1) need not be an integer.

We also use $\mathcal{P}$ to denote the set of primes.

As usual $\pi(X)$ denotes the number of primes $p \le X$ and, for integers $k \ge 1$ and a, $\pi(X;k,a)$ denotes the number of primes $p \le X$ with $p \equiv a \pmod k$.

For an integer $m \ge 2$, $\tau(m)$ denotes the number of positive integer divisors of m, $\varphi(m)$ denotes the *Euler function* of m, $\nu(m)$ denotes the number of prime divisors of the integer m.

For an integer $m \ge 1$, $\mu(m)$ denotes the Möbius function. We recall that $\mu(1) = 1$, $\mu(m) = 0$ if $m \ge 2$ is not square-free and $\mu(m) = (-1)^{\nu(m)}$ otherwise.

We say that an integer $k \ge 1$ is Y-smooth if it is divisible only by primes $p \le Y$.

Throughout this book, for a polynomial f over a ring $\mathcal{R}$, $\operatorname{spr} f$ denotes its *sparsity*, which is defined as the total number of its non-zero coefficients.

For an integer $m \ge 1$ we define the *Carmichael function* $\lambda(m)$ as the largest possible order of elements of the unit group in the residue ring modulo m. More explicitly, for a prime power p^k we define

$$\lambda\left(p^k\right) = \begin{cases} p^{k-1}(p-1), & \text{if } p \ge 3 \text{ or } k \le 2; \\ 2^{k-2}, & \text{if } p = 2 \text{ and } k \ge 3; \end{cases}$$

and finally,

$$\lambda(m) = \operatorname{lcm}\left(\lambda(p_1^{k_1}), \ldots, \lambda(p_\nu^{k_\nu})\right),$$

where

$$m = p_1^{k_1} \ldots p_\nu^{k_\nu}$$

is the prime number factorisation of m.

Let us define the *additive complexity* $\mathrm{C}_{\pm}(f)$ of a multivariate polynomial f over a ring $\mathcal{R}$ as the smallest number of '$+$' and '$-$' signs necessary to write down a polynomial [88, 237, 293, 473, 474, 479, 480]. Obviously, for any univariate polynomial f,

$$\mathrm{C}_{\pm}(f) \le \operatorname{spr}(f) - 1 \le \deg f$$

but neither the sparsity $\operatorname{spr}(f)$ nor the degree $\deg f$ can be estimated in terms of $\mathrm{C}_{\pm}(f)$.

For example, the polynomials

$$\begin{aligned} f_1(X) &= \left((X+1)^k - (X+2)^m\right)^l, \\ f_2(X,Y) &= \left((X+1)^k(Y-2)^m + X\right)^l + (Y-1)^m\left((X+1)^k + Y^l\right)\end{aligned}$$

are of additive complexity $C_{\pm}(f_1) \le 3$ and $C_{\pm}(f_2) \le 7$ respectively but neither of low degree nor sparse (for k, l, m large enough). It is easy to see that, using repeated squaring, values of polynomials of low additive complexity can be computed very quickly at any point, thus they belong to the family of "easily computable" functions regardless of their degree or sparsity.

Following [366], for a function

$$f : \mathbb{R} \to \{0,1\}$$

we define its $M_f(r)$*-invariant* as the largest integer M such that for any $\lambda < M$ there are two r-bit integers $0 \le x_1 < x_2 \le 2^r - 1$, both divisible by λ and such that $f(x_1) \ne f(x_2)$; see also [81, 365, 366, 380] for applications to complexity theory.

We recall that a sequence $(u(x))$ is called a *linear recurrence sequence* of order n over a ring $\mathcal{R}$, if it satisfies the relation

$$u(x+n) = c_{n-1}u(x+n-1) + \ldots + c_0u(x), \qquad x = 1,2,\ldots, \tag{1.2}$$

for some $c_{n-1}, \ldots, c_1, c_0 \in \mathcal{R}$ and $c_0 \ne 0$. The polynomials

$$\psi(T) = T^n - c_{n-1}T^{n-1} - \ldots - c_0 \in \mathcal{R}[T]$$

is called the *characteristic polynomial* of this sequence.

We also recall, that the *linear complexity profile* $L(N)$ of an infinite sequence (s_x) over a ring $\mathcal{R}$ is the function which for every integer $N \ge 2$ is defined as the shortest length k of linear recurrence relations of the form

$$s_{x+k} = a_{k-1}s_{x+k-1} + \ldots + a_0s_x, \qquad 0 \le x \le N-k-1, \tag{1.3}$$

which are satisfied by this sequence, see [138, 389, 439, 483, 484].

The largest value

$$L = \max_{N \ge 2} L(N)$$

is called the *linear complexity* of the sequence.

The notion of *linear complexity* is closely related with another important characteristic which is the largest s for which the sequence passes the s-dimensional lattice test introduced by Marsaglia [368], see [444] for a quantitative form of this relation and some applications.

Obviously, for some sequences the linear complexity can be equal to infinity. However, for the linear complexity of any periodic sequence of period t one easily verifies that

$$L = L(2t) \le t.$$

Similarly, the *non-linear complexity profile* of order d of an infinite sequence (s_x) over a ring $\mathcal{R}$ is the function $NL_d(N)$ which for every integer $N \ge 2$ is defined as the shortest length k of non-linear recurrence relations of the form

$$s_{x+k} = \Psi\left(s_{x+k-1}, \dots, s_x\right), \qquad 0 \le x \le N-k-1,$$

with polynomials $\Psi(T_1, \dots, T_k) \in \mathcal{R}[T_1, \dots, T_k]$ of degree $\deg \Psi \le d$ which are satisfied by this sequence. Also, the largest value

$$NL_d = \max_{N \ge 2} NL_d(N)$$

is called the *non-linear complexity* of order d of the sequence.

It is useful to note that generally speaking $NL_1(N) \ne L(N)$ because in the definition of $L(N)$ one can use only homogeneous linear polynomials, but it is easy to see that $NL_1(N) + 1 \ge L(N)$. We also have

$$L(N) \ge NL_1(N) \ge NL_2(N) \ge \dots.$$

Let $H(\gamma)$ denote the *binary entropy function*

$$H(\gamma) = -\gamma \log \gamma - (1-\gamma)\log(1-\gamma), \qquad 0 < \gamma < 1, \tag{1.4}$$

where $\log z$ denotes the logarithm to base 2.

We also denote by $\mathfrak{B}_r$ the r-dimensional Boolean cube,

$$\mathfrak{B}_r = \{0,1\}^r.$$

More generally, we define

$$\mathfrak{B}_r(h) = \{\mathbf{x} = (x_1, \dots, x_r) \in \mathbb{Z}^r \mid 0 \le x_j \le h-1,\ j = 1, \dots, r\}. \tag{1.5}$$

We also define $\mathfrak{B}_{r,k}(h)$ as the subset of $\mathfrak{B}_r(h)$ consisting of vectors with precisely k non-zero components. Hence

$$|\mathfrak{B}_r(h)| = h^r \qquad \text{and} \qquad |\mathfrak{B}_{r,k}(h)| = (h-1)^k \binom{r}{k}.$$

In the special case $h = 2$ we put

$$\mathfrak{B}_{r,k}(2) = \mathfrak{B}_{r,k}, \tag{1.6}$$

and of course $\mathfrak{B}_r(2) = \mathfrak{B}_r$.

The *Hamming weight* $\|a\|$ of a vector $a \in \mathfrak{B}_r$ is the number of its non-zero components.

For a Boolean function $B(U_1, \dots, U_r)$ we define its *Fourier coefficients* $\widehat{B}(a)$, where $a = (a_1, \dots, a_r) \in \mathfrak{B}_r$, as

$$\widehat{B}(a) = \sum_{u \in \mathfrak{B}_r} (-1)^{B(u)+a \cdot u}, \tag{1.7}$$

where

$$a \cdot u = a_1 u_1 + \ldots + a_r u_r.$$

Following [556], for a Boolean function $B(U_1, \dots, U_r)$ we define the *sensitivity*, $\sigma(B)$, which is also known as the *critical complexity*, as the largest integer $s \le r$ such that there is a binary vector $x = (u_1, \dots, u_r) \in \mathfrak{B}_r$ for which $B(x) \neq B(x^{(i)})$ for s values of i, $1 \le i \le r$, where $x^{(i)}$ is the vector obtained from x by flipping its ith coordinate,

$$\sigma(B) = \max_{x \in \mathfrak{B}_r} \sum_{i=1}^{r} \left| B(x) - B(x^{(i)}) \right| .$$

In other words, $\sigma(B)$ is the maximum, over all binary vectors

$$x = (u_1, \dots, u_r) \in \mathfrak{B}_r,$$

of the number of points $y \in \mathfrak{B}_r$ on the unit *Hamming sphere* around x with $B(y) \neq B(x)$.

This function gives lower bounds for several other complexity characteristics of B including its CREW PRAM complexity; see [152] or Section 20.4.1 of [179] or [453] or Chapter 13 of [556].

We recall that *CREW PRAM* complexity is the complexity on a *parallel random access machine* (PRAM) with an unlimited number of all-powerful processors. More precisely, we consider the modification which is known as CREW (concurrent read, exclusive write) PRAM. Such a machine has an infinite shared memory, each cell of which can hold an integer number and such that simultaneous reads of a single cell by several processors are permitted, but simultaneous writes are not [151, 152, 179, 453, 556].

Following [33, 35, 36, 67] one can also define the *average sensitivity* $\sigma_{av}(B)$ of a Boolean function $B(U_1, \dots, U_r)$ as

$$\sigma_{av}(B) = 2^{-r} \sum_{x \in \mathfrak{B}_r} \sum_{i=1}^{r} \left| B(x) - B(x^{(i)}) \right| .$$

We note that an explicit expression for $\sigma_{av}(B)$ in terms of the Fourier coefficients is given in [36]; see also [33] and the proof of Theorem 29.5 below.

Generalisations of this notion, introduced in [35], can be studied similarly.

The sensitivity and the average sensitivity are also related to the complexity of so called *private computation*, see [317].

Further we mention the following definitions from [447]: for a Boolean function $B(X_1, \dots, X_r)$ let the *real degree* of f, denoted by $\Delta(B)$, be the smallest possible degree of a real polynomial $f(X_1, \dots, X_r)$ for which

$$B(x_1, \dots, x_r) = f(x_1, \dots, x_r)$$

holds for every $(x_1, \dots, x_r) \in \mathfrak{B}_r$.

We also define the *real approximate degree* of B, denoted by $\delta(B)$, as the smallest possible degree of a multilinear real polynomial $f(X_1, \dots, X_r)$ for which

$$|B(x_1, \dots, x_r) - f(x_1, \dots, x_r)| \le 1/3$$

holds for every $(x_1, \dots, x_r) \in \mathfrak{B}_r$. Certainly, in all our results $1/3$ can be replaced by any constant $\alpha < 1/2$.

Clearly,

$$\delta(B) \le \Delta(B) \le r$$

for any B.

Now we recall the notion of communication complexity. Assume that two collaborating parties $\mathcal{A}$ and $\mathcal{B}$ want to compute a Boolean function $B(X, Y)$ depending on two arguments X and Y of r bits each. The party $\mathcal{A}$ knows only the first argument x, the party $\mathcal{B}$ knows only the second argument y. Parties $\mathcal{A}$ and $\mathcal{B}$ may exchange messages. The *communication complexity* $\mathrm{Comm}\,(B)$ of B is the smallest number of bits to be exchanged in the worst case for any communication protocol [14, 15, 148, 184, 211, 278, 316, 448, 469, 503].

The *combinatorial discrepancy* $\Gamma(B)$ of a Boolean function $B(X, Y)$ depending on two arguments X and Y of r bits each is defined as

$$\Gamma(B) = 2^{-2r} \max_{\mathcal{X}, \mathcal{Y}} |N_1(\mathcal{X}, \mathcal{Y}) - N_0(\mathcal{X}, \mathcal{Y})|,$$

where the maximum is taken over all sets $\mathcal{X}, \mathcal{Y} \subseteq \mathcal{B}$ and $N_\mu(\mathcal{X}, \mathcal{Y})$ is the number of pairs $(x, y) \in \mathcal{X} \times \mathcal{Y}$ with $B(x, y) = \mu$.

The notion of combinatorial discrepancy is a very powerful tool in the study of communication complexity of B, see [15, 316]. The link between the combinatorial discrepancy, and communication complexity is given by Lemma 7.7 below.

Now we define several complexity classes which we consider in this book.

We consider the class $\mathrm{BC}(d)$ of *bounded fan-in Boolean circuits* of depth d. Given a sequence of bits $u_1, \dots, u_r$, such a circuit $C \in \mathrm{BC}(d)$ computes the values of some Boolean function $B(u_1, \dots, u_r)$ in the following way.

Each circuit $C \in \mathrm{BC}(d)$ has one special *starting* level and d levels of Boolean processors which are called *gates*. Levels are numbered from 0 for the starting level to d for the last level. Each level may have an unlimited number of gates, with only one gate on the last level.

Each gate of the starting level accepts either some constant or the value of one of the input variables $u_1, \dots, u_r$ for which we want to compute the function B.

Each gate of level $k \geq 1$ accepts two values from gates of previous levels, φ_1, φ_2, and then computes and outputs the value of $\varphi_1 \# \varphi_2$, where $\#$ stands for one of the arithmetic operations over $\mathbb{F}_2$, that is, $\# \in \{+, \times\}$.

Finally, the unique gate of the last level outputs the result of the computation, that is, $C(u_1, \dots, u_r)$.

The *combinatorial complexity* $CC(B)$ of a Boolean function B is defined as the smallest number of gates over all possible circuits of the same kind as circuits of the class $\mathrm{BC}(d)$ with the only restriction that the fun-out is 1. That is, the result of the computation of each gate can be used only once; see [252, 492, 493] for more details.

The class $\mathrm{AC}_p(d)$ of *bounded fan-in arithmetic circuits* of depth d is defined quite analogously, the only difference being that the gates of the starting level accept either constants or the value of one of the input variables, and each gate of level $k \geq 1$ accepts two values from gates of previous levels, φ_1, φ_2, and then computes and outputs the value of $\varphi_1 \# \varphi_2$, where $\#$ stands for an arithmetic operation modulo p, that is, $\# \in \{+, -, \times, /\}$.

We also consider the subclass $\mathrm{DFAC}_p(d) \subseteq \mathrm{AC}_p(d)$ of *division-free arithmetic circuits* of depth d which may use addition, subtraction or multiplication but do not use division modulo p.

We consider, in addition, the following class $\mathrm{PAC}_q(d)$ of *probabilistic branching arithmetic circuits* of depth d over $\mathbb{F}_q$.

Each circuit $C \in \mathrm{PAC}_q(d)$ has one special *starting* level and d levels of arithmetic processors which are called arithmetic *gates*. Levels are numbered from 0 for the starting level to d for the last level. Each level may have an unlimited number of gates, with only one gate on the last level.

Each gate of the starting level accepts either some constant or the value of one of the input variables.

Each gate of level $d - 1 \geq k \geq 1$ accepts four values from some gates of previous levels, $\varphi_1, \varphi_2, \varphi_3, \varphi_4$, and one special *branching* value ψ which is either some value computed on a previous level or the value of a random variable taking values 0 and 1 with probability $1/2$.

Then it computes and outputs the value of $\varphi_1 \# \varphi_2$ if $\psi \neq 0$ and $\varphi_3 \# \varphi_4$ otherwise, where $\#$ stands for one of the arithmetic operations over $\mathbb{F}_q$, that is, $\# \in \{+, -, \times, /\}$ (not necessarily the same each time).

The gate of the last level accepts only one value ψ (which is either some value computed on a previous level or the value of a random variable taking values 0 and 1 with probability $1/2$) and outputs $C(u,v) = \textbf{YES}$ if $\psi = 0$ and $C(u,v) = \textbf{NO}$ otherwise.

Finally, we define the class $\mathrm{UBC}(d,S)$ of *unbounded fan-in Boolean circuits* of depth d and of size S. This class incorporates the more standard classes Σ_d^S and Π_d^S; see Section 3.1 of [68].

Given a sequence of bits $u_1, \dots, u_r$, a circuit $C \in \mathrm{UBC}(d,S)$ computes the values of some Boolean function $B(u_1, \dots, u_r)$ in the following way.

Each circuit $C \in \mathrm{UBC}(d,S)$ has one special *starting* level and d levels of at most S unbounded fan-in Boolean processors which are called *unbounded fan-in gates*. Levels are numbered from 0 for the starting level to d for the last level. Each level may have any number of gates, with only one gate on the last level. The total number of gates is at most S.

Each gate of the starting level accepts either some constant or the value of one of the input variables $u_1, \dots, u_r$ for which we want to compute the function B.

Each gate of level $k \geq 1$ accepts any number s of values from gates of previous levels, $\varphi_1, \dots, \varphi_s$, and then computes and outputs the value of $\varphi_1 \# \dots \# \varphi_s$, where $\#$ stands for one of the Boolean operations $\# \in \{\vee, \wedge, \neg\}$ (the same for each gate but possibly different for different gates). If $\#$ is $\neg$, then $s = 1$ of course.

Finally, the unique gate of the last level outputs the result of the computation $C(u_1, \dots, u_r)$.

A *decision tree* with input variables $X_1, \dots, X_r$ is a rooted binary tree in which each edge is labeled with a variable or a negated variable in such a way that labels of edges leaving the same inner node are negations of each other. Further each leaf v of the tree is labeled with some value $\lambda(v) \in \{1, -1\}$.

A decision tree T defines a Boolean function B_T as follows: Given an input $x = (x_1, \dots, x_r)$, replace each edge label X_i by the induced value, that is, replace each X_i by x_i and each $\neg X_i$ by $\neg x_i$. After the replacement there will be exactly one path from the root to some leaf v whose edges are all labeled 1 which is called the *computation path of the input* x. Define $B_T(x) \in \{1, -1\}$ to be $\lambda(v)$.

The number of leaves is called the *size* of the decision tree.

Given a decision tree with input variables $X_1, \dots, X_r$, we denote by l_i the number of inputs whose computation paths have length i, $i = 1, \dots, r$. We call $l_0, \dots, l_r$ the *leaf numbers* of the tree. Observe that the size M of the decision tree is given by the formula

$$M = \sum_{i=0}^{r} 2^{i-r} l_i.$$

We denote by DT (B) the smallest possible size of a decision tree which computes B.

We say that a Boolean function $B(U_1, \ldots, U_r)$ belongs to the *class* $\mathcal{P}^r_{k,s}$ if for any choice of k integers $1 \le i_1 < \ldots < i_k \le r$ there are at least s distinct functions obtainable by making all 2^k possible assignments to $U_{i_1}, \ldots, U_{i_k}$.

In particular, functions of the class $\mathcal{P}^r_{1,2}$ depend on all their variables.

A *restriction* ρ of the set $\{1, \ldots, r\}$ is a mapping of this set to the set $\{0, 1, \star\}$, where

- $\rho(i) = 0$ means that we substitute the value 0 for x_i;
- $\rho(i) = 1$ means that we substitute the value 1 for x_i;
- $\rho(i) = \star$ means that X_i remains a variable.

Given a Boolean function $B(X_1, \ldots, X_r)$ of r variables, we will denote by B_ρ the function obtained from B by applying the restriction ρ; B_ρ will be a function of the variables X_i for which $\rho(X_i) = \star$, $1 \le i \le r$. We call B_ρ a *subfunction* of B generated by the restriction ρ.

The subscripts i and the corresponding variables x_i are called *fixed* if $\rho(i) = 0, 1$, and *free* if $\rho(i) = \star$.

A Boolean function $B(X_1, \ldots, X_r)$ of r variables is called δ*-approximated at level* k if

$$\left| \frac{1}{2^l} \widehat{B}_\rho(0) - \frac{1}{2^r} \widehat{B}(0) \right| \le \delta$$

for any subfunction B_ρ depending on $l \ge k$ variables.

In case $\delta = 0$ this notion has been introduced in [33, 34] and extended to arbitrary δ in [39, 41] where it proved to be an effective tool in studying Boolean functions associated to number-theoretic problems.

For a real z and an integer m we use the notation

$$\mathbf{e}(z) = \exp(2\pi i z) \qquad \text{and} \qquad \mathbf{e}_m(z) = \exp(2\pi i z/m).$$

For an integer vector $\mathbf{a} = (a_1, \ldots, a_N) \in \mathbb{Z}^N$ we define

$$|\mathbf{a}| = \max_{i=1,\ldots,N} |a_i|, \qquad r(\mathbf{a}) = \prod_{i=1}^{N} \max\{|a_i|, 1\}. \tag{1.8}$$

For a sequence of H points

$$\Gamma = (\gamma_{0,x}, \dots, \gamma_{N-1,x})_{x=1}^{H} \tag{1.9}$$

in the N-dimensional unit cube, denote its *discrepancy* by Δ_Γ. That is,

$$\Delta_\Gamma = \sup_{B \subseteq [0,1)^N} \left| \frac{T_\Gamma(B)}{H} - |B| \right|,$$

where $T_\Gamma(B)$ is the number of points of the sequence Γ in the box

$$B = [\alpha_0, \beta_0) \times \dots \times [\alpha_{N-1}, \beta_{N-1}) \subseteq [0,1)^N$$

and the supremum is taken over all such boxes.

We recall that the *multiplicative order* $\mathrm{ord}_m \vartheta$ of an integer ϑ modulo an integer $m \geq 1$ with $\gcd(\vartheta, m) = 1$ is the smallest positive integer τ for which

$$\vartheta^\tau \equiv 1 \pmod m.$$

In particular τ divides any integer t with $\vartheta^t \equiv 1 \pmod m$.

Throughout the book

- $A = O(B)$ and $A \ll B$ mean that $|A| \leq cB$;
- $A = \Omega(B)$ and $A \gg B$ mean that $|A| \geq cB$;

where c is a positive constant, which may occasionally, where obvious, depend on the small positive parameter ε (or some other relevant parameters) and is absolute otherwise.

We also denote by $\log x$ the binary logarithm of x and by $\ln x$ the natural logarithm of x.

Chapter 2

Polynomials and Recurrence Sequences

Here we collect some lemmas we will need later.

Lemma 2.1. *Let a sequence $(u(x))$ satisfy a linear recurrence relation of order $n \geq 1$ of the form (1.2) over a field $\mathbb{F}$. Then for any $K \geq n+1$ pairwise distinct non-negative integers $j_1, \ldots, j_K$ there exist $c_1, \ldots, c_K \in \mathbb{F}$, not all equal to zero, such that*

$$\sum_{i=1}^{K} c_i u(x + j_i) = 0, \qquad x = 1, 2, \ldots .$$

Proof. It is easy to see that the set of all solutions of any linear recurrence relation over any field $\mathbb{F}$ is a linear space of dimension n over $\mathbb{F}$, for example, see Chapter 8 of [343]. Therefore, any $T \geq n+1$ solutions are linearly dependent. In particular, the sequences $(u(x+j_1)), \ldots, (u(x+j_K))$ are linearly dependent. □

We also need the following variant of Lemma 2.1 which can be proved using similar arguments.

Lemma 2.2. *Let an N-element sequence w_n, $n = 0, \ldots, N-1$, over a field $\mathbb{K}$ be of linear complexity L. Then for any integers $m \geq 1$, $h \geq 1$ and $0 \leq e_0, \ldots, e_L \leq h$ there are some elements $c_0, \ldots, c_L \in \mathbb{K}$ not all equal to zero and such that*

$$\sum_{j=0}^{L} c_j w_{mk+e_j} = 0$$

for any integer k with $0 \leq mk + h \leq N-1$.

Proof. Using the corresponding linear recurrence relation of order L, we can extend the sequence w_n to all integers $n \geq 0$.

For $i = 0, \ldots, L$ we consider $L + 1$ sequences

$$w_{mk+e_j}, \qquad k = 0, 1, \ldots .$$

Obviously, they all satisfy the same linear recurrence relation of order $l \leq L$. For example, this follows from the explicit expression for solutions of linear recurrence relations via polynomials and exponential functions, see [343]. It is easy to see that the set of solutions of a linear recurrence relation of order l forms a vector space of dimension l over $\mathbb{K}$. Therefore the above $L + 1$ sequences are linearly dependent. □

Lemma 2.3. *Let a sequence $(u(x))$ satisfy a linear recurrence relation of the form (1.2) over a ring $\mathcal{R}$ and let $(w_i(x))$ be the linear recurrence sequences which satisfy the same linear recurrence relation as $(u(x))$ and with initial values*

$$w_i(j) = \begin{cases} 1, & \text{if } i = j, \\ 0, & \text{if } i \neq j, \end{cases} \qquad 1 \leq i, j \leq n.$$

Then

$$u(x + y) = \sum_{i=1}^{n} u(x + i) w_i(y)$$

for any integers $x, y \geq 1$.

Proof. Indeed, for each fixed x the expressions on the left- and right-hand sides of the last equation satisfy the same linear recurrence relation (with respect to the index y) and have the same initial values. □

Lemma 2.4. *Let $(u(x))$ be a linear recurrence sequence of order n satisfying an equation of the form (1.2) over a field $\mathbb{K}$ with $c_0 \neq 0$ and let $u(x) \neq 0$ for at least one integer $x \geq 1$. Then for any integer $H \geq 1$ there are at least $\lfloor H/n \rfloor$ values of $x = 1, \ldots, H$ with $u(x) \neq 0$.*

Proof. It is obvious that if n consecutive values

$$u(x) = \ldots = u(x + n - 1) = 0,$$

then all further values $u(x + n), u(x + n + 1), \ldots$ are zeros as well.

Moreover, using the fact that $c_0 \neq 0$ one easily derives that all previous values $u(x - 1), u(x - 2), \ldots$ are zeros as well. So among every n consecutive terms of the sequence there is at least one non-zero term and the bound follows. □

For small n an essentially better estimate is known (at least for linear recurrence sequences whose characteristic polynomials have no multiple roots).

Indeed, let $a_1, \dots, a_n, \vartheta_1, \dots, \vartheta_n \in \mathbb{F}^*$ be $2n$ arbitrary non-zero elements of a field $\mathbb{F}$.

Let ρ denote the smallest integer $r > 0$ with $\vartheta_i^r = \vartheta_j^r$ for at least one pair $1 \le i < j \le n$ (and formally set $\rho = \infty$ if such i and j do not exist).

Combining the result of [506] (for $n = 2, 3, 4$) with the estimates of [531, 532] (for $n \ge 5$) we derive the following estimate (see also Section 3.3 of [508]).

The number of solutions $T_n(H)$ of the equation

$$a_1\vartheta_1^x + \dots + a_n\vartheta_n^x = 0, \qquad 1 \le x \le H,$$

does not exceed

$$T_n(H) \le 2nH(H^{-\delta_n} + \rho^{-\delta_n}), \tag{2.1}$$

where the sequence δ_n is defined as:

$$\delta_2 = 1, \qquad \delta_3 = 1/2, \qquad \delta_4 = 1/4,$$

and

$$\delta_n = \frac{\delta_{\lfloor n/2 \rfloor + 1}}{\lfloor n/2 \rfloor + 1}, \qquad n \ge 5.$$

One sees that

$$\delta_n \ge \exp(-c \log^2 n),$$

where $c > 0$ is some absolute constant.

The method of [506] has been used in [189] to obtain an upper bound on the number of solutions of exponential congruences modulo a composite.

It is useful to recall that such a linear recurrence sequence $(u(x))$ can be uniquely represented in the form

$$u(x) = \sum_{i=1}^{m} \lambda_i^x f_i(x), \tag{2.2}$$

where $\lambda_1, \dots, \lambda_m$ are the roots of its characteristic polynomial with multiplicities $k_1, \dots, k_m$, respectively, and $f_1(X), \dots, f_m(X)$ are polynomials (over an algebraic extension of $\mathbb{K}$) of degrees at most $k_1 - 1, \dots, k_m - 1$, respectively.

The inverse statement is also true: any sequence having a representation of the form (2.2) is a linear recurrence sequence of order $n = m + \deg f_1 + \dots + \deg f_m$.

In particular, a polynomial $f(X) \in \mathbb{Z}[X]$ of degree n satisfies a linear recurrence equation of order $n + 1$,

$$f(x + n + 1) = \sum_{k=0}^{n} (-1)^{n-k} \binom{n+1}{k} f(x + k), \tag{2.3}$$

with characteristic polynomial $\psi(T) = (T - 1)^{n+1}$.

The following lemma is based on very similar elementary considerations.

Lemma 2.5. *Let $f(X) \in \mathbb{F}_q[X]$ be a non-zero polynomial of degree* $\deg f \le q-2$ *and of sparsity* $\operatorname{spr} f = t \ge 1$. *Then there are at least $(q-1)/t$ values of $x \in \mathbb{F}_q^*$ with $f(x) \ne 0$.*

Proof. Let g be a primitive root of $\mathbb{F}_q$. Noticing that $f(g^x) = f(g^{x+q-1})$, we see that the number N of $x = 0, \ldots, q-2$ with $f(g^x) \ne 0$ is t times less than the number T of pairs (y,i), $y = 0, \ldots, q-2$, $i = 0, \ldots, t-1$, with $f(g^{y+i}) \ne 0$. Using properties of Vandermonde matrices and recalling that $\deg f \le q-2$, one can easily see that for any $y = 0, \ldots, q-2$, $f(g^{y+i}) \ne 0$ for at least one $i = 0, \ldots, t-1$. Thus $tN = T \ge q-1$ and the estimate follows. □

In the next lemma, obtained in [94], we obtain an essentially stronger estimate but under some conditions on the set exponents occurring in the sparse polynomial. This restricts the area of applications of the result. Nevertheless it is an essential ingredient of the proof of Lemma 3.21 and, therefore, Theorems 13.6, 13.8. This lemma has also been used

- in [190], for some coding theory problems;
- in [191], to derive first non-trivial results about the distribution of the power generator;
- in [519], to estimate the number of singular matrices of the form

$$\left(\lambda_i^{u_j}\right)_{i,j=1}^{m} \qquad 0 \le u_1, \ldots, u_m \le q-2$$

  where $\lambda_1, \ldots, \lambda_m \in \mathbb{F}_q^*$;
- in [221, 520], for certain sparse polynomial interpolations and approximation problems.

Finally, it is also used in the proof of Theorems 12.8 (see also [518, 521]). It has also been used to show that the conjecture of [224] holds for almost all values of d, see [225]. We hope it may find several other applications.

Lemma 2.6. *For $t \ge 2$ elements $a_1, \ldots, a_t \in \mathbb{F}_q^*$ and integers $\tau_1, \ldots, \tau_t \in \mathbb{Z}$ let us denote by Q the number of solutions of the equation*

$$\sum_{i=1}^{t} a_i x^{\tau_i} = 0, \qquad x \in \mathbb{F}_q^*.$$

Then

$$Q \le 2q^{1-1/(t-1)} D^{1/(t-1)} + O(q^{1-2/(t-1)} D^{2/(t-1)}),$$

where

$$D = \min_{1 \le i \le t} \max_{j \ne i} \gcd(\tau_j - \tau_i, q-1).$$

Proof. Assume that

$$D = \max_{2 \le j \le t} \gcd(\tau_j - \tau_1, q-1).$$

If $D = q-1$, then the bound is trivial. Thus we may assume that $D \le (q-1)/2$. Let g be a primitive root of $\mathbb{F}_q$. Putting $r_i = \tau_i - \tau_t$ we see that Q equals the number of solutions of the equation

$$\sum_{i=1}^{t-1} a_i g^{r_i y} + a_t = 0, \qquad 0 \le y \le q-2.$$

Put

$$L = (q-1)/D, \qquad K = \left\lceil L^{1/(t-1)} \right\rceil - 1, \qquad M = \lfloor (q-1)/K \rfloor .$$

From the Dirichlet pigeonhole principle we see that there exists l with $1 \le l \le L-1$ and such that the remainders of $s_i \equiv r_i l \pmod{q-1}$, taken in the interval $-(q-1)/2 < s_i \le (q-1)/2$, are all

$$|s_i| \le M, \qquad i = 1, \dots, t-1.$$

Indeed, for each $l = 1, \dots, L$ the corresponding vector $(s_1, \dots, s_{t-1})$ represents a point in the $(n-1)$-dimensional cube with side length $q-1$. This cube can be split into K^{t-1} cubes with side length $h = (q-1)/K$. Since $K^{t-1} < L$, then at least one subcube contains at least two vectors corresponding to some $1 \le l_1 < l_2 \le L$. Putting $l = l_2 - l_1$ we obtain the claim.

Let $d = \gcd(l, q-1)$. Now one easily verifies that for any y, $0 \le y \le q-2$, there is a unique representation of the form

$$y = dz + \nu, \qquad 0 \le z \le (q-1)/d - 1, \ 0 \le \nu \le d-1,$$

and therefore of the form

$$y \equiv lz + \nu \pmod{q-1}, \qquad 0 \le z \le (q-1)/d - 1, \ 0 \le \nu \le d-1.$$

Then

$$Q \le \sum_{\nu=0}^{d-1} Q_\nu,$$

where Q_ν, $\nu = 0, \dots, d-1$, is the number of solutions of the equation

$$\sum_{i=1}^{t-1} a_i g^{r_i(lz+\nu)} + a_t = 0, \qquad 0 \le z \le (q-1)/d - 1.$$

It is obvious that

$$Q_\nu = \frac{1}{d} R_\nu, \qquad \nu = 0, \dots, d-1,$$

where R_ν is the number of solutions of the equation

$$\sum_{i=1}^{t-1} a_i g^{r_i \nu} g^{s_i z} + a_t = 0, \qquad 0 \le z \le q-2,$$

or, equivalently, of the polynomial equation

$$\sum_{i=1}^{t-1} a_i g^{r_i \nu} x^{s_i+M} + a_t x^M = 0, \qquad x \in \mathbb{F}_q^*.$$

Using the inequality $dD < LD = q-1$, one also easily verifies that for $j = 2, \dots, t-1$,

$$s_1 - s_j \equiv r_1 l - r_j l \equiv (r_1 - r_j) l \equiv (\tau_1 - \tau_j) l \not\equiv 0 \pmod{q-1}$$

and

$$s_1 \equiv r_1 l \equiv (\tau_1 - \tau_t) l \not\equiv 0 \pmod{q-1}.$$

Therefore R_ν does not exceed the number of zeros of a non-zero polynomial (in particular it contains x^{s_1+M} with a non-zero coefficient) of degree at most $2M$. Hence

$$Q \le \sum_{\nu=0}^{d-1} Q_\nu = \frac{1}{d} \sum_{\nu=0}^{d-1} R_\nu \le 2M \tag{2.4}$$

$$2M \le 2q^{1-1/(t-1)} D^{1/(t-1)} + O(q^{1-2/(t-1)} D^{2/(t-1)})$$

and the bound follows. □

In a more explicit form the bound of Lemma 2.6 can be written down as

$$Q \le 3(q-1)^{1-1/(t-1)} D^{1/(t-1)}. \tag{2.5}$$

Indeed, if $L \le 3^{t-1}$, then we obviously have

$$Q \le q-1 \le 3(q-1)L^{-1/(t-1)} \le 3(q-1)^{1-1/(t-1)} D^{1/(t-1)}.$$

Otherwise

$$K \ge \frac{2}{3} L^{1/(t-1)}$$

and from (2.4) we derive

$$Q \le 2M \le 2(q-1)/K \le 3(q-1)L^{-1/(t-1)} \le 3(q-1)^{1-1/(t-1)} D^{1/(t-1)}.$$

In [473, 474] the number of zeros of a real polynomial has been estimated in terms of its additive complexity; see also [88, 237, 293, 479, 480]. We present that estimate in the following form.

Lemma 2.7. *For a non-zero polynomial* $f(X) \in \mathbb{R}[X]$ *having* N *real zeros the bound*

$$C_{\pm}(f) \geq \left(\frac{1}{5}\log N\right)^{1/2}$$

holds.

In [479], using [480],the following sharper bound has been given on the number of rational roots.

Lemma 2.8. *For a non-zero polynomial* $f(X) \in \mathbb{Z}[X]$ *having* N *rational zeros the bound*

$$\log N \ll C_{\pm}(f)\log C_{\pm}(f)$$

holds.

We also need some results about vanishing linear combinations of polynomials and rational functions.

Let us consider the following sequence of rational functions over $\mathbb{F}_p$:

$$H_0(X) = X, \qquad H_i(X) = H_{i-1}\left(aX^{-1} + b\right), \quad i = 1, 2, \ldots .$$

It is obvious that this sequence is purely periodic. Denote by T the smallest period.

Lemma 2.9. *If for some* $a_0, \ldots, a_k \in \mathbb{F}_q$ *we have an identity*

$$H_0(X) = a_0 + \sum_{i=1}^{k} a_i H_i(X),$$

then $k \geq T$.

Proof. The result is immediate for $b = 0$, so we can assume that $b \neq 0$. We see that $H_i(X) = f_i(X)/g_i(X)$ are non-constant rational functions, where $f_i, g_i \in \mathbb{F}_p[X]$ with $\max\{\deg f_i, \deg g_i\} = 1$. Now we write $g_i(X) = c_iX + d_i$ with $c_i, d_i \in \mathbb{F}_q$. Clearing denominators in the identity of the lemma, we have:

$$H_0(X)\prod_{i=1}^{k}(c_iX + d_i) = a_0\prod_{i=1}^{k}(c_iX + d_i) + \sum_{i=1}^{k} a_i f_i(X)\prod_{\substack{j=1\\ j\neq i}}^{k}(c_jX + d_j).$$

Comparing the coefficient of X^{k+1} of the left-hand side and the right-hand side, we see that $c_1 \ldots c_k = 0$. Therefore $c_i = 0$ for some $1 \leq i \leq k$. It follows by straightforward induction that

$$H_i(X) = \frac{(bc_i + d_i)X + ac_i}{c_iX + d_i},$$

and so $c_i = 0$ implies that $H_i(X) = X$. Therefore $k \geq i \geq T$. □

In [247] Lemma 2.9 has been extended to polynomial relations

$$H_k(X) = \Psi\left(H_{k-1}(X), \dots, H_0(X)\right)$$

with $\Psi(T_{k-1}, \dots, T_0) \in \mathbb{F}_q[T_{k-1}, \dots, T_0]$.

Lemma 2.10. *Let $f(X) \in \mathbb{Z}[X]$ be a polynomial whose degree satisfies $p > \deg f \geq 3$, and whose leading coefficient does not vanish modulo p. Suppose that integers s and m satisfy the inequality*

$$s(m+1) < \log p.$$

Let $L = 2^s - 1$ and $e_0 = 0$, $e_i = 2^{si}$, $i = 1, \dots, m$. Then for any vector $\mathbf{a} \in \mathbb{Z}^{m+1}$, with $0 < |\mathbf{a}| \leq L$, where $|\mathbf{a}|$ is defined by (1.8) with $N = m+1$, the linear combination

$$F(X) = \sum_{i=0}^{m} a_i f(X + e_i)$$

is neither a constant nor a linear function modulo p.

Proof. Without loss of generality we may assume that $a_m \neq 0$. Let $n = \deg f$ and $f(X) = c_n X^n + c_{n-1} X^{n-1} + \ldots + c_0$. Then the leading coefficient of F equals

$$C_n = c_n \sum_{i=0}^{m} a_i,$$

and the second leading coefficient of $F(X)$ equals

$$C_{n-1} = n c_n \sum_{i=0}^{m} a_i e_i + c_{n-1} \sum_{i=0}^{m} a_i.$$

By the condition of the theorem we have $c_n \not\equiv 0 \pmod p$. Hence if $C_n = 0$, then

$$\sum_{i=0}^{m} a_i = 0 \pmod p$$

and

$$C_{n-1} \equiv n c_n \sum_{i=0}^{m} a_i e_i \pmod p.$$

One easily verifies that

$$\left| \sum_{i=0}^{m} a_i e_i \right| \leq L \sum_{i=1}^{m} e_i < p,$$

and that

$$\left| \sum_{i=0}^{m} a_i e_i \right| \geq e_m - L \sum_{i=1}^{m-1} e_i > 0.$$

Therefore $C_{n-1} \not\equiv 0 \pmod p$, thus F is of degree at least $n - 1 \geq 2$ modulo p. □

Finally we need a similar result for rational functions.

Lemma 2.11. *Let $f(X), g(X) \in \mathbb{Z}[X]$ be polynomials such that the rational function $h(X) = f(X)/g(X)$ is not a polynomial modulo p. Then for any integer $k < \log p$ there exist at least*

$$M \geq k - \frac{\deg g(\deg g - 1)}{2}$$

integers

$$0 \leq s_i \leq k, \qquad i = 1, \ldots, M,$$

such that any non-trivial modulo p linear combination

$$H(X) = \sum_{i=0}^{M} a_i h(X + e_i),$$

where $e_0 = 0$, $e_i = 2^{s_i}$, $i = 1, \ldots, M$, is neither a constant nor a linear function modulo p.

Proof. From the condition of the theorem we conclude that $g(X)$ is not constant modulo p. Obviously it is enough to show that there exist at least M integers $0 \leq s_1 \leq \ldots \leq s_M \leq k$ such that the polynomials $g(X + e_i)$, $i = 0, \ldots, M$, are pairwise relatively prime in the residue ring modulo p.

Let $\lambda_1, \ldots, \lambda_N$ be the $N \leq \deg g$ distinct roots of g in the algebraic closure of $\mathbb{Q}$.

Let D be the set of integers δ, $0 \leq \delta \leq p - 1$, such that

$$\lambda_l - \lambda_r \equiv \delta \pmod p.$$

for some pair (l, r), with $1 \leq l < r \leq N$.

We define the 2-adic order $\nu_2(\delta)$ of an integer δ as the largest power of 2 which divides δ, that is, $\nu_2(\delta) = \nu$ if and only if

$$\delta \equiv 0 \pmod{2^\nu}, \qquad \delta \not\equiv 0 \pmod{2^{\nu+1}},$$

and put

$$T = \{\nu_2(\delta) \mid \delta \in D\}.$$

We define the sequence $s_1, \ldots, s_M$ by the relation

$$\{s_1, \ldots, s_M\} = \{1, \ldots, k\} \setminus T.$$

It is obvious that if

$$\lambda_l - \lambda_r \equiv 2^{s_i} - 2^{s_j} \pmod p$$

or

$$\lambda_l - \lambda_r \equiv 2^{s_i} \pmod{p}$$

for some pair (l, r) with $1 \le l < r \le N$ and some pair (i, j) with $0 \le i, j \le M$, $i \ne j$, then

$$\delta \equiv 2^{s_i} - 2^{s_j} \pmod{p}$$

or

$$\delta \equiv 2^{s_i} \pmod{p}$$

for some $\delta \in D$. Because $k < \log p$ these congruences imply that

$$\delta = 2^{s_i} - 2^{s_j}$$

or

$$\delta = 2^{s_i}$$

which is impossible because of the choice of the sequence $s_1, \dots, s_M$.

Thus the corresponding polynomials $g(x + e_i)$, $i = 0, \dots, M$, are pairwise relatively prime in the residue ring modulo p. It is easy to see that $M \ge k - |T| \ge k - N(N-1)/2$. □

Chapter 3

Exponential Sums

We need some bounds of exponential sums as well.

To simplify our formulas we always assume that $q \geq 7$ (or in the case of prime fields, that $p \geq 7$). In particular, in this case we have the inequality

$$\ln(q-1) + 1 \leq \log q. \tag{3.1}$$

First of all we recall the *Weil bound* in its classical form given in Example 12 of Appendix 5 of [557]; see also Theorem 3 of Chapter 6 in [341] and Theorem 5.41 and comments to Chapter 5 of [343].

Lemma 3.1. *Let χ be a non-trivial multiplicative character modulo p of order d. For any polynomial $f(X) \in \mathbb{Z}[X]$ of degree m and any $n \geq 0$ linear forms $a_i x + b_i$ with $a_i \neq 0$ and $b_i/a_i \not\equiv b_j/a_j \pmod p$, $i, j = 1 \ldots, n$, $i \neq j$, the bound*

$$\left| \sum_{x=1}^{p} \chi\left((a_1 x + b_1)^{e_1} \ldots (a_n x + b_n)^{e_n}\right) \mathbf{e}_p\left(f(x)\right) \right| \leq (m+n) p^{1/2}$$

holds provided that either f is not a constant polynomial modulo p or, if $n \geq 1$, at least one of the integers $e_1, \ldots, e_n$ is not divisible by d.

Similar results for rational functions are known as well. For example, we make use of the following estimate which can be found in [397].

Lemma 3.2. *For any polynomials $g(X), h(X) \in \mathbb{F}_p[X]$ such that the rational function $F(X) = h(X)/g(X)$ is not constant on $\mathbb{F}_p$, the bound*

$$\left| \sum_{\substack{x \in \mathbb{F}_p \\ g(x) \neq 0}} \mathbf{e}_p\left(F(x)\right) \right| \leq \left(\max\{\deg g, \deg h\} + r - 2\right) p^{1/2} + \delta$$

holds, where

$$(r,\delta) = \begin{cases} (s,1), & \text{if } \deg h \le \deg g, \\ (s+1,0), & \text{if } \deg h > \deg g, \end{cases}$$

and s *is the number of distinct zeros of* $g(X)$ *in the algebraic closure of* $\mathbb{F}_p$.

We repeatedly use the following well-known elementary facts about exponential sums with linear functions.

First of all we need the identity (see Exercise 11.a in Chapter 3 of [553])

$$\sum_{c=0}^{m-1} \mathbf{e}_m(cu) = \begin{cases} 0, & \text{if } u \not\equiv 0 \pmod m, \\ m, & \text{if } u \equiv 0 \pmod m, \end{cases} \tag{3.2}$$

and the inequality (see Exercise 11.b in Chapter 3 of [553])

$$\left| \sum_{u=1}^{H} \mathbf{e}_p(au) \right| \le \frac{p}{2|a|}, \tag{3.3}$$

which holds for any integer a with $1 \le |a| \le (p-1)/2$ and any integer H. Both of these statements follow from the formula for the sum of a geometric progression.

Now, from (3.3) one easily derives the estimate

$$\sum_{c=0}^{m-1} \left| \sum_{y=H+1}^{H+N} \mathbf{e}_m(cy) \right| \le m(\ln m + 1) \tag{3.4}$$

which holds for any integers $m \ge 2$, H and $1 \le N \le m$ (see Exercise 11.c in Chapter 3 of [553]).

The next four statements are slightly more general forms of the Weil bound as it is usually known. They are combinations of Lemma 3.1 and Lemma 3.2 with the standard method of estimation of incomplete sums via complete ones [107, 279, 553] (which makes use of the bound (3.4)).

We present a proof of Lemma 3.3 only; the other three can be obtained quite similarly (and apparently can be found in the literature as well).

Lemma 3.3. *Let* g *be a primitive root of* $\mathbb{F}_q$. *For any non-trivial multiplicative character* χ *of* $\mathbb{F}_q$ *of order* d *and any* $n \ge 1$ *integers* $c_0, c_1, \ldots, c_n$ *which are not all divisible by* d, *the bound*

$$\left| \sum_{x=N+1}^{N+H} \chi\left(\prod_{i=0}^{n} (a_i g^x + b_i)^{c_i} \right) \right| \le (n+1) q^{1/2} \log q$$

holds for any integers N *and* $H \le q-1$ *and any linear forms* $a_i x + b_i$ *with* $a_i, b_i \ne 0$ *and* $b_i/a_i \not\equiv b_j/a_j \pmod p$, $i,j = 0,\ldots,n$, $i \ne j$.

Proof. From (3.2) we obtain

$$\begin{aligned}
&\sum_{x=N+1}^{N+H} \chi\left(\prod_{i=0}^{n}(a_i g^x + b_i)^{c_i}\right) \\
&\quad = \frac{1}{q-1}\sum_{x=0}^{q-2} \chi\left(\prod_{i=0}^{n}(a_i g^x + b_i)^{c_i}\right) \sum_{c=0}^{q-2}\sum_{y=N+1}^{N+H} \psi(c(x-y)) \\
&\quad = \frac{1}{q-1}\sum_{c=0}^{q-2}\sum_{x=0}^{q-2} \chi\left(\prod_{i=0}^{n}(a_i g^x + b_i)^{c_i}\right) \psi(cx) \sum_{y=N+1}^{N+H} \psi(-cy),
\end{aligned}$$

where

$$\psi(x) = \mathbf{e}_{q-1}(x).$$

Let

$$\rho(x) = \mathbf{e}_{q-1}(\operatorname{ind} x)$$

be a primitive multiplicative character of $\mathbb{F}_q$, where $\operatorname{ind} x$ means the discrete logarithm of $x \in \mathbb{F}_q$ to base g. Therefore χ is some power of ρ,

$$\chi(u) = \rho(u)^e,$$

where the exponent e satisfies

$$1 \le e \le q-2, \qquad \gcd(e, q-1) = (q-1)/d,$$

and

$$\psi(x) = \rho(g^x).$$

Applying the Weil bound in the form of Theorem 5.41 of [343] (which is a modification of Lemma 3.1 for the case of arbitrary finite fields) we obtain

$$\begin{aligned}
\left|\sum_{x=0}^{q-2} \chi\left(\prod_{i=0}^{n}(a_i g^x + b_i)^{c_i}\right) \rho(g^{cx})\right| &= \left|\sum_{x=0}^{q-2} \rho\left(g^{xc}\prod_{i=0}^{n}(a_i g^x + b_i)^{ec_i}\right)\right| \\
&= \left|\sum_{x\in\mathbb{F}_q} \rho\left(x^c\prod_{i=0}^{n}(a_i x + b_i)^{ec_i}\right)\right| \\
&\le (n+1)q^{1/2}
\end{aligned}$$

(one easily verifies that the condition of that theorem is satisfied). Taking into account the inequality (3.1), from (3.4) we derive the desired bound. □

In a similar way, from Lemma 3.1 we derive the following two statements.

Lemma 3.4. *For any non-trivial multiplicative character χ modulo p of order d and any $n \geq 1$ integers $e_1, \dots, e_n$ which are not all divisible by d, the bound*

$$\left| \sum_{x=N+1}^{N+H} \chi\left((a_1x+b_1)^{e_1} \dots (a_nx+b_n)^{e_n}\right) \right| \leq np^{1/2} \log p$$

holds for any integers N and $H \leq p$ and any linear forms $a_ix + b_i$ with $a_i \not\equiv 0$ and $b_i/a_i \not\equiv b_j/a_j \pmod p$, $i, j = 1, \dots, n$, $i \neq j$.

Lemma 3.5. *For any non-trivial multiplicative character χ modulo p and any polynomial $f(X) \in \mathbb{Z}[X]$ of degree $n = \deg f \geq 1$, the bound*

$$\left| \sum_{x=N+1}^{N+H} \chi(x)\, \mathbf{e}_p\left(f(x)\right) \right| \leq np^{1/2} \log p$$

holds for any integers N and $H \leq p$.

Accordingly, using Lemma 3.2 we also obtain a more general bound.

Lemma 3.6. *For any polynomials $f(X), g(X) \in \mathbb{Z}[X]$ such that the rational function $h(X) = f(X)/g(X)$ is neither a constant nor a linear function modulo p, the bound*

$$\left| \sum_{\substack{x=N+1 \\ g(x) \not\equiv 0 \pmod p}}^{N+H} \mathbf{e}_p\left(h(x)\right) \right| \leq \left(\max\{\deg f,\, \deg g\} + s\right) p^{1/2} \log p$$

holds for any integers N and $H \leq p$, where s is the number of distinct zeros of $g(X)$ in the algebraic closure of $\mathbb{F}_p$.

For a divisor $s|m$, we denote by $\mathcal{N}_{m,s}(p)$ the multiplicative group of elements $z \in \mathbb{F}_{p^m}^*$ with the norm in $\mathbb{F}_{p^{m/s}}$ equal to 1, that is,

$$\mathcal{N}_{m,s}(p) = \left\{ z \in \mathbb{F}_{p^m}^* \mid \mathrm{Nm}_{\mathbb{F}_{p^m}/\mathbb{F}_{p^{m/s}}}(z) = 1 \right\}. \tag{3.5}$$

Thus

$$\mathcal{N}_{m,s}(p) = \frac{p^m - 1}{p^{m/s} - 1}.$$

The results of [342] depend on the following estimate conjectured by Deligne in [150] (and proved in the case of the trivial character $\chi = \chi_0$). In the full generality it has been proved by Katz [292] (the proof follows from Theorem 4.1.1 of [292] after some standard transformations). In fact, for the cases relevant to the XTR and LUC key exchange protocols, simpler and more explicit statements are given in [340] and in Chapter 6 of [341].

Lemma 3.7. *For any divisor $s|m$, any $\gamma \in \mathbb{F}_{p^m}^*$, and any multiplicative character χ of $\mathbb{F}_{p^m}$, the bound*

$$\left| \sum_{t \in \mathcal{N}_{m,s}(p)} \chi(t)\, \mathbf{e}_p\left(\operatorname{Tr}_{\mathbb{F}_{p^m}/\mathbb{F}_p}(\gamma t)\right) \right| \le s p^{(m-m/s)/2}$$

holds.

We recall the following standard property of the group of characters of an abelian group. In fact, the identity (3.2) is a special partial case of the next statement.

Lemma 3.8. *Let $\mathcal{H}$ be an abelian group and let $\widehat{\mathcal{H}} = \mathrm{Hom}(\mathcal{H}, \mathbb{C}^*)$ be its dual group. Then for any character χ of $\mathcal{H}$, we have*

$$\frac{1}{|\widehat{\mathcal{H}}|} \sum_{\chi \in \widehat{\mathcal{H}}} \chi(h) = \begin{cases} 1, & \text{if } h = \mathrm{id}, \\ 0, & \text{if } h \ne \mathrm{id}, \end{cases}$$

where id *is the identity element of $\mathcal{H}$.*

Lemma 3.9. *For any divisor $s|m$, any subgroup $\mathcal{G}$ of $\mathcal{N}_{m,s}(p)$ and any $\gamma \in \mathbb{F}_{p^m}^*$, the bound*

$$\left| \sum_{t \in \mathcal{G}} \mathbf{e}_p\left(\operatorname{Tr}_{\mathbb{F}_{p^m}/\mathbb{F}_p}(\gamma t)\right) \right| \le s p^{(m-m/s)/2}$$

holds.

Proof. Let $\Omega_{\mathcal{G}}$ be the set of all multiplicative characters of $\mathcal{N}_{m,s}(p)$ trivial on $\mathcal{G}$. Using Lemma 3.8, we write

$$\begin{aligned} \sum_{t \in \mathcal{G}} \mathbf{e}_p\left(\operatorname{Tr}_{\mathbb{F}_{p^m}/\mathbb{F}_p}(\gamma t)\right) &= \frac{1}{|\Omega_{\mathcal{G}}|} \sum_{t \in \mathcal{N}_{m,s}(p)} \mathbf{e}_p\left(\operatorname{Tr}_{\mathbb{F}_{p^m}/\mathbb{F}_p}(\gamma t)\right) \sum_{\chi \in \Omega_{\mathcal{G}}} \chi(t) \\ &= \frac{1}{|\Omega_{\mathcal{G}}|} \sum_{\chi \in \Omega_{\mathcal{G}}} \sum_{t \in \mathcal{N}_{m,s}(p)} \chi(t)\, \mathbf{e}_p\left(\operatorname{Tr}_{\mathbb{F}_{p^m}/\mathbb{F}_p}(\gamma t)\right). \end{aligned}$$

Applying the inequality of Lemma 3.7, we obtain the desired estimate. □

A combination of the bounds of Lemmas 3.9, 3.15 and 3.16 has been used in [526] to obtain new bounds of Gauss sums in finite fields.

We also recall the *Burgess bound*, see Theorem 2.3 of [407].

Lemma 3.10. *For any $\varepsilon > 0$ there exists $\gamma > 0$ such that for any non-trivial multiplicative character χ modulo p*

$$\left| \sum_{x=N+1}^{N+H} \chi(x) \right| \le H p^{-\gamma}$$

for $H \ge p^{1/4+\varepsilon}$ and sufficiently large p.

It is known that the *Extended Riemann Hypothesis* implies non-trivial upper bounds for much shorter sums. In particular, we use a bound which follows from one of the results of [230].

Lemma 3.11. *Let*

$$v = \frac{\ln H}{\ln \ln p} \to \infty.$$

Then, assuming the Extended Riemann Hypothesis, the bound

$$\left| \sum_{x=N+1}^{N+H} \chi(x) \right| \le H v^{-v/2+o(v)}$$

holds.

Proof. We recall that an integer $n \ge 1$ is called Y-smooth if all primes dividing it are $\le Y$. As before, let $\Psi(X,Y)$ denote the total number of Y-smooth integers which do not exceed X. It has been proved in Theorem 2 of [230] that

$$\sum_{x=N+1}^{N+H} \chi(x) = O\left(\Psi\left(H, \ln^2 p \ln^{20} \ln p\right)\right),$$

provided that $u \to \infty$. One easily verifies that the bound (5.7) can be applied to the last function with $u = v/2 + o(v)$, yielding the desired result. □

The following estimate is a combination of Lemma 3.5 (or Lemma 3.6) and the bound (3.3).

Lemma 3.12. *Let $p \ge 3$ be a prime, $\rho = (1-p)/2$ and let integers N, m, λ and L satisfy the conditions*

$$1 \le L < \lambda, \qquad 2^{m+1} L \lambda^N < p.$$

Then for any integer vector $\mathbf{a} = (a_1, \dots, a_N) \in \mathbb{Z}^N$, with $0 < |\mathbf{a}| \le L$, where $|\mathbf{a}|$ is defined by (1.8), the bound

$$\left| \sum_{x=1}^{H} \mathbf{e}_p \left(a_1 \rho 2^{2m} x^2 + \sum_{i=1}^{N-1} a_{i+1} \rho (2^m x + \lambda^i)^2 \right) \right|$$
$$\ll \max\{p^{1/2} \log p,\ p/2^m L\lambda\}$$

holds.

Proof. If

$$\sum_{i=1}^{N} a_i \not\equiv 0 \pmod p,$$

then the estimate follows from Lemmas 3.5 or 3.6.

Assume that

$$\sum_{i=1}^{N} a_i \equiv 0 \pmod p; \tag{3.6}$$

then

$$\sum_{x=1}^{H} \mathbf{e}_p \left(a_1 \rho 2^{2m} x^2 + \sum_{i=1}^{N-1} a_{i+1} \rho (2^m x + \lambda^i)^2 \right) = \sum_{x=1}^{H} \mathbf{e}_p (Ax + B),$$

where

$$A = 2^m \sum_{i=1}^{N-1} a_{i+1} \lambda^i \equiv \rho 2^{m+1} \sum_{i=1}^{N-1} a_{i+1} \lambda^i \pmod p, \quad B = \rho \sum_{i=1}^{N-1} a_{i+1} \lambda^{2i}.$$

We see that

$$|a_1 + \ldots + a_N| \le NL < p$$

for $|\mathbf{a}| \le L$. Thus the congruence (3.6) implies the equality $a_1 + \ldots + a_N = 0$. Hence, there exists at least one i with $N \ge i \ge 2$ such that $a_i \ne 0$. Therefore,

$$2^m L\lambda \le |A| \le 2^m L \lambda^N \le p/2.$$

From the estimate (3.3) the desired result follows. □

We need various estimates of exponential sums with exponential functions from [306, 308, 309, 423, 426] (as well as their simple modifications).

Lemma 3.13. *Let g and $m \ge 1$ be integers with* $\gcd(g, m) = 1$ *and let T be the order of g modulo m. Then for any integer c we have*

$$\left| \sum_{x=1}^{T} \mathbf{e}_m (cg^x) \right| \le \delta^{1/2} m^{1/2},$$

where $\gcd(c, m) = \delta$.

Proof. If $\delta = 1$, then this bound is essentially Theorem 10 of Chapter 1 of [309]; see also the proof of Lemma 2 of [308]. For $\delta > 1$, we denote by T_μ the order of g modulo $\mu = m/\delta$, and we also put $c/\delta = \gamma$. Thus $\gcd(\gamma, \mu) = 1$ and we obtain

$$\left| \sum_{x=1}^{T} \mathbf{e}_m (cg^x) \right| \le \frac{T}{T_\mu} \left| \sum_{x=1}^{T} \mathbf{e}_m (\gamma g^x) \right| \le \frac{T}{T_\mu} \mu^{1/2}.$$

To finish the proof, we remark that

$$T \le \frac{\varphi(m)}{\varphi(\mu)} T_\mu \le \frac{m}{\mu} T_\mu = \delta T_\mu,$$

by Lemma 5.3. □

Let $g \in \mathbb{F}_p^*$ be of multiplicative order T modulo p. In the next several statements we estimate exponential sums

$$S(a,c) = \sum_{x=0}^{T-1} \mathbf{e}_T(ax)\,\mathbf{e}_p(cg^x). \tag{3.7}$$

For $a = 0$ these sums have been estimated in [306]. Accordingly we follow very closely the arguments of that work.

For an integer $m \geq 1$ we denote by N_m the number of solutions of the congruence

$$g^{x_1} + \ldots + g^{x_m} \equiv g^{y_1} + \ldots + g^{y_m} \pmod p,$$

where $0 \leq x_1, \ldots, x_m, y_1, \ldots, y_m \leq T-1$.

Lemma 3.14. *For any integers $m, l \geq 1$ and any integers a and c with* $\gcd(c,p) = 1$ *for sums $S(a,c)$ given by (3.7), the inequality*

$$|S(a,c)| \leq p^{1/2ml} N_m^{1/2ml} N_l^{1/2ml} T^{1-1/m-1/l}$$

holds.

Proof. For an integer $l \geq 1$ and $\lambda \in \mathbb{F}_p$ denote by $\mathcal{Q}_l(\lambda)$ the set

$$\mathcal{Q}_l(\lambda) = \left\{(x_1, \ldots, x_l) \in [0, T-1]^l \mid g^{x_1} + \ldots + g^{x_l} \equiv \lambda \pmod p\right\}$$

and define the exponential sum

$$Q_l(a,\lambda) = \sum_{(x_1,\ldots,x_l) \in \mathcal{Q}_l(\lambda)} \mathbf{e}_T\left(a(x_1 + \ldots + x_l)\right).$$

One easily verifies that for any integer u,

$$\begin{aligned} Q_l(a, g^u\lambda) &= \sum_{(x_1,\ldots,x_l) \in \mathcal{Q}_l(g^u\lambda)} \mathbf{e}_T\left(a(x_1 + \ldots + x_l)\right) \\ &= \sum_{(x_1,\ldots,x_l) \in \mathcal{Q}_l(\lambda)} \mathbf{e}_T\left(a((x_1+u) + \ldots + (x_l+u))\right) \\ &= \mathbf{e}_T(alu) Q_l(a,\lambda). \end{aligned}$$

Thus

$$Q_l(a,\lambda) = Q_l(a, g^u\lambda)\,\mathbf{e}_T(-alu), \qquad u = 0, \ldots, T-1.$$

Obviously,

$$\sum_{\lambda \in \mathbb{F}_p} |Q_l(a,\lambda)|^2 \leq \sum_{\lambda \in \mathbb{F}_p} Q_l(0,\lambda)^2 = N_l$$

and

$$\sum_{\lambda \in \mathbb{F}_p} |Q_l(a, \lambda)| \le \sum_{\lambda \in \mathbb{F}_p} Q_l(0, \lambda) = T^l.$$

Therefore,

$$\begin{aligned} |S(a,c)|^m &= \left| \sum_{\lambda \in \mathbb{F}_p} Q_m(a,\lambda)\, \mathbf{e}_p(c\lambda) \right| \\ &= \frac{1}{T} \left| \sum_{u=0}^{T-1} \sum_{\lambda \in \mathbb{F}_p} Q_m(a, \lambda g^u)\, \mathbf{e}_p(c\lambda g^u)\, \mathbf{e}_T(-alu) \right| \\ &\le \frac{1}{T} \sum_{\lambda \in \mathbb{F}_p} |Q_m(a,\lambda)| \left| \sum_{u=0}^{T-1} \mathbf{e}_p(c\lambda g^u)\, \mathbf{e}_T(-alu) \right| . \end{aligned}$$

Hence, from the Hölder inequality we obtain

$$|S(a,c)|^{2ml} \le \frac{1}{T^{2l}} \left[\sum_{\lambda \in \mathbb{F}_p} |Q_m(a,\lambda)|^{2l/(2l-1)} \right]^{2l-1} \sum_{\lambda \in \mathbb{F}_p} \left| \sum_{u=0}^{T-1} \mathbf{e}_p(c\lambda g^u)\, \mathbf{e}_T(-alu) \right|^{2l} .$$

For the last sum we have

$$\begin{aligned} &\sum_{\lambda \in \mathbb{F}_p} \left| \sum_{u=0}^{T-1} \mathbf{e}_p(c\lambda g^u)\, \mathbf{e}_T(-alu) \right|^{2l} \\ &\quad = \sum_{0 \le u_1, \ldots, u_l, v_1, \ldots, v_l \le T-1} \mathbf{e}_T(-al(u_1 + \ldots + u_l - v_1 - \ldots - v_l)) \\ &\qquad \times \sum_{\lambda \in \mathbb{F}_p} \mathbf{e}_p\left(c\lambda\left(g^{u_1} + \ldots + g^{u_l} - g^{v_1} - \ldots - g^{v_l}\right)\right) \le pN_l \end{aligned}$$

because the sum over λ vanishes if

$$g^{u_1} + \ldots + g^{u_l} \not\equiv g^{v_1} + \ldots + g^{v_l} \pmod p$$

and is equal to p otherwise. Therefore, applying the Hölder inequality again, we derive

$$\begin{aligned} |S(a,c)|^{2ml} &= \frac{pN_l}{T^{2l}} \left[\sum_{\lambda \in \mathbb{F}_p} |Q_m(a,\lambda)|^{2l/(2l-1)} \right]^{2l-1} \\ &\le \frac{pN_l}{T^{2l}} \left[\sum_{\lambda \in \mathbb{F}_p} |Q_m(a,\lambda)| \right]^{2l-2} \sum_{\lambda \in \mathbb{F}_p} |Q_m(a,\lambda)|^2 \\ &\le pN_m N_l T^{2ml-2m-2l}, \end{aligned}$$

and the desired estimate follows. □

Now we can use Lemma 3.3 of [306], see also [259], in the same way as in the proof of Theorem 3.4 of [306], to estimate the sums $S(a,c)$.

Lemma 3.15. *Let $g \in \mathbb{F}_p^*$ be of multiplicative order T modulo p. Then for sums $S(a,c)$ given by (3.7) the bound*

$$\max_{0\le a\le T-1}\ \max_{\gcd(c,p)=1} |S(a,c)| \ll B(T,p),$$

holds, where

$$B(T,p) = \begin{cases} p^{1/2}, & \text{if } T \ge p^{2/3}; \\ p^{1/4}T^{3/8}, & \text{if } p^{2/3} > T \ge p^{1/2}; \\ p^{1/8}T^{5/8}, & \text{if } p^{1/2} > T \ge p^{1/3}. \end{cases} \tag{3.8}$$

Proof. The bound

$$\max_{0\le a\le T-1}\ \max_{\gcd(c,p)=1} |S(a,c)| \le p^{1/2}$$

is well known, see Lemma 3.13. It also follows from the first inequality of Lemma 3.14 if one takes $m = l = 1$ and uses that $N_1 = T$. We may now assume that $T < 0.7p^{2/3}$ because otherwise the first estimate is stronger than the other two. For such T we have the estimate

$$N_2 \ll T^{5/2},$$

see [259] and Lemma 3.3 of [306]. Substituting this bound in the inequalities of Lemma 3.14 with $(m,l) = (2,1)$ and with $(m,l) = (2,2)$ we obtain the desired statement. □

We remark that the implicit constant in the first estimate is just 1, see Lemma 3.13. The bound of Lemma 3.15 is non-trivial for $T \ge p^{1/3+\varepsilon}$ with any fixed $\varepsilon > 0$. Namely for any $\varepsilon > 0$ there exists some $\delta > 0$ such that

$$\max_{0\le a\le T-1}\ \max_{\gcd(c,p)=1} |S(a,c)| = O\left(Tp^{-\delta}\right).$$

This result has recently been improved in [304], where the same bound is established already for $T \ge p^{1/4+\varepsilon}$. Accordingly one can get the corresponding improvements to all the results which are based on Lemma 3.15.

Our next result shows that for almost all primes p even much shorter sums admit a non-trivial estimate. It readily follows from Theorem 5.5 of [306].

Lemma 3.16. *Let Q be a sufficiently large integer. For any $\varepsilon > 0$ there exists $\delta > 0$ such that for all primes $p \in [Q, 2Q]$, except at most $Q^{5/6+\varepsilon}$ of them, and any element $g \in \mathbb{F}_p$ of multiplicative order $T \ge p^{\varepsilon}$ the bound*

$$\max_{\gcd(c,p)=1} \left| \sum_{x=0}^{T-1} \mathbf{e}_p\left(cg^x\right) \right| \le T^{1-\delta}$$

holds.

Proof. For each integer $T \geq 1$ and for each prime $p \equiv 1 \pmod T$ we fix an element $g_{p,T}$ of multiplicative order T. Then Theorem 5.5 of [306] claims that for any $U > 1$ and any integer $\nu \geq 2$, for all primes $p \equiv 1 \pmod T$ except at most $O(U/\log U)$ of them, the bound

$$\max_{\gcd(c,p)=1} \left| \sum_{x=0}^{T-1} \mathbf{e}_p\left(c g_{p,T}^x\right) \right| = O\left(T p^{1/2\nu^2}\left(T^{-1/\nu} + U^{-1/\nu^2}\right)\right),$$

holds. We remark that the value of the above exponential sum does not depend on the particular choice of the element $g_{p,T}$.

Taking

$$\nu = \left\lfloor \frac{1}{\varepsilon} \right\rfloor + 1 \qquad \text{and} \qquad U = Q^{1/2+\varepsilon/3},$$

after simple computation we obtain that there exists some $\delta > 0$, depending only on ε, such that for any fixed $T \geq Q^\varepsilon$ the bound

$$\max_{\gcd(c,p)=1} \left| \sum_{x=0}^{T-1} \mathbf{e}_p\left(c g_{p,T}^x\right) \right| \leq T^{1-\delta},$$

holds for all except $O(Q^{1/2+\varepsilon/3})$ primes $p \equiv 1 \pmod T$ in the interval $p \in [Q, 2Q]$. As it follows from Lemma 3.15, a similar bound also holds for $T \geq Q^{1/3+\varepsilon/3}$. So the total number of exceptional primes p for which the bound of the lemma does not hold for at least one $T \geq p^\varepsilon \geq Q^\varepsilon$ is $O\left(Q^{5/6+2\varepsilon/3}\right)$. Thus for sufficiently large Q we obtain the desired result. □

This immediately implies that the size of the exceptional set in Lemma 3.16 and all its applications can be reduced to $Q^{3/4+\varepsilon}$.

Combining Lemmas 3.15 and 3.16 with the identity

$$\sum_{u \in \mathbb{F}_p^*} \mathbf{e}_p\left(c u^n\right) = \frac{p-1}{T} \sum_{x=0}^{T-1} \mathbf{e}_p\left(c g^x\right),$$

where $g \in \mathbb{F}_p^*$ is of multiplicative order

$$T = \frac{p-1}{\gcd(p-1, n)} \geq (p-1)/n,$$

we obtain the following bound of *Gauss sums*.

Lemma 3.17. *Let Q be a sufficiently large integer. The following statement holds with $\vartheta = 1/3$ for all primes $p \in [Q, 2Q]$, and with $\vartheta = 0$ for all primes $p \in [Q, 2Q]$*

except at most $Q^{5/6+\varepsilon}$ *of them. For any* $\varepsilon > 0$ *there exists* $\delta > 0$ *such that for* $n \le p^{1-\vartheta-\varepsilon}$ *the bound*

$$\max_{\gcd(c,p)=1} \left| \sum_{u \in \mathbb{F}_p^*} \mathbf{e}_p(cu^n) \right| \le p^{1-\delta}.$$

holds.

We also need several more bounds of exponential sums which follows from Lemmas 3.13 and 3.15.

Lemma 3.18. *Let* $g \in \mathbb{F}_p^*$ *be of multiplicative order* T *modulo* p. *Then for any* b *with* $\gcd(b, p-1) = 1$ *the bound*

$$\max_{\gcd(c,p)=1} \left| \sum_{\substack{x=0 \\ \gcd(bx+1,p-1)=1}}^{p-1} \mathbf{e}_p(cg^x) \right| = O\left(2^{\nu(p-1)} p^{3/2} T^{-1} \log p\right)$$

holds.

Proof. Using the Möbius function $\mu(d)$ over the divisors of $p-1$ to detect the co-primality condition and interchanging the order of summation we obtain (see Section 3.d of Chapter 2 of [553])

$$\sum_{\substack{x=0 \\ \gcd(bx+1,p-1)=1}}^{p-2} \mathbf{e}_p(cg^x) = \sum_{d|p-1} \mu(d) \sum_{\substack{x \in [0,T-1] \\ bx+1 \equiv 0 \pmod d}} \mathbf{e}_p(cg^x).$$

Since $\gcd(b, p-1) = 1$, we can find for each $d|p-1$ the condition $bx+1 \equiv 0 \pmod d$ can be written in the form $x = dz + \alpha_d$ with some integer α_d, $1 \le \alpha_d \le d-1$. Therefore

$$\sum_{\substack{x=0 \\ bx+1 \equiv 0 \pmod d}}^{p-2} \mathbf{e}_p(cg^x) = \sum_{0 \le dz+\alpha_d \le p-2} \mathbf{e}_p\left(cg^{dz+\alpha_d}\right).$$

Denoting by τ_d the multiplicative order of $g_d = g^d$ and remarking that $\tau_d \ge T/d$, we derive from Lemma 3.13

$$\begin{aligned}
\sum_{0 \le dz+\alpha_d \le p-2} \mathbf{e}_p\left(cg^{dz+\alpha_d}\right) &= \sum_{0 \le z \le (p-2-\alpha_d)/d} \mathbf{e}_p(cg^{\alpha_d} g_d^z) \\
&= O\left(\left(\left\lfloor \frac{p-1}{\tau_d d} \right\rfloor + 1\right) p^{1/2} \log p\right) \\
&= O\left(\left(\left\lfloor \frac{p-1}{T} \right\rfloor + 1\right) p^{1/2} \log p\right) \\
&= O\left(p^{3/2} T^{-1} \log p\right).
\end{aligned}$$

Taking into account that

$$\sum_{d|p-1} |\mu(d)| = 2^{\nu(p-1)}$$

(see Section 3.b of Chapter 2 of [553]), we obtain the desired result. □

Lemma 3.19. *Let g be a primitive root modulo p and $\gcd(k, p-1) = d$. Then the bounds*

$$\max_{\gcd(c,p)=1} \left| \sum_{x=1}^{p-1} \mathbf{e}_p\left(cg^{kx}\right) \right| \le dp^{1/2}$$

hold.

Proof. It is easy to verify that g^k is of multiplicative order $T = (p-1)/d$. Thus

$$\left| \sum_{x=1}^{p-1} \mathbf{e}_p\left(cg^{kx}\right) \right| = d \left| \sum_{x=1}^{T} \mathbf{e}_p\left(cg^{kx}\right) \right|$$

and by Lemma 3.13 the stated bound follows. □

Lemma 3.20. *Let g be a primitive root modulo p. Then the bound*

$$\max_{\gcd(c,p)=1} \left| \sum_{x \in \mathbb{Z}_{p-1}^*} \mathbf{e}_p\left(cg^{x}\right) \right| \le 2^{\nu(p-1)} p^{1/2}$$

holds.

Proof. Using the Möbius function $\mu(d)$ over the divisors of $p-1$ to detect the co-primality condition, we obtain (see Section 3.d of Chapter 2 of [553])

$$\begin{aligned} \sum_{x \in \mathbb{Z}_{p-1}^*} \mathbf{e}_p\left(cg^{x}\right) &= \sum_{d|p-1} \mu(d) \sum_{\substack{x=1 \\ x \equiv 0 \pmod d}}^{p-1} \mathbf{e}_p\left(cg^{x}\right) \\ &= \sum_{d|p-1} \mu(d) \sum_{x=1}^{(p-1)/d} \mathbf{e}_p\left(cg^{dx}\right). \end{aligned}$$

Because g^d is of multiplicative order $(p-1)/d$ we can apply Lemma 3.13, getting

$$\left| \sum_{x \in \mathbb{Z}_{p-1}^*} \mathbf{e}_p\left(cg^{x}\right) \right| \le p^{1/2} \sum_{d|p-1} |\mu(d)|,$$

provided that $\gcd(c, p) = 1$. Remarking that

$$\sum_{d|p-1} |\mu(d)| = 2^{\nu(p-1)},$$

we obtain the desired statement. □

Let g be a primitive root modulo a prime number $p \geq 3$. For integers a, b, c we define the exponential sum

$$S_{a,b,c} = \sum_{x,y=1}^{p-1} \mathbf{e}_p\left(ag^x + bg^y + cg^{xy}\right). \tag{3.9}$$

Our results on the distribution of the Diffie–Hellman triples are based on an upper bound on these sums; see [94, 95].

In fact, instead of sums $S_{a,b,c}$ we study slightly different sums

$$W_{a,c} = \sum_{y=1}^{p-1} \left| \sum_{x=1}^{p-1} \mathbf{e}_p\left(ag^x + cg^{xy}\right) \right|$$

for which obviously $|S_{a,b,c}| \leq W_{a,c}$.

The following bound is a partial case of a more general result from [94]. It improves the previous estimate from [95] which is based on the upper bound (2.1) rather than on Lemma 2.6. Nevertheless, for bounds of incomplete sums analogous to $W(a,c)$ the results of [95] are still the best known. Analogous sums with an arbitrary composite denominator have been estimated in [189].

Lemma 3.21. *Let g be a primitive root modulo a sufficiently large prime p, then for any integers a, c such that $\gcd(a,c,p) = 1$, the bound*

$$W_{a,c} \leq \begin{cases} p^{3/2}\tau(p-1)), & \text{if } a \equiv 0 \pmod p, \\ 2.5p^{23/12}, & \text{otherwise}. \end{cases}$$

holds, where $\tau(m)$ denotes the number of integer divisors of $m \geq 1$.

Proof. If $c \equiv 0 \pmod p$, then $a \not\equiv 0 \pmod p$ and the bound follows trivially from Lemma 3.19 with $d = 1$.

If $a \equiv 0 \pmod p$, then $c \not\equiv 0 \pmod p$ and from Lemma 3.19 we derive

$$\begin{aligned} W_{a,c} &= \sum_{y=1}^{p-1} \left| \sum_{x=1}^{p-1} \mathbf{e}_p\left(cg^{xy}\right) \right| \\ &\leq p^{1/2} \sum_{y=1}^{p-1} \gcd(y, p-1) \leq p^{1/2} \sum_{d|p-1} d \sum_{\substack{y=1 \\ \gcd(y,p-1)=d}}^{p-1} 1 \\ &\leq p^{1/2} \sum_{d|p-1} d \left(\frac{p-1}{d}\right) \leq (p-1)p^{1/2}\tau(p-1). \end{aligned}$$

Now let us consider the case $\gcd(a,p) = 1$.

We apply the Hölder inequality, getting

$$W_{a,c}^4 \le (p-1)^3 \sum_{y=1}^{p-1} \left| \sum_{x=1}^{p-1} \mathbf{e}_p \left(a g^x + c g^{xy} \right) \right|^4 . \tag{3.10}$$

Making a change of variable we obtain

$$\begin{aligned}
\sum_{y=1}^{p-1} \left| \sum_{x=1}^{p-1} \mathbf{e}_p \left(a g^x + c g^{xy} \right) \right|^4 &= \frac{1}{p-1} \sum_{y=1}^{p-1} \sum_{z=1}^{p-1} \left| \sum_{x=1}^{p-1} \mathbf{e}_p \left(a g^{x+z} + c g^{(x+z)y} \right) \right|^4 \\
&= \frac{1}{p-1} \sum_{y=1}^{p-1} \sum_{z=1}^{p-1} \left| \sum_{x=1}^{p-1} \mathbf{e}_p \left(a g^z g^x + c g^{zy} g^{xy} \right) \right|^4 \\
&\le \frac{1}{p-1} \sum_{y=1}^{p-1} \sum_{\lambda,\mu=0}^{p-1} \left| \sum_{x=1}^{p-1} \mathbf{e}_p \left(\lambda g^x + \mu g^{xy} \right) \right|^4 ,
\end{aligned}$$

because for each fixed $y = 1, \ldots, p-1$ the pairs (ag^z, cg^{zy}), $z = 1, \ldots, p-1$, are all distinct modulo p.

Using (3.2) we obtain

$$\sum_{y=1}^{p-1} \left| \sum_{x=1}^{p-1} \mathbf{e}_p \left(a g^x + c g^{xy} \right) \right|^4 \le \frac{p^2}{p-1} T, \tag{3.11}$$

where T is the number of solutions of the system of congruences

$$\begin{aligned}
g^{x_1} + g^{x_2} &\equiv g^{x_3} + g^{x_4} \pmod p, \\
g^{x_1 y} + g^{x_2 y} &\equiv g^{x_3 y} + g^{x_4 y} \pmod p, \\
x_1, x_2, x_3, x_4, y &= 1, \ldots, p-1.
\end{aligned}$$

Now one easily verifies that for each fixed quadruple (x_1, x_2, x_3, x_4) the number of solutions of the second congruence equals the number of solutions of the polynomial congruence

$$z^{x_1} + z^{x_2} - z^{x_3} - z^{x_4} \equiv 0 \pmod p, \qquad z = 1, \ldots, p-1.$$

Using the bound of Lemma 2.6 with $t = 4$, we see that for each quadruple (x_1, x_2, x_3, x_4), there are at most

$$N_d \le 2p^{2/3} d^{1/3} + O(p^{1/3} d^{2/3})$$

values of $y = 1, \ldots, p-1$ which satisfy that system, where

$$d = \min_{1 \le i \le 4} \max_{j \ne i} \gcd(x_i - x_j, p-1). \tag{3.12}$$

We claim that the number of quadruples (x_1, x_2, x_3, x_4) with (3.12) and which satisfy the first congruence of the above system of congruences does not exceed $p^3 d^{-2} + p^2$.

Indeed, it is easy to see that any (x_1, x_2, x_3, x_4) for which (3.12) holds belongs to one of the following two types of configuration.

In the first case they can be separated into two pairs (x_{i_1}, x_{i_2}) and (x_{i_3}, x_{i_4}) with $\{i_1, i_2, i_3, i_4\} = \{1, 2, 3, 4\}$ and such that

$$\gcd(x_{i_1} - x_{i_2}, p-1) \geq d, \qquad \gcd(x_{i_3} - x_{i_4}, p-1) \geq d.$$

There are three choices of a partner for x_1 and this determines the pairings.

In the second case there is a permutation $\{i_1, i_2, i_3, i_4\} = \{1, 2, 3, 4\}$ such that

$$\gcd(x_{i_1} - x_{i_k}, p-1) \geq d, \qquad k = 2, 3, 4.$$

Here there are four choices for i_1.

Let us consider the congruence

$$A_1 g^{u_1} + A_2 g^{u_2} + A_3 g^{u_3} + A_4 g^{u_4} \equiv 0 \pmod p$$

with $A_1 A_2 A_3 A_4 \not\equiv 0 \pmod p$ and let us count the number of solutions $1 \leq u_1, u_2, u_3, u_4 \leq p-1$ with $u_1 \equiv u_2 \pmod{d_1}$ and $u_3 \equiv u_4 \pmod{d_2}$ for two divisors d_1, d_2 of $p-1$. We first consider the solutions for which

$$A_1 g^{u_1} + A_2 g^{u_2} \equiv 0 \pmod p.$$

In this case, each value of u_1 gives rise to at most one value of u_2 and then each value of u_3 gives rise to at most one value of u_4. There are thus at most $(p-1)^2$ of these diagonal solutions. For the other solutions, each choice of u_1, u_2 (there are at most $(p-1)^2/d_1$ such pairs), determines a non-zero class for $g^{u_3}(A_3 + A_4 g^{u_4 - u_3})$. Once we then specify $u_4 - u_3$ (there are $(p-1)/d_2$ ways) the rest is determined. Therefore the total number of solutions does not exceed $(p-1)^3/d_1 d_2 + (p-1)^2$.

This proves our claim for the first configuration.

For the second configuration we merely ignore the congruence condition on x_{i_4} and remark that it is uniquely defined for each choice of $x_{i_1}, x_{i_2}, x_{i_3}$ for which we have at most $(p-1)^3/d_1 d_2$ possibilities.

Therefore,

$$\begin{aligned} T &\leq 14 p^{11/3} \sum_{d \mid p-1} d^{-5/3} + O\left(p^{8/3} \sum_{d \mid p-1} d^{1/3} + p^{10/3} \sum_{d \mid p-1} d^{-4/3}\right) \\ &\leq 14\zeta(5/3) p^{11/3} + O(p^3 \tau(p-1) + p^{10/3}), \end{aligned}$$

where

$$\zeta(5/3) = \sum_{k=1}^{\infty} k^{-5/3} = 2.123\ldots.$$

Substituting this estimate in (3.11) we obtain

$$\sum_{y=1}^{p-1} \left| \sum_{x=1}^{p-1} \mathbf{e}_p \left(a g^x + c g^{xy} \right) \right|^4 \le 14\zeta(5/3) p^{14/3} + O(p^{13/3}). \tag{3.13}$$

Now from (3.10) we derive the desired result. □

In fact, a more general estimate has been obtained in [94] which applies to elements of $\lambda \in \mathbb{F}_p$ of multiplicative order $T \ge p^{3/4+\varepsilon}$. It is based on the following generalisation of (3.13),

$$\sum_{y=1}^{T} \left| \sum_{x=1}^{T} \mathbf{e}_p \left(a\lambda^x + c\lambda^{xy} \right) \right|^4 = O\left(pT^{11/3} \right). \tag{3.14}$$

We fix an element $g \in \mathbb{F}_p^*$ of multiplicative order t and consider double exponential sums

$$S_a(\mathcal{X}, \mathcal{Y}) = \sum_{x \in \mathcal{X}} \sum_{y \in \mathcal{Y}} \mathbf{e}_p \left(a g^{xy} \right),$$

where $a \in \mathbb{F}_p$ and $\mathcal{X}, \mathcal{Y} \subseteq \mathbb{Z}_t$. We also put $S_a(\mathcal{X}) = S_a(\mathcal{X}, \mathcal{X})$.

We need the result of [195] which gives an upper bound of the sums $S_a(\mathcal{X}, \mathcal{Y})$ for arbitrary sets $\mathcal{X}$ and $\mathcal{Y}$. In the special case that both sets are of the same cardinality $|\mathcal{X}| = |\mathcal{Y}| = N$ and $t = p-1$, that is, g is a primitive root of $\mathbb{F}_p$, this estimate is non-trivial for $N \ge p^{15/16+\delta}$ with any fixed $\delta > 0$. Further examples are given below.

Lemma 3.22. *For any sets $\mathcal{X}, \mathcal{Y} \subseteq \mathbb{Z}_t$, the bound*

$$\max_{a \in \mathbb{F}_p^*} |S_a(\mathcal{X}, \mathcal{Y})| \ll |\mathcal{X}|^{1/2} |\mathcal{Y}|^{5/6} t^{1/2} p^{1/8+\varepsilon}$$

holds.

Proof. For a divisor $d|t$ we denote by $\mathcal{Y}(d)$ the subset of $y \in \mathcal{Y}$ with $\gcd(y,t) = d$. Then

$$|S_a(\mathcal{X}, \mathcal{Y})| \le \sum_{d|t} |\sigma_d|,$$

where

$$\sigma_d = \sum_{x \in \mathcal{X}} \sum_{y \in \mathcal{Y}(d)} \mathbf{e}_p \left(a g^{xy} \right).$$

Using the Cauchy inequality, we derive

$$\begin{aligned}|\sigma_d|^2 &\le |\mathcal{X}| \sum_{x\in\mathcal{X}} \left| \sum_{y\in\mathcal{Y}(d)} \mathbf{e}_p\left(ag^{xy}\right)\right|^2 \le |\mathcal{X}| \sum_{x\in\mathbb{Z}_t} \left| \sum_{y\in\mathcal{Y}(d)} \mathbf{e}_p\left(ag^{xy}\right)\right|^2 \\ &= |\mathcal{X}| \sum_{y.z\in\mathcal{Y}(d)} \sum_{x\in\mathbb{Z}_t} \mathbf{e}_p\left(a\left(g^{xy}-g^{xz}\right)\right).\end{aligned}$$

By the Hölder inequality we have

$$\begin{aligned}|\sigma_d|^8 &\le |\mathcal{X}|^4|\mathcal{Y}(d)|^6 \sum_{y.z\in\mathcal{Y}(d)} \left|\sum_{x\in\mathbb{Z}_t} \mathbf{e}_p\left(a\left(g^{xy}-g^{xz}\right)\right)\right|^4 \\ &\le |\mathcal{X}|^4|\mathcal{Y}(d)|^6 \sum_{y\in\mathcal{Y}(d)} \sum_{u\in\mathbb{Z}_{t/d}} \left|\sum_{x\in\mathbb{Z}_t} \mathbf{e}_p\left(a\left(g^{xy}-g^{xud}\right)\right)\right|^4.\end{aligned}$$

Because each element $y \in \mathcal{Y}(d)$ can be represented in the form $y = dv$ with $\gcd(v, t/d) = 1$ and $g_d = g^d$ is of multiplicative order t/d, we see that the double sum over u and x does not depend on y. Therefore,

$$\begin{aligned}|\sigma_d|^8 &\le |\mathcal{X}|^4|\mathcal{Y}(d)|^7 \sum_{u\in\mathbb{Z}_{t/d}} \left|\sum_{x\in\mathbb{Z}_t} \mathbf{e}_p\left(a\left(g_d^{x}-g_d^{xu}\right)\right)\right|^4 \\ &= |\mathcal{X}|^4|\mathcal{Y}(d)|^7 d^4 \sum_{u\in\mathbb{Z}_{t/d}} \left|\sum_{v\in\mathbb{Z}_{t/d}} \mathbf{e}_p\left(a\left(g_d^{v}-g_d^{vu}\right)\right)\right|^4.\end{aligned}$$

By (3.14) we obtain

$$|\sigma_d|^8 \ll |\mathcal{X}|^4|\mathcal{Y}(d)|^7 p t^{11/3} d^{1/3}.$$

Using the bound $|\mathcal{Y}(d)| \le |\mathcal{Y}|$ for $d \le t/|\mathcal{Y}|$ and the bound $|\mathcal{Y}(d)| \le t/d$ for $d > t/|\mathcal{Y}|$, we obtain that

$$|\sigma_d| \ll |\mathcal{X}|^{1/2}|\mathcal{Y}|^{5/6}t^{1/2}p^{1/8}$$

for any divisor $d|t$. Applying (5.1), we derive the desired result. □

Of course the sets $\mathcal{X}$ and $\mathcal{Y}$ can be interchanged in the above estimate. Combining those two bounds we obtain the symmetric estimate

$$\max_{a\in\mathbb{F}_p^*} |S_a(\mathcal{X},\mathcal{Y})| \ll (|\mathcal{X}|^{-1/3} + |\mathcal{Y}|^{-1/3})^{-1}|\mathcal{X}|^{1/2}|\mathcal{Y}|^{1/2}t^{1/2}p^{1/8+\varepsilon}$$

or, more simply, multiplying the two bounds and taking the square root we obtain the somewhat weaker result

$$\max_{a\in\mathbb{F}_p^*} |S_a(\mathcal{X},\mathcal{Y})| \ll |\mathcal{X}|^{2/3}|\mathcal{Y}|^{2/3}t^{1/2}p^{1/8+\varepsilon}.$$

In particular, if $|\mathcal{X}| = |\mathcal{Y}| = N$ and $t = p-1$, then the bound takes the form

$$\max_{a\in\mathbb{F}_p^*} |S_a(\mathcal{X},\mathcal{Y})| \ll N^{4/3}p^{5/8+\varepsilon}.$$

We remark that if some non-trivial information about the size of $\mathcal{Y}(d)$ is available, then the result can be refined. The bounds (5.8) and (5.9) provide such information for sets of integers with a given number of non-zero bits in their bit representation.

A slight generalization of Lemma 3.22, and its analogue of [189] for a composite denominator m, has been used in [20] to estimate exponential sums

$$\sum_{l\le N} \mathbf{e}_p\left(ag^l\right) \qquad \text{and} \qquad \sum_{l\le N} \mathbf{e}_m\left(ag^l\right),$$

taking over primes $l \le N$. In particular, nontrivial upper bounds of exponential sums with *Mersenne numbers* $M_l = 2^l - 1$ have been obtained. It is interesting to remark, that the results of [187, 188, 189, 194], see also Theorem 23.6, which are based on similar ideas and techniques, yield an upper bound for exponential sums with *Fermat numbers* $F_n = 2^{2^n} + 1$.

Assume that $r = \lfloor \log p \rfloor$ is the bit length of p. We denote by $\mathcal{W}_k$ the set of r-bit integers which have precisely k non-zero bits in their bit representation.

We define the functions

$$G(\gamma) = \begin{cases} H(\gamma) & \text{if } \gamma < 1/2, \\ 1 & \text{if } \gamma \ge 1/2, \end{cases}$$

and

$$F(\alpha,\gamma) = 12H(\alpha) - 14/3 - 7(1-\gamma)G\left(\alpha/(1-\gamma)\right) - \gamma/3,$$

where $H(\gamma)$ is the binary entropy function given by (1.4). Finally, we put

$$E(\alpha) = \inf_{\gamma\in[0,1]} F(\alpha,\gamma). \tag{3.15}$$

One easily verifies that $E(\alpha)$ monotonically increases for $0 \le \alpha \le 1/2$. Therefore we can define $\alpha_0 = 0.349\ldots$ as the unique root of the equation $E(\alpha) = 0$, $0 \le \alpha \le 1/2$.

Unfortunately the bound (5.8) is not enough for our purposes but using (5.9) one can derive the following estimate from [196] which improves Lemma 3.22 in the case $\mathcal{X} = \mathcal{Y} = \mathcal{W}_k$ (we remark that in the case $c = 0$ some other arguments are required, see [196]).

Lemma 3.23. *Let $g \in \mathbb{F}_p^*$ be a primitive root. For any fixed $\alpha > \alpha_0$ there exists $\delta > 0$ such that for $n/2 \ge k \ge \alpha n$ the bound*

$$\max_{\gcd(a,b,c,p)=1} \left| \sum_{x\in\mathcal{W}_k} \sum_{y\in\mathcal{W}_k} \mathbf{e}_p\left(ag^x + bg^y + cg^{xy}\right) \right| \ll |\mathcal{W}_k|^{2-\delta}$$

holds.

We note that such sparse exponents $x, y \in \mathcal{W}_k$ are of cryptographic interest because for them the computation of g^x, g^y, g^{xy} takes less time than for arbitrary x and y. Bounds of some other exponential sums and sums of multiplicative characters with numbers with restricted (in several different ways) g-ary digits have been given in [21].

The following result is a special case of Theorem 6 of [187].

Lemma 3.24. *Let $g \in \mathbb{F}_p^*$ be of multiplicative order T modulo p. Then the bound*

$$\max_{0 \le a \le T-1} \max_{\gcd(c,p)=1} \left| \sum_{x=1}^{T} \mathbf{e}_p\left(cg^{x^2+2ax}\right) \right| \ll T^{3/4} p^{1/8+\varepsilon}$$

holds.

Proof. We have

$$\sum_{x=1}^{T} \mathbf{e}_p\left(cg^{x^2+2ax}\right) = \sum_{x=1}^{T} \mathbf{e}_p\left(c\left(g^{(x+a)^2-a^2}\right)\right) = \sum_{x=1}^{T} \mathbf{e}_p\left(cg^{-a^2}g^{x^2}\right).$$

Using Theorem 6 of [187], we obtain the desired result. □

The following statement shows how to reduce exponential sums with polynomials and arbitrary denominators to exponential sums with prime power denominators (see for example Problem 12.d to Chapter 3 of [553]).

Lemma 3.25. *Let $m = m_1 m_2$ where $m_1 \ge m_2 \ge 2$ and $\gcd(m_1, m_2) = 1$ and let k_1, k_2 be such that*

$$k_1 m_2 \equiv 1 \pmod{m_1} \qquad \text{and} \qquad k_2 m_1 \equiv 1 \pmod{m_2}.$$

Then for any polynomial $f(x)$ with integer coefficients,

$$\sum_{\substack{x=0 \\ \gcd(x,m)=1}}^{m-1} \mathbf{e}_m\left(f(x)\right) = \sum_{\substack{x_1=0 \\ \gcd(x_1,m_1)=1}}^{m_1-1} \mathbf{e}_{m_1}\left(k_1 f(x_1)\right) \sum_{\substack{x_2=0 \\ \gcd(x_2,m_2)=1}}^{m_2-1} \mathbf{e}_{m_2}\left(k_2 f(x_2)\right).$$

Indeed, this statement follows from Problem 12.d to Chapter 3 of [553] if one remarks that

$$k_1 m_2 + k_2 m_1 \equiv 1 \pmod{m}.$$

Now we give several more applications of the bound (3.13) which are used in Chapter 21 to establish some results of [514] about the uniformity of distribution of RSA pairs.

For an integer $m \geq 2$ let $\mathcal{E}$ be the set of integers e, $1 \leq e \leq \varphi(m)$, with $\gcd(e, \varphi(m)) = 1$ We define exponential sums

$$W_m(a,b) = \sum_{e \in \mathcal{E}_m} \left| \sum_{x \in \mathbb{Z}_m^*} \mathbf{e}_m\left(ax + bx^e\right) \right| .$$

Lemma 3.26. *Let $m = pl$ where p and l are two distinct primes. Then the bound*

$$\max_{\gcd(a,b,m)=1} W_m(a,b) = O\left(m^{23/12}\right)$$

holds.

Proof. Lemma 3.25 implies that there exist some integer numbers k_p and k_l with $\gcd(p, k_p) = \gcd(l, k_l) = 1$ and such that

$$\sum_{x \in \mathbb{Z}_m^*} \mathbf{e}_m\left(ax + bx^e\right) = \sum_{x_1=1}^{p-1} \mathbf{e}_p\left(k_p\left(ax_1 + bx_1^e\right)\right) \sum_{x_2=1}^{l-1} \mathbf{e}_l\left(k_l\left(ax_2 + bx_2^e\right)\right).$$

From the previous equation and the Cauchy inequality we derive

$$\begin{aligned}
W_m(a,b) &\leq \sum_{e \in \mathcal{E}_m} \left| \sum_{x_1=1}^{p-1} \mathbf{e}_p\left(k_p\left(ax_1 + bx_1^e\right)\right)\right| \left| \sum_{x_2=1}^{l-1} \mathbf{e}_l\left(k_l\left(ax_2 + bx_2^e\right)\right)\right| \\
&\leq \varphi(m)^{1/2} \left(\sum_{e \in \mathcal{E}_m} \left| \sum_{x_1=1}^{p-1} \mathbf{e}_p\left(k_p\left(ax_1 + bx_1^e\right)\right)\right|^4 \right)^{1/4} \\
&\quad \times \left(\sum_{e \in \mathcal{E}_m} \left| \sum_{x_2=1}^{l-1} \mathbf{e}_l\left(k_l\left(ax_2 + bx_2^e\right)\right)\right|^4 \right)^{1/4} \\
&\leq \varphi(m)^{1/2} \left(\frac{\varphi(m)}{p-1} \sum_{e=1}^{p-1} \left| \sum_{x_1=1}^{p-1} \mathbf{e}_p\left(k_p\left(ax_1 + bx_1^e\right)\right)\right|^4 \right)^{1/4} \\
&\quad \times \left(\frac{\varphi(m)}{l-1} \sum_{e=1}^{l-1} \left| \sum_{x_2=1}^{l-1} \mathbf{e}_l\left(k_l\left(ax_2 + bx_2^e\right)\right)\right|^4 \right)^{1/4} .
\end{aligned}$$

Using the bound (3.13), we obtain

$$W_m(a,b) \ll m^{1/2} \left(mp^{11/3}\right)^{1/4} \left(ml^{11/3}\right)^{1/4} = m^{23/12}$$

and the desired result follows. □

We also remark that the same (and even somewhat simpler) considerations imply the bounds

$$\max_{\gcd(a,b,m)=p} W_m(a,b) = O\left(m^2 l^{-1/12}\right) \tag{3.16}$$

and

$$\max_{\gcd(a,b,m)=l} W_m(a,b) = O\left(m^2 p^{-1/12}\right). \tag{3.17}$$

Our next result is an analogue of Lemma 3.26 for sums whose argument x runs through certain subset products.

For a given r-dimensional vector $\mathbf{z} = (z_1, \dots, z_r) \in \left(\mathbb{Z}_m^*\right)^r$ and a binary vector $\mathbf{u} = (u_1, \dots, u_r) \in \mathfrak{B}_r$ we put

$$x_{\mathbf{z}}(\mathbf{u}) = \prod_{j=1}^{r} z_j^{u_j}.$$

For an integer $m \ge 2$ define

$$S_{m,k,r}(a,b) = \sum_{\mathbf{z}\in(\mathbb{Z}_m^*)^r} \sum_{e\in\mathcal{E}_m} \left| \sum_{\mathbf{u}\in\mathfrak{B}_{r,k}} \mathbf{e}_m\left(a x_{\mathbf{z}}(\mathbf{u}) + b x_{\mathbf{z}}^e(\mathbf{u})\right) \right|$$

where $\mathfrak{B}_{r,k}$ is defined by (1.5) and (1.6).

Lemma 3.27. *Let $m = pl$ where p and l are two distinct primes. Then the bound*

$$\max_{\gcd(a,b,m)=1} S_{m,k,r}(a,b) = O\left(m|\mathbb{Z}_m^*|^r \left(|\mathfrak{B}_{r,k}|^{1/2} + |\mathfrak{B}_{r,k}| m^{-1/12}\right)\right)$$

holds.

Proof. Using the Cauchy inequality and changing the order of summation, we derive

$$\begin{aligned} &S_{m,k,r}(a,b)^2 \\ &\le |\mathbb{Z}_m^*|^r |\mathcal{E}_m| \sum_{\mathbf{z}\in(\mathbb{Z}_m^*)^r} \sum_{e\in\mathcal{E}_m} \left| \sum_{\mathbf{u}\in\mathfrak{B}_{r,k}} \mathbf{e}_m\left(a x_{\mathbf{z}}(\mathbf{u}) + b x_{\mathbf{z}}^e(\mathbf{u})\right) \right|^2 \\ &= |\mathbb{Z}_m^*|^r |\mathcal{E}_m| \sum_{\mathbf{u},\mathbf{v}\in\mathfrak{B}_{r,k}} \sum_{\mathbf{z}\in(\mathbb{Z}_m^*)^r} \\ &\qquad\qquad \times \sum_{e\in\mathcal{E}_m} \mathbf{e}_m\left(a x_{\mathbf{z}}(\mathbf{u}) + b x_{\mathbf{z}}^e(\mathbf{u}) - a x_{\mathbf{z}}(\mathbf{v}) - b x_{\mathbf{z}}^e(\mathbf{v})\right). \end{aligned}$$

For each pair with $\mathbf{u} = \mathbf{v}$ the inner double sums is equal to $|\mathbb{Z}_m^*|^r |\mathcal{E}_m|$. Therefore, the total contribution from such pairs is $|\mathbb{Z}_m^*|^{2r} |\mathcal{E}_m|^2 |\mathfrak{B}_{r,k}|$.

For each pair $\mathbf{u}, \mathbf{v} \in \mathfrak{B}_{r,k}$ with $\mathbf{u} \neq \mathbf{v}$ we can find i and j, $1 \le i < j \le r$, with $u_i = v_j = 1$ and $u_j = v_i = 0$. Without loss of generality we may assume that $i = 1$, $j = 2$. In this case $x_{\mathbf{z}}(\mathbf{u}) = Az_1$ and $x_{\mathbf{z}}(\mathbf{v}) = Bz_2$ where A and B do not depend on z_1 and z_2. Therefore

$$\begin{aligned}\sum_{\mathbf{z}\in(\mathbb{Z}_m^*)^r} \sum_{e\in\mathcal{E}_m} &\mathbf{e}_m\left(ax_{\mathbf{z}}(\mathbf{u}) + bx_{\mathbf{z}}^e(\mathbf{u}) - ax_{\mathbf{z}}(\mathbf{v}) - bx_{\mathbf{z}}^e(\mathbf{v})\right)\\ &= \sum_{z_3,\dots,z_r\in\mathbb{Z}_m^*} \sum_{e\in\mathcal{E}_m} \sum_{z_1\in\mathbb{Z}_m^*} \mathbf{e}_m\left(aAz_1 + bA^e z_1^e\right)\\ &\qquad \times \sum_{z_2\in\mathbb{Z}_m^*} \mathbf{e}_m\left(-aBz_2 - bB^e z_2^e\right),\end{aligned}$$

where A and B depend only on $\mathbf{u}$, $\mathbf{v}$ and $z_3, \dots, z_r$. Because $\gcd(A, m) = \gcd(B, m) = 1$ we have

$$\sum_{e\in\mathcal{E}_m} \sum_{z_1\in\mathbb{Z}_m^*} \mathbf{e}_m\left(aAz_1 + bA^e z_1^e\right) = \sum_{e\in\mathcal{E}_m} \sum_{z_1\in\mathbb{Z}_m^*} \mathbf{e}_m\left(az_1 + bz_1^e\right)$$

and

$$\sum_{z_2\in\mathbb{Z}_m^*} \mathbf{e}_m\left(-aBz_2 - bB^e z_2^e\right) = \sum_{z_2\in\mathbb{Z}_m^*} \mathbf{e}_m\left(-az_2 - bz_2^e\right).$$

Therefore

$$\begin{aligned}\sum_{\mathbf{z}\in(\mathbb{Z}_m^*)^r} \sum_{e\in\mathcal{E}_m} &\mathbf{e}_m\left(ax_{\mathbf{z}}(\mathbf{u}) + bx_{\mathbf{z}}^e(\mathbf{u}) - ax_{\mathbf{z}}(\mathbf{v}) - bx_{\mathbf{z}}^e(\mathbf{v})\right)\\ &= |\mathbb{Z}_m^*|^{r-2} \sum_{e\in\mathcal{E}_m} \sum_{z_1\in\mathbb{Z}_m^*} \mathbf{e}_m\left(az_1 + bz_1^e\right) \sum_{z_2\in\mathbb{Z}_m^*} \mathbf{e}_m\left(-az_2 - bz_2^e\right)\\ &= |\mathbb{Z}_m^*|^{r-2} \sum_{e\in\mathcal{E}_m} \left|\sum_{z\in\mathbb{Z}_m^*} \mathbf{e}_m\left(az + bz^e\right)\right|^2.\end{aligned}$$

As in the proof of Lemma 3.26 we obtain that the double sum in the above expression is $O(m^{17/6})$. Indeed, in the notations of the proof of Lemma 3.26, using the Cauchy inequality, we derive

$$\begin{aligned}
&\sum_{e\in\mathcal{E}_m}\left|\sum_{z\in\mathbb{Z}_m^*}\mathbf{e}_m\left(az+bz^e\right)\right|^2\\
&\quad=\sum_{e\in\mathcal{E}_m}\left|\sum_{z_1=1}^{p-1}\mathbf{e}_p\left(k_p(az_1+bz_1^e)\right)\right|^2\left|\sum_{z_2=1}^{l-1}\mathbf{e}_l\left(k_l(az_2+bz_2^e)\right)\right|^2\\
&\quad\le\left(\sum_{e\in\mathcal{E}_m}\left|\sum_{z_1=1}^{p-1}\mathbf{e}_p\left(k_p(az_1+bz_1^e)\right)\right|^4\right)^{1/2}\\
&\qquad\qquad\times\left(\sum_{e\in\mathcal{E}_m}\left|\sum_{z_2=1}^{l-1}\mathbf{e}_l\left(k_l(az_2+bz_2^e)\right)\right|^4\right)^{1/2}\\
&\quad=\left(\frac{\varphi(m)}{p-1}\sum_{e=1}^{p-1}\left|\sum_{z_1=1}^{p-1}\mathbf{e}_p\left(k_p(az_1+bz_1^e)\right)\right|^4\right)^{1/2}\\
&\qquad\qquad\times\left(\frac{\varphi(m)}{l-1}\sum_{e=1}^{l-1}\left|\sum_{z_2=1}^{l-1}\mathbf{e}_l\left(k_l(az_2+bz_2^e)\right)\right|^4\right)^{1/2}.
\end{aligned}$$

Using the bound (3.13), we obtain

$$\sum_{e\in\mathcal{E}_m}\left|\sum_{z\in\mathbb{Z}_m^*}\mathbf{e}_m\left(az+bz^e\right)\right|^2\ll\left(mp^{11/3}\right)^{1/2}\left(ml^{11/3}\right)^{1/2}=mp^{11/6}l^{11/6}=m^{17/6}.$$

Thus,

$$S_{m,k,r}(a,b)^2\ll|\mathbb{Z}_m^*|^{2r}|\mathcal{E}_m|^2|\mathfrak{B}_{r,k}|+|\mathbb{Z}_m^*|^{2r-2}|\mathcal{E}_m||\mathfrak{B}_{r,k}|^2m^{17/6}.$$

Now from the inequality $|\mathbb{Z}_m^*|=\varphi(m)=(p-1)(l-1)\ge 0.25m$ and the trivial bound $|\mathcal{E}_m|\le m$ we derive the desired result. □

As after Lemma 3.26, we also remark that

$$\max_{\gcd(a,b,m)=p}S_{m,k,r}(a,b)=O\left(m|\mathbb{Z}_m^*|^r\left(|\mathfrak{B}_{r,k}|^{1/2}+|\mathfrak{B}_{r,k}|l^{-1/12}\right)\right)\tag{3.18}$$

and

$$\max_{\gcd(a,b,m)=l}S_{m,k,r}(a,b)=O\left(m|\mathbb{Z}_m^*|^r\left(|\mathfrak{B}_{r,k}|^{1/2}+|\mathfrak{B}_{r,k}|p^{-1/12}\right)\right).\tag{3.19}$$

Chapter 4

Distribution and Discrepancy

Here we collect some results about the distribution properties of various sequences. Some of them are quite generic, while the others are rather specialised and tailored to our applications.

One of our basic tools to study the uniformity of distribution is the celebrated *Koksma–Szüsz inequality* [302, 545] (see also Theorem 1.21 of [161]) for the discrepancy of a sequence of points of the N-dimensional unit cube. This statement provides a very important link between the discrepancy and exponential sums. We present it in the following form.

Lemma 4.1. *There exists an absolute constant $C > 0$ such that, for any integer $L > 1$ and any sequence Γ of H points (1.9) the bound*

$$\Delta_\Gamma < C^N \left(\frac{2}{L+1} + \frac{1}{H} \sum_{0<|\mathbf{a}|\le L} \frac{1}{r(\mathbf{a})} \left| \sum_{x=1}^{H} \mathbf{e}\left(\sum_{j=0}^{N-1} a_j \gamma_{j,x} \right) \right| \right)$$

on the discrepancy Δ_Γ holds, where $|\mathbf{a}|$, $r(\mathbf{a})$ are defined by (1.8) and the sum is taken over all nonzero integer vectors

$$\mathbf{a} = (a_0, \ldots, a_{N-1}) \in \mathbb{Z}^N$$

with $|\mathbf{a}| \le L$.

According to Theorem 1.21 of [161], one can take $C = 3/2$.

We remark that the sequence Γ hits any box

$$B = [\beta_0, \alpha_0] \times \ldots \times [\beta_{N-1}, \alpha_{N-1}] \subseteq [0,1]^N$$

of size $|B| > \Delta_\Gamma$.

We also need several results about the distribution of various sequences in intervals.

To study the bit security of the XTR [332, 333, 334, 335, 551] and LUC [47, 536] key exchange protocol we also need the following direct corollary of Lemma 3.9.

For a subgroup $\mathcal{G} \subseteq \mathbb{F}_{p^m}^*$ of the multiplicative group $\mathbb{F}_{p^m}^*$ an element $\gamma \in \mathbb{F}_{p^m}$ and integers r and h we denote by $N_\gamma(\mathcal{G}, r, h)$ the number of solutions of the congruence

$$\mathrm{Tr}_{\mathbb{F}_{p^m}/\mathbb{F}_p}(\gamma t) \equiv r + y \pmod p, \qquad t \in \mathcal{G}, \ y = 0, \ldots, h-1.$$

From Lemmas 3.9 and 4.1 one immediately derives an upper bound on $N_\gamma(\mathcal{G}, r, h)$.

Lemma 4.2. *For any divisor $s|m$, any $\gamma \in \mathbb{F}_{p^m}^*$ and any subgroup $\mathcal{G} \subseteq \mathcal{N}_{m,s}(p)$, where $\mathcal{N}_{m,s}(p)$ given by (3.5), the bound*

$$N_\gamma(\mathcal{G}, r, h) = \frac{h}{p}|\mathcal{G}| + O(p^{(m-m/s)/2} \log p)$$

holds.

Let $H_p(T)$ be the largest integer h such that for some subgroup $\mathcal{G} \subseteq \mathbb{F}_p^*$ of the multiplicative group $\mathbb{F}_p^*$ of size at least T there exist integers $a \not\equiv 0 \pmod p$ and r such that the congruence

$$at \equiv r + y \pmod p, \qquad t \in \mathcal{G}, \ y = 0, \ldots, h-1,$$

has no solution.

Theorem 7.10 of [306] claims the following inequality.

Lemma 4.3. *For any $\varepsilon > 0$ and $T \geq p^{1/2}$ the bound*

$$H_p(T) \ll p^{34/37+\varepsilon}$$

holds.

We remark, that it is known, see [172, 281, 452], that if an integer $g > 1$ is fixed that for any function $\varepsilon(p) \to 0$, for almost all primes it generates a subgroup of $\mathbb{F}_p^*$ of order at least $p^{1/2+\varepsilon(p)}$, thus Lemma 4.3 applies. An brief outline of some other related results on multiplicative orders of a fixed integer modulo different prime and integer numbers is given in Chapter 5.

Lemma 4.4. *Let $\tau = \mathrm{ord}_M e$ be the multiplicative order of an integer e modulo an integer $M \geq 1$ with $\gcd(e, M) = 1$. Then for any fixed $\delta > 0$ and any integer $h \geq M^\delta$ there exists an integer r with*

$$\gcd(r, M) = 1$$

and such that the congruence

$$re^k \equiv y \pmod{M}, \qquad 1 \le k \le \tau,\ 0 \le y \le h-1,$$

has

$$L_r(h) \gg \frac{\tau h}{M}$$

solutions.

Proof. For each $k = 1, \dots, \tau$ and every y, $0 \le y \le h-1$ with $\gcd(y, M) = 1$, the integer r, $1 \le r \le M$, such that the above congruence holds is uniquely determined and satisfies $\gcd(r, M) = 1$ since $\gcd(e, M) = 1$. Hence we have

$$\sum_{\substack{r=1 \\ \gcd(r,M)=1}}^{M} L_r(h) \gg \tau \sum_{\substack{y=0 \\ \gcd(y,M)=1}}^{h-1} 1.$$

Since $h \ge M^\delta$, by Lemma 5.4 and by the inequalities (5.1) and (5.2) below we have

$$\sum_{\substack{y=0 \\ \gcd(y,M)=1}}^{h-1} 1 \gg \frac{\varphi(M)h}{M}$$

and substituting this we get

$$\sum_{\substack{r=1 \\ \gcd(r,M)=1}}^{M} L_r(h) \gg \frac{\tau\varphi(M)h}{M}.$$

Hence for some choice of r we have the required lower bound for $L_r(h)$. □

For the attack on the Digital Signature Algorithm described in Chapter 16, we need the following result from [417] about the of distribution residues of exponential functions after two consecutive modular reductions.

As in Chapter 16 we assume that p and q are primes with $q|p-1$ and $g \in \mathbb{F}_p$ is a fixed element of multiplicative order q. For an integer k we also define

$$r(k) = \left\lfloor \left\lfloor g^k \right\rfloor_p \right\rfloor_q.$$

Finally, for an integer $\rho \in [0, q-1]$, we denote by $N(\rho)$ the number of solutions of the equation

$$r(k) = \rho, \qquad k \in [1, q-1].$$

Lemma 4.5. *Let Q be a sufficiently large integer. The following statement holds with $\vartheta = 1/3$ for all primes $p \in [Q, 2Q]$, and with $\vartheta = 0$ for all primes $p \in [Q, 2Q]$ except at most $Q^{5/6+\varepsilon}$ of them. For any $\varepsilon > 0$ there exists $\delta > 0$ such that for any element $g \in \mathbb{F}_p$ of multiplicative order $q \geq p^{\vartheta+\varepsilon}$ the bound*

$$N(\rho) = O\left(q^{1-\delta}\right), \qquad \rho \in [0, q-1],$$

holds.

Proof. Let

$$L = \left\lfloor \frac{p-\rho-1}{q} \right\rfloor.$$

We remark that $N(\rho)$ is the number of solutions of the congruence

$$g^k \equiv qx + \rho \pmod p, \qquad k \in [1, q-1], \ x \in [0, L].$$

Using the identity (3.2) we obtain

$$\begin{aligned} N(\rho) &= \frac{1}{p}\sum_{k=1}^{q-1}\sum_{x=0}^{L}\sum_{c=0}^{p-1} \mathbf{e}_p\left(c\left(g^k - qx - \rho\right)\right) \\ &= \frac{1}{p}\sum_{c=0}^{p-1} \mathbf{e}_p(-c\rho)\sum_{k=1}^{q-1}\mathbf{e}_p\left(cg^k\right)\sum_{x=0}^{L}\mathbf{e}_p(-cqx). \end{aligned}$$

Separating the term

$$\frac{(q-1)(L+1)}{p} \leq \frac{(q-1)(p/q+1)}{p} \leq 2$$

corresponding to $c = 0$, we derive

$$\begin{aligned} N(\rho) &\leq 2 + \frac{1}{p}\sum_{c=1}^{p-1}\left|\sum_{k=1}^{q-1}\mathbf{e}_p\left(cg^k\right)\right|\left|\sum_{x=0}^{L}\mathbf{e}_p(-cqx)\right| \\ &\leq 2 + \frac{1}{p}\sum_{c=1}^{p-1}\left|\sum_{k=1}^{q-1}\mathbf{e}_p\left(cg^k\right)\right|\left|\sum_{x=0}^{L}\mathbf{e}_p(cqx)\right|. \end{aligned}$$

Combining Lemmas 3.15 and 3.16 to estimate the sum over $k \in [1, q-1]$ (certainly the missing term corresponding to $k = 0$ does not change the order of magnitude of this sum) with (3.4), we obtain the desired upper bound. □

Using the result of [304] one can show that the statement of Lemma 4.5 holds with $\vartheta = 1/4$ for all primes $p \in [Q, 2Q]$.

Finally we need a certain statement about the distribution subset products.

Denote by $\mathfrak{B}_m^*$ the set of all $2^m - 1$ non-zero binary vectors of length m. As in [528], we need information about the distribution in $\mathbb{F}_l^*$ of products of the form

$$\mathbf{b}^{\mathbf{z}} = b_1^{z_1} \dots b_m^{z_m}, \qquad \mathbf{z} = (z_1, \dots, z_m) \in \mathfrak{B}_m^*,$$

for m-tuples $\mathbf{b} \in (\mathbb{F}_l^*)^m$,

For an integer $h \geq 1$ and a vector $\mathbf{b} \in (\mathbb{F}_l^*)^m$, we denote by $T_{\mathbf{b}}(h)$ the number of distinct values taken by the products $\mathbf{b}^{\mathbf{z}}$, $\mathbf{z} \in \mathfrak{B}_m^*$ among the elements of the set $\{1, \dots, h\}$.

The following statement has been proved in [528].

Lemma 4.6. *For any integers $m \geq 2$, $l - 1 \geq h \geq 1$ and any $\Delta > 0$ with $h \geq \Delta^{-1}\left(l 2^{-m/2} + 1\right) \log l$, for all except at most $O\left(\Delta(l-1)^m\right)$ vectors $\mathbf{b} \in (\mathbb{F}_l^*)^m$ the bound*

$$T_{\mathbf{b}}(h) \geq \frac{\Delta h 2^m}{l + 2^m}$$

holds.

Throughout the book we also consistently use a combination of Lemma 4.1 with bounds of various exponential sums with polynomial, rational and exponential functions, given in Chapter 3, to study the distribution of these functions. In fact, in many cases we could simply quote some results of [229], which apply to a very general setting of joint distribution of values of several rational functions "along a curve" in arbitrary convex sets.

Chapter 5

Arithmetic Functions

We use the well-know estimates

$$\frac{m}{\varphi(m)} = O\left(\log\log(m+1)\right), \qquad \log\tau(m) = O\left(\frac{\log m}{\log\log(m+1)}\right), \qquad (5.1)$$

see Theorem 5.1 and Theorem 5.2 of Chapter 1 of [468]. Also, from the inequality $\nu(m)! \le m$, we derive

$$\nu(m) = O\left(\frac{\log m}{\log\log(m+1)}\right). \qquad (5.2)$$

The above bounds are tight. Thus although sometimes the values of $\tau(m)$ are rather big, "on average", much stronger bounds hold. Namely, for any $X \ge 2$,

$$\sum_{k\le X} \tau(k) \ll X\log X \qquad \text{and} \qquad \sum_{k\le X} \tau^2(k) \ll X\left(\log X\right)^3, \qquad (5.3)$$

see Theorems 5.3 and 5.4 of Chapter 1 of [468].

Besides the *Prime Number Theorem*

$$\pi(x) \sim \frac{x}{\ln x},$$

we also recall the *Brun–Titchmarsh* theorem. Namely, for any integers $k, a \ge 1$ with $1 \le k \le X^{1/2}$ and $\gcd(a,k) = 1$ the bound

$$\pi(X;k,a) \ll \frac{X}{\varphi(k)\log X} \qquad (5.4)$$

holds, see Theorem 4.1 of Chapter 2 of [468].

From the Brun–Titchmarsh theorem we derive estimates of various sums with prime numbers. For example, for any integers $X \ge 3$ and $k \ge \log X$,

$$\sum_{\substack{q \in \mathcal{P},\ q \le X \\ q \equiv 1 \pmod k}} \frac{1}{q} \ll \frac{\log k}{\varphi(k)}. \tag{5.5}$$

Indeed, for $k \ge X^{1/2}$ we have

$$\sum_{\substack{q \in \mathcal{P},\ q \le X \\ q \equiv 1 \pmod k}} \frac{1}{q} \le \sum_{\substack{k \le n \le X \\ n \equiv 1 \pmod k}} \frac{1}{n} \ll \frac{\log X}{k} \ll \frac{\log k}{k}.$$

For $k < X^{1/2}$ we use the same bound, but also the Brun–Titchmarsh estimate (5.4), to deduce that

$$\begin{aligned} \sum_{\substack{q \in \mathcal{P},\ q \le X \\ q \equiv 1 \pmod k}} \frac{1}{q} &\ll \sum_{s=\lceil \log k \rceil}^{\lceil \log X \rceil} \frac{\pi(\exp(s); k, 1)}{\exp(s)} \\ &\ll \frac{\log k}{k} + \sum_{s=\lfloor 2\log k \rfloor}^{\lceil \log X \rceil} \frac{\pi(\exp(s); k, 1)}{\exp(s)} \\ &\ll \frac{\log k}{k} + \frac{1}{\varphi(k)} \sum_{s=1}^{\lceil \log X \rceil} \frac{1}{s} \ll \frac{\log k}{k} + \frac{\log\log X}{\varphi(k)} \ll \frac{\log k}{\varphi(k)}. \end{aligned}$$

We also require a bound for smaller k. For any fixed $\alpha > 0$, there exists $X_0(\alpha)$ such that the following holds. For all $X \ge X_0(\alpha)$ and all $k \le \log X$,

$$\sum_{\substack{q \in \mathcal{P},\ X^\alpha \le q \le X \\ q \equiv 1 \pmod k}} \frac{1}{q} \ll \frac{\log(1/\alpha)}{\varphi(k)}. \tag{5.6}$$

This follows from (5.4) by partial summation.

Let $\Psi(X, Y)$ denote the total number of Y-smooth numbers $k \le X$. The following estimate is a substantially relaxed and simplified version of (for example) Corollary 1.3 of [262], see also [96]. Let $X = Y^u$; then for any $u \to \infty$ with $u \le Y^{1/2}$ we have the bound

$$\Psi(X, Y) \ll X u^{-u+o(u)}. \tag{5.7}$$

Now we are prepared to derive the lower bound from [192] on the number of small values of the Carmichael function. Various upper and lower bounds for $\lambda(n)$ have been obtained in [174]. In particular, it follows from Theorem 2 of [174] that for all except $o(N)$ positive integers $n \le N$,

$$\lambda(n) = n \exp\left(-\log\log n \log\log\log n - C \log\log n + o(\log\log n)\right)$$

for some explicitly given constant C. However for our applications we need to say a little more about the size of the exceptional set (even although we have to lower the bound a little bit).

Lemma 5.1. *For sufficiently large numbers N and for $\Delta \geq (\log\log N)^3$, the number of positive integers $n \leq N$ with*

$$\lambda(n) \leq n \exp(-\Delta)$$

is at most $N \exp\left(-0.69\,(\Delta \log \Delta)^{1/3}\right)$.

Proof. Fix $\Delta \geq (\log\log N)^3$ and let us define K from the equation

$$\frac{(\log K)^3}{\log\log K} = \Delta - \Delta^{1/2};$$

thus $K = \exp\left(\left(3^{-1/3} + o(1)\right)(\Delta \log \Delta)^{1/3}\right)$. Note that $K \geq (\log N)^{\alpha(N)}$ for some $\alpha(N) \to \infty$ as $N \to \infty$.

Let $\mathcal{S}_1$ denote the set of $n \leq N$ with $p^2 | \varphi(n)$ for some prime $p > K$. There are four possibilities for $n \in \mathcal{S}_1$.

- There exists a prime $p > K$ with $p^3 | n$. There are at most

$$\sum_{K \leq p \leq N^{1/3}} \left\lfloor \frac{N}{p^3} \right\rfloor \leq N \sum_{K \leq k \leq N} \frac{1}{k^3} \ll NK^{-2}$$

  such $n \leq N$. Here, and in a number of places below, we are a little inefficient by not saving all possible logarithmic factors.

- There exists a prime $p > K$ with $p^2 | n$ and also there exists a prime $q | n$ with $q \equiv 1 \pmod{p}$. From (5.5) and partial summation we derive that the number of such $n \leq N$ is at most

$$\sum_{K \leq p \leq N^{1/3}} \sum_{\substack{q \in \mathcal{P},\ p < q \leq N/p^2 \\ q \equiv 1 \pmod{p}}} \left\lfloor \frac{N}{p^2 q} \right\rfloor \ll N \sum_{K \leq p \leq N} \frac{\log p}{p^3} \ll NK^{-2}.$$

- There exists a prime $p > K$ and there exists a prime $q | n$ with $q \equiv 1 \pmod{p^2}$. As before we see that the number of such $n \leq N$ is bounded by

$$N \sum_{K \leq p \leq N^{1/3}} \frac{\log p}{p^2} \ll NK^{-1}.$$

- There exists a prime $p > K$ and there exist two distinct primes $q_1 q_2 | n$ with $q_1 \equiv q_2 \equiv 1 \pmod p$. In this, the most frequently occurring case, we see that the number of such $n \le N$ is majorised by

$$N \sum_{K \le p \le N^{1/3}} \frac{(\log p)^2}{p^2} \ll NK^{-1} \log K.$$

So the cardinality of $\mathcal{S}_1$ satisfies $|\mathcal{S}_1| \ll NK^{-1} \log K \le NK^{-1+o(1)}$.

We put

$$v = \frac{\log K}{\log \log K} \qquad \text{and} \qquad L = K^v.$$

and denote by $\mathcal{Q}$ the set of primes $p \le N$ such that the contribution to $p-1$ from primes $q < K$ is at least L. Let $\mathcal{L}$ denote the set of K-smooth integers k with $N \ge k \ge L$. Then, from (5.5) we see that

$$\sum_{p \in \mathcal{Q}} \frac{1}{p} \le \sum_{k \in \mathcal{L}} \sum_{\substack{p \le N \\ p \equiv 1 \pmod k}} \frac{1}{p} \ll \sum_{k \in \mathcal{L}} \frac{\log k}{\varphi(k)}.$$

For the last sum, using (5.1) we derive

$$\sum_{k \in \mathcal{L}} \frac{\log k}{\varphi(k)} \ll \sum_{s = \lceil \log L \rceil}^{\lceil \log N \rceil} \frac{s \log s}{\exp(s)} \Psi(\exp(s), K).$$

It is easy to verify that if for $\lceil \log L \rceil \le s \le \lceil \log N \rceil$ we define u by the equation $\exp(s) = K^u$, then $v \le u \le \log N + 1 \le K^{1/2}$ so the bound (5.7) applies. Therefore

$$\begin{aligned} \sum_{p \in \mathcal{Q}} \frac{1}{p} &\le v^{-v+o(v)} \sum_{s = \lceil \log L \rceil}^{\lceil \log N \rceil} s \log s \\ &\le v^{-v+o(v)} (\log N)^2 \log \log N \le K^{-1+o(1)}. \end{aligned}$$

Denote by $\mathcal{S}_2$ the set of $n \le N$ that are divisible by a prime from $\mathcal{Q}$. Then

$$|\mathcal{S}_2| \le \sum_{p \in \mathcal{Q}} \left\lfloor \frac{N}{p} \right\rfloor \le N \sum_{p \in \mathcal{Q}} \frac{1}{p} \le NK^{-1+o(1)}.$$

Similarly, the cardinality of the set $\mathcal{S}_3$ of $n \le N$, such that the contribution to n itself from primes $q < K$ is at least L, satisfies

$$|\mathcal{S}_3| \le \sum_{k \in \mathcal{L}} \left\lfloor \frac{N}{k} \right\rfloor \le N \sum_{k \in \mathcal{L}} \frac{1}{k} \ll NK^{-1+o(1)}.$$

Let $\omega(n)$ denote the number of distinct prime divisors of an integer $n \geq 1$. Denote by $\mathcal{S}_4$ the set of $n \leq N$ with $\omega(n) \geq \log K - 1$. Because, for every $n \in \mathcal{S}_4$, $\tau(n) \geq 2^{\omega(n)} \geq \frac{1}{2}K^{\log 2} > K^{1/2}$, we derive from the second part of (5.3) that

$$|\mathcal{S}_4| \leq K^{-1} \sum_{n \leq N} \tau^2(n) \ll NK^{-1+o(1)}.$$

Finally, we define the set

$$\mathcal{N} = \{NK^{-1} \leq n \leq N\} \setminus (\mathcal{S}_1 \cup \mathcal{S}_2 \cup \mathcal{S}_3 \cup \mathcal{S}_4).$$

Combining the above results one verifies that $|\mathcal{N}| = N + O\left(NK^{-1+o(1)}\right)$.

For $n \in \mathcal{N}$ we write $\varphi(n) = mM$, where m is the contribution of primes $q < K$ and M is the contribution of primes $q \geq K$. We see that for any $n \in \mathcal{N}$,

$$m \leq L^{\omega(n)+1} \leq L^{\log K} = \exp\left(v\,(\log K)^2\right) = \exp\left(\Delta - \Delta^{1/2}\right)$$

because of our choice of K and v. We also remark that M is square-free. Therefore,

$$\begin{aligned} \lambda(n) \quad &\geq \quad M = \frac{\varphi(n)}{m} \gg \frac{n}{m \log\log n} \\ &\gg \quad \frac{NK^{-1}}{\log\log N} \exp\left(-\Delta + \Delta^{1/2}\right) \gg N \exp(-\Delta) \end{aligned}$$

for sufficiently large N. Taking into account that $3^{-1/3} > 0.69$ we obtain the desired statement. □

Our results about the period of the power generator depend on some general statements from [192] about multiplicative orders of integers modulo $M \geq 2$. They could be of independent interest and can probably find other applications. For example, Lemma 5.2 is a direct improvement of Proposition 1 from [255], see also [149, 216]. It can also be useful for the purposes of the works [350, 467], and possibly some others.

First of all we estimate the proportion of numbers whose multiplicative order modulo $M \geq 2$ is much smaller than the Carmichael function $\lambda(M)$.

Lemma 5.2. *Let M be a positive integer and j a divisor of $\lambda(M)$. Let $N_j(M)$ be the number of integers g with $1 \leq g \leq M$, $\gcd(g, M) = 1$ and $\mathrm{ord}_M g$ dividing $\lambda(M)/j$. Then*

$$N_j(M) \leq \varphi(M)/j.$$

Moreover, for any real $K \geq 1$ the number $S_K(M)$ of integers g, $1 \leq g \leq M$, with $\gcd(g, M) = 1$ and $\mathrm{ord}_M g \leq \lambda(M)/K$ satisfies

$$S_K(M) \leq \varphi(M)\tau(\lambda(M))/K.$$

Proof. Let

$$M = p_1^{m_1} \dots p_\nu^{m_\nu} \qquad \text{and} \qquad \lambda(M) = q_1^{l_1} \dots q_\mu^{l_\mu}$$

be the prime number factorisations of M and $\lambda(M)$. For a divisor j of $\lambda(M)$, let $j = q_1^{j_1} \dots q_\mu^{j_\mu}$ be its prime factorisation, where each exponent j_t satisfies $0 \le j_t \le l_t$. For each $p_i^{m_i}$ in the prime factorisation of M, let d_i be the product of those $q_t^{j_t}$ in the prime factorisation of j for which $q_t^{l_t}$ (in the prime factorisation of $\lambda(M)$) divides $\lambda(p_i^{m_i})$. Note that for each $q_t^{j_t}$ there is at least one d_i divisible by $q_t^{j_t}$, so that $j|d_1 \dots d_\nu$.

For $\mathrm{ord}_M g$ to divide $\lambda(M)/j$ it is necessary that for each $p_i^{m_i}$ we have that $g^{\lambda(p_i^{m_i})/d_i} \equiv 1 \pmod{p_i^{m_i}}$. The number of residues g modulo $p_i^{m_i}$ which satisfy this congruence is at most $\varphi(p_i^{m_i})/d_i$. (It is equal to this bound except when $p_i = 2$, $m_i \ge 3$, $d_i = 2^{m_i - 2}$, in which case it is half of this bound.) Thus, by the Chinese Remainder Theorem, we have

$$N_j(M) \le \varphi(M)/d_1 \dots d_\nu \le \varphi(M)/j,$$

as required. The second statement follows from the first simply by summing over $j|\lambda(M)$, $j \ge K$ (certainly this last part could be sharpened). □

We also need the following elementary statement.

Lemma 5.3. *Assume $d \ge 1$ is a divisor of an integer $n > 1$. Then for any integer g with $gcd(g, n) = 1$ we have that $\varphi(d)/\mathrm{ord}_d g$ divides $\varphi(n)/\mathrm{ord}_n g$. We also have $\varphi(d)/\lambda(d)$ divides $\varphi(n)/\lambda(n)$ and, as a consequence,*

$$\frac{\lambda(d)}{d} \ge \frac{\lambda(n)}{n}.$$

Proof. The natural projection of multiplicative groups

$$(\mathbb{Z}/n\mathbb{Z})^* \to (\mathbb{Z}/d\mathbb{Z})^*$$

gives rise to the projection

$$(\mathbb{Z}/n\mathbb{Z})^* / \langle g \rangle \to (\mathbb{Z}/d\mathbb{Z})^* / \langle g \rangle,$$

and so $\varphi(d)/\mathrm{ord}_d g$ divides $\varphi(n)/\mathrm{ord}_n g$ as claimed. Next, taking any g satisfying $\mathrm{ord}_n g = \lambda(n)$, we deduce that, for this g, $\varphi(d)/\mathrm{ord}_d g$ divides $\varphi(n)/\lambda(n)$. But since $\varphi(d)/\lambda(d)$ divides $\varphi(d)/\mathrm{ord}_d g$ the second statement follows. Finally, using the inequality $d/\varphi(d) \le n/\varphi(n)$, we obtain the third statement from the second one. □

We also need the following well-known consequence of the sieve of Eratosthenes.

Lemma 5.4. *For any integers $m, J \geq 1$,*

$$\sum_{\substack{j=1 \\ \gcd(j,m)=1}}^{J} 1 = \frac{\varphi(m)}{m} J + O(2^{\nu(m)}).$$

Proof. Using the Möbius function $\mu(d)$ over the divisors of q to detect the co-primality condition (see Section 3.d of Chapter 2 of [553]) and interchanging the order of summation, we obtain the Legendre formula

$$\sum_{\substack{j=1 \\ \gcd(j,m)=1}}^{J} 1 = \sum_{d|m} \mu(d) \left\lfloor \frac{J}{d} \right\rfloor = J \sum_{d|m} \frac{\mu(d)}{d} + O\left(\sum_{d|m} |\mu(d)|\right)$$

from which the result follows at once (see Section 4 of Chapter 2 of [553]). □

Let $T_{r,k}(d)$ be the number of r-bit integers $x \in [1, 2^r - 1]$ with exactly k non-zero digits and such that $x \equiv 0 \pmod d$. It follows from Theorem 2 of [370] that there exists an absolute constant $c > 0$ such that

$$T_{r,k}(d) = \frac{1}{d} \binom{r}{k} \left(1 + O\left(\exp(-ck/\log d)\right)\right) \tag{5.8}$$

uniformly for $d \leq K$ where $K = \exp(ck^{1/2})$.

To estimate $T_{r,k}(d)$ for larger values of d we remark that if d is in the range $2^s \leq d \leq 2^{s+1} - 1$ and if $d|x$, then those bits of x in the s rightmost positions are uniquely determined by the bits at the other $r - s$ positions. Therefore

$$T_{r,k}(d) \leq \sum_{j=0}^{k} \binom{r-s}{j}. \tag{5.9}$$

We need a result about the uniformity of distribution of odd square-free numbers with some fixed binary digits.

Let ρ be a restriction on the set $\{1, \ldots, r\}$. We denote by $\mathcal{M}_\rho(r)$ the set of integers x, $0 \leq x \leq 2^r - 1$ such that $x_i = \rho(i)$ for all fixed subscripts $i \in \{1, \ldots, r\}$, where $x = x_1 \ldots x_r$ is the bit representation of x. We also denote by $Q_\rho(r)$ the number of $x \in \mathcal{M}_\rho(r)$ for which $2x + 1$ is square-free.

Lemma 5.5. *For any restriction ρ with $l \leq (r/3)^{1/2} - 1$ fixed positions,*

$$Q_\rho(r) = \frac{8}{\pi^2} 2^{r-l} + O\left(2^{r-l-r/3(l+1)}\right).$$

Proof. Let $R_\rho(r,m)$ be the number of $x \in \mathcal{M}_\rho(r)$ with $m^2|2x+1$. From the inclusion-exclusion principle it follows that

$$Q_\rho(r) = \sum_{\substack{1\le m\le 2^{(n+1)/2}\\ m\equiv 1 \pmod 2}} \mu(m) R_\rho(r,m).$$

Let t be the length of the largest substring of free positions. It is obvious that the elements of $\mathcal{M}_\rho(r)$ can be separated into 2^{r-l-t} groups such that in each group the numbers are of the form $2^s z + a$, $0 \le z \le 2^t - 1$, for some integers s and a. For an odd integer $m \ge 1$, each such group contains $2^t/m^2 + O(1)$ numbers which are congruent to zero modulo m^2. Taking into account that $t \ge r/(l+1)$, we then obtain

$$R_\rho(r,m) = 2^{r-l}/m^2 + O\left(2^{r-l-r/(l+1)}\right).$$

Put $K = 2^{2r/3(l+1)}$. Applying the above asymptotic formula for $m \le K$ and the trivial bound

$$R_\rho(r,m) \le 2^r/m^2 + 1$$

for $m > K$, we obtain

$$\begin{aligned} Q_\rho(r) &= \sum_{\substack{1\le m\le K\\ m\equiv 1 \pmod 2}} \mu(m)\left(\frac{2^{r-l}}{m^2} + O\left(2^{r-l-r/(l+1)}\right)\right) \\ &\quad + O\left(\sum_{\substack{K< m\le 2^{(n+1)/2}\\ m\equiv 1 \pmod 2}} \frac{2^r}{m^2}\right). \end{aligned}$$

We also have

$$\sum_{\substack{1\le m\le K\\ m\equiv 1 \pmod 2}} \frac{\mu(m)}{m^2} = \sum_{m\equiv 1 \pmod 2} \frac{\mu(m)}{m^2} + O(K^{-1}).$$

Since

$$\begin{aligned} \sum_{m\equiv 1 \pmod 2} \frac{\mu(m)}{m^2} &= \sum_{m=1}^{\infty} \frac{\mu(m)}{m^2} - \sum_{m\equiv 0 \pmod 2} \frac{\mu(m)}{m^2} \\ &= \sum_{m=1}^{\infty} \frac{\mu(m)}{m^2} + \frac{1}{4} \sum_{m\equiv 1 \pmod 2} \frac{\mu(m)}{m^2}, \end{aligned}$$

from Exercises 13.a, 13.c and 15 in Chapter 7 of [553] we derive

$$\sum_{m\equiv 1 \pmod 2} \frac{\mu(m)}{m^2} = \frac{4}{3}\sum_{m=1}^{\infty} \frac{\mu(m)}{m^2} = \frac{8}{\pi^2}.$$

Therefore,

$$Q_\rho(r) = \frac{8}{\pi^2} 2^{r-l} + O\left(2^{r-l-r/(l+1)} K + 2^r K^{-1}\right).$$

Finally, since for $l \le (r/3)^{1/2} - 1$ the first term in the 'O'-symbol dominates, the result follows. □

We now need a result about the number of pairs of odd co-prime numbers with some fixed binary digits.

Let $r = 2m$ be an even integer and let ρ be a restriction on the set $\{1, \dots, r\}$. We denote by $\mathcal{N}_\rho(r)$ the set of pairs (x, y) of integers $0 \le x, y \le 2^m - 1$ such that for all fixed subscripts i we have $x_i = \rho(i)$ if $1 \le i \le m$ and $y_{(i-m)} = \rho(i)$ if $m+1 \le i \le 2m$, where $x = x_1 \dots x_m$ and $y = y_1 \dots y_m$ are the bit representations of x and y.

We also denote by $S_\rho(r)$ the number of pairs $(x, y) \in \mathcal{N}_\rho(r)$ for which $\gcd(2x + 1, 2y + 1) = 1$.

Lemma 5.6. *For any restriction ρ with $l \le (r/6)^{1/2} - 1$ fixed positions,*

$$S_\rho(r) = \frac{8}{\pi^2} 2^{r-l} + O\left(2^{r-l-r/6(l+1)}\right).$$

Proof. Let $T_\rho(r, d)$ be the number of pairs $(x, y) \in \mathcal{N}_\rho(r)$ with $d | 2x + 1$ and $d | 2y + 1$. From the inclusion-exclusion principle it follows that

$$S_\rho(r) = \sum_{\substack{1 \le d \le 2^{m+1} \\ d \equiv 1 \pmod 2}} \mu(d) T_\rho(r, d).$$

Let s be the length of the largest substring of free positions i with $1 \le i \le m$ and let t be the length of the largest substring of free positions i with $m+1 \le i \le 2m$. It is obvious that the elements of $\mathcal{N}_\rho(r)$ can be separated into $2^{2m-l-s-t}$ groups such that in each group the pairs are of the form

$$\left(2^\alpha z + a, 2^\beta w + b\right), \qquad 0 \le z \le 2^s - 1, \ 0 \le w \le 2^t - 1,$$

for some integers α, β, a, b.

For an odd integer $d \ge 1$, each such group contains

$$\left(\frac{2^s}{d} + O(1)\right)\left(\frac{2^t}{d} + O(1)\right) = \frac{2^{s+t}}{d^2} + O\left(2^s d^{-1} + 2^t d^{-1} + 1\right)$$

pairs whose both components are divisible by d.

Put $K = 2^{r/3(l+1)}$ and let $d \le K$. Taking into account that s and t are greater than or equal to $m/(l+1)$, we obtain

$$\begin{aligned} T_\rho(r, d) &= 2^{r-l}/d^2 + O\left(2^{r-l-t} d^{-1} + 2^{r-l-s} d^{-1} + 2^{r-l-s-t}\right) \\ &= 2^{r-l}/d^2 + O\left(2^{r-l-m/(l+1)}\right). \end{aligned}$$

Applying the above asymptotic formula for $d \le K$ and the trivial bound

$$T_\rho(r,d) \le 2^r/d^2$$

for $d > K$, we then obtain

$$\begin{aligned} S_\rho(r) &= \sum_{\substack{1\le d\le K\\ d\equiv 1 \pmod 2}} \mu(d)\left(\frac{2^{r-l}}{d^2} + O\left(2^{r-l-r/2(l+1)}\right)\right) \\ &\quad + O\left(\sum_{\substack{K<d\le 2^{m+1}\\ d\equiv 1 \pmod 2}} \frac{2^r}{d^2}\right). \end{aligned}$$

We also have

$$\sum_{\substack{1\le d\le K\\ d\equiv 1 \pmod 2}} \frac{\mu(d)}{d^2} = \sum_{d\equiv 1 \pmod 2} \frac{\mu(d)}{d^2} + O(K^{-1}).$$

As in the proof of Lemma 5.5 we derive

$$S_\rho(r) = \frac{8}{\pi^2}2^{r-l} + O\left(2^{r-l-r/2(l+1)}K + 2^r K^{-1}\right).$$

Finally, since for $l \le (r/6)^{1/2} - 1$ the first term in the 'O'-symbol dominates, the result follows. □

Apparently the results of Lemmas 5.5 and 5.6 can be sharpened by means of some more sophisticated sieve methods (see for instance [257]). However this would not improve our main result.

We define the constant

$$\gamma = \frac{8}{\pi^2} - 2\prod_{p\in\mathcal{P}}\left(1 - \frac{2}{p^2}\right) = 0.165\ldots. \tag{5.10}$$

For an r-bit integer x we denote by x^i the integer obtained by flipping the ith leftmost bit of x, $1 \le i \le r$.

Let i be an integer with $1 \le i \le r$ and let $\mathcal{N}_i(r)$ denote the set of integers x, $0 \le x \le 2^r - 1$ such that $2x+1$ is not square-free but $2x^{(i)}+1$ is. Let $N_i(r)$ be the number of elements in $\mathcal{N}_i(r)$.

Similarly, let $\mathcal{M}_i(r)$ denote the set of integers x, $0 \le x \le 2^r - 1$ such that $2x+1$ is square-free but $2x^{(i)}+1$ is not. Let $M_i(r)$ be the number of elements in $\mathcal{M}_i(r)$.

Lemma 5.7. *For any* i, $1 \le i \le r$, *the bound*

$$N_i(r) = M_i(r) = \gamma 2^r + O\left(2^{7r/8}\right)$$

holds, where γ *is defined by (5.10).*

Proof. The identity $N_i(r) = M_i(r)$ is obvious.

Let $T_i(r,m)$ be the number of integers x, $0 \le x \le 2^r - 1$, such that $2x+1$ is not square-free and

$$2x^{(i)} + 1 \equiv 0 \pmod{(2m-1)^2}.$$

Applying the inclusion-exclusion principle we derive that

$$N_i(r) = \sum_{m=1}^{\infty} \mu(2m-1)\, T_i(r,m). \tag{5.11}$$

Let $R_i(r,k,m)$ be the number of integers x, $0 \le x \le 2^r - 1$, such that $2x+1$ is not square-free and

$$2x + 1 \equiv 0 \pmod{(2k-1)^2}, \quad 2x^{(i)} + 1 \equiv 0 \pmod{(2m-1)^2}. \tag{5.12}$$

Again, applying the inclusion-exclusion principle we derive that

$$T_i(r,m) = -\sum_{k=2}^{\infty} \mu(2k-1)\, R_i(r,k,m).$$

It is easy to see that $2x+1$ and $2x^{(i)}+1$ are relatively prime because they differ by a power of 2. Therefore $R_i(r,k,m) = 0$ if $\gcd(2k-1, 2m-1) > 1$.

Let us now estimate $R_i(r,k,m)$ for k and m with $\gcd(2k-1, 2m-1) = 1$.

The flipping position splits the bits of x into two parts. Let us denote by $t = \max\{i-1,\ r-i\}$ the length of the longer part. Then it is obvious that for any fixing of the ith binary digit and all digits of the shortest part we obtain that the system of congruences (5.12) can be replaced by 2^{r-t} systems of congruences, each of them of the form

$$2^s z + a \equiv 0 \pmod{(2k-1)^2}, \quad 2^s z + b \equiv 0 \pmod{(2m-1)^2}$$

with $0 \le z \le 2^t - 1$, for some integers s, a and b.

Since $\gcd(2k-1, 2m-1) = 1$, applying the Chinese Remainder Theorem we see that these congruences define z uniquely modulo $(2k-1)^2(2m-1)^2$. Thus there are $2^t/(2k-1)^2(2m-1)^2 + O(1)$ such values of z in the interval $0 \le z \le 2^t - 1$.

Putting everything together we obtain

$$\begin{aligned} R_i(r,k,m) &= 2^{r-t}\left(\frac{2^t}{(2k-1)^2(2m-1)^2}+O(1)\right) \\ &= \frac{2^r}{(2k-1)^2(2m-1)^2}+O(2^{r-t}). \end{aligned}$$

From the inequality $t \geq (r-1)/2$ we conclude that

$$R_i(r,k,m) = \frac{2^r}{(2k-1)^2(2m-1)^2}+O(2^{r/2}). \tag{5.13}$$

It is also clear that

$$R_i(r,k,m) \leq 2^r/(2k-1)^2. \tag{5.14}$$

Let $K \geq 2$ be an integer. Using (5.13) for $k \leq K$ and (5.14) for $k > K$ and taking into account that

$$\begin{aligned} \sum_{k>K}\frac{1}{(2k-1)^2} &< \frac{1}{4}\sum_{k>K}\frac{1}{(k-1)^2} < \frac{1}{4}\sum_{k=K}^{\infty}\frac{1}{k(k-1)} \\ &= \frac{1}{4}\sum_{k=K}^{\infty}\left(\frac{1}{k-1}-\frac{1}{k}\right) = \frac{1}{4(K-1)}, \end{aligned}$$

we obtain

$$\begin{aligned} T_i(r,m) &= -\sum_{k=2}^{\infty}\mu(2k-1)\,R_i(r,k,m) \\ &= -\frac{2^r}{(2m-1)^2}\sum_{\substack{2\leq k\leq K\\ \gcd(2k-1,2m-1)=1}}\frac{\mu(2k-1)}{(2k-1)^2} \\ &\qquad +O\left(2^{r/2}\sum_{2\leq k\leq K}1+2^r\sum_{k>K}\frac{1}{(2k-1)^2}\right) \\ &= -\frac{2^r}{(2m-1)^2}\sum_{\substack{2\leq k\leq K\\ \gcd(2k-1,2m-1)=1}}\frac{\mu(2k-1)}{(2k-1)^2}+O\left(2^{r/2}K+2^rK^{-1}\right). \end{aligned}$$

Extending the summation range in the sum to all integers k introduces an additional error of order 2^rK^{-1}. Thus

$$T_i(r,m) = -\frac{2^r}{(2m-1)^2}\sum_{\substack{k=2\\ \gcd(2k-1,2m-1)=1}}^{\infty}\frac{\mu(2k-1)}{(2k-1)^2}+O\left(2^{r/2}K+2^rK^{-1}\right).$$

It is easy to verify that

$$
\begin{aligned}
-\sum_{\substack{k>1\\ \gcd(2k-1,2m-1)=1}} \frac{\mu(2k-1)}{(2k-1)^2} &= 1-\sum_{\gcd(2k-1,2m-1)=1} \frac{\mu(2k-1)}{(2k-1)^2}\\
&= 1-\prod_{\substack{p\in\mathcal{P},\ p\ge 3\\ \gcd(p,2m-1)=1}} \left(1-\frac{1}{p^2}\right).
\end{aligned}
$$

Thus, selecting $K = \lfloor 2^{r/4} \rfloor$ we obtain

$$
T_i(r,m) = \frac{2^r}{(2m-1)^2}\left(1-\prod_{\substack{p\in\mathcal{P},\ p\ge 3\\ \gcd(p,2m-1)=1}} \left(1-\frac{1}{p^2}\right)\right) + O\left(2^{3r/4}\right).
$$

We also have the inequality.

$$
T_i(r,m) \le \frac{2^r}{(2m-1)^2}.
$$

As before we obtain from these bounds and from (5.11) that for any integer M the following asymptotic formula

$$
\begin{aligned}
N_i(r) &= 2^r\sum_{m=1}^{\infty} \frac{\mu(2m-1)}{(2m-1)^2}\left(1-\prod_{\substack{p\in\mathcal{P},\ p\ge 3\\ \gcd(p,2m-1)=1}} \left(1-\frac{1}{p^2}\right)\right)\\
&\qquad + O\left(2^{3r/4}M + 2^r M^{-1}\right)
\end{aligned}
$$

holds

Thus selecting $M = \lfloor 2^{r/8} \rfloor$ we obtain

$$
\begin{aligned}
N_i(r) = 2^r &\sum_{m=1}^{\infty} \frac{\mu(2m-1)}{(2m-1)^2}\\
&-2^r\sum_{m=1}^{\infty} \frac{\mu(2m-1)}{(2m-1)^2}\prod_{\substack{p\in\mathcal{P},\ p\ge 3\\ \gcd(p,2m-1)=1}} \left(1-\frac{1}{p^2}\right) + O\left(2^{7r/8}\right).
\end{aligned}
$$

As in the proof of Lemma 5.5, for the first sum we derive

$$
\sum_{m=1}^{\infty} \frac{\mu(2m-1)}{(2m-1)^2} = \frac{4}{3}\sum_{m=1}^{\infty} \frac{\mu(m)}{m^2} = \frac{8}{\pi^2}.
$$

For the second sum we have

$$
\begin{aligned}
\sum_{m=1}^{\infty} \frac{\mu(2m-1)}{(2m-1)^2} & \prod_{\substack{p\in\mathcal{P},\ p\ge 3\\ \gcd(p,2m-1)=1}} \left(1-\frac{1}{p^2}\right) \\
&= \prod_{\substack{p\in\mathcal{P}\\ p\ge 3}} \left(1-\frac{1}{p^2}\right) \sum_{m=1}^{\infty} \frac{\mu(2m-1)}{(2m-1)^2} \prod_{\substack{p\in\mathcal{P}\\ p\mid 2m-1}} \left(1-\frac{1}{p^2}\right)^{-1} \\
&= \prod_{\substack{p\in\mathcal{P}\\ p\ge 3}} \left(1-\frac{1}{p^2}\right) \sum_{m=1}^{\infty} \frac{\mu(2m-1)}{(2m-1)^2} \prod_{\substack{p\in\mathcal{P}\\ p\mid 2m-1}} \left(\frac{p^2}{p^2-1}\right) \\
&= \prod_{\substack{p\in\mathcal{P}\\ p\ge 3}} \left(1-\frac{1}{p^2}\right) \sum_{m=1}^{\infty} \mu(2m-1) \prod_{\substack{p\in\mathcal{P}\\ p\mid 2m-1}} \left(\frac{1}{p^2-1}\right) \\
&= \prod_{\substack{p\in\mathcal{P}\\ p\ge 3}} \left(1-\frac{1}{p^2}\right) \prod_{\substack{p\in\mathcal{P}\\ p\ge 3}} \left(1-\frac{1}{p^2-1}\right) \\
&= 2\prod_{p\in\mathcal{P}} \left(1-\frac{1}{p^2}\right) \prod_{p\in\mathcal{P}} \left(1-\frac{1}{p^2-1}\right) \\
&= 2\prod_{p\in\mathcal{P}} \left(1-\frac{2}{p^2}\right)
\end{aligned}
$$

and we obtain the desired result. □

Many of the results of this book are non-trivial only under the condition that the base ϑ of the relevant exponential function ϑ^x is of sufficiently large order modulo m. Thus a natural question arises how common this property is.

It is certainly relevant to recall the famous result of [272] proving *Artin's conjecture* under the Extended Riemann Hypothesis and the results of [242, 258, 401, 403] showing (unconditionally) that *Artin's conjecture* may fail only for a very "thin" set of integers.

We also recall the widely known elementary fact that for any fixed integer $\vartheta > 1$, for almost all primes p the multiplicative order t_p of ϑ modulo p is at least $p^{1/2}/\log p$. This simple result has been improved in a series of works, see [172, 281, 404, 452]. In particular, it is shown in [172] that for any function $\varepsilon(p) \to 0$, $t_p \ge p^{1/2+\varepsilon(p)}$ for almost all primes p.

Either of the above bounds (even the elementary $t_p \ge p^{1/2}/\log p$) can be combined with currently known record lower bounds on the largest prime divisor of $p-1$, see [18], showing that for sufficiently many primes p, t_p is of order substantially larger than $p^{1/2}$. In particular, it follows from [18] that $t_p \ge p^{0.677}$ for at least $\Omega(x/\log^2 x)$ primes $p \le x$. It is interesting to remark that similar arguments constitute an essential ingredient of the unconditional deterministic polynomial time primality testing algorithm of [5].

One can also use very similar arguments in a combination with Lemma 1 of [183] to obtain lower bounds on the period of the inversive congruential pseudorandom numbers. More precisely, let us fix an irreducible polynomial $f(X) = X^2 - bX - a \in \mathbb{Z}[X]$. Then for any function $\varepsilon(p) \to 0$, for almost all primes p, for which $f(X)$ remains irreducible (that is, for $(0.5 + o(1))\pi(x)$ primes $p \le x$), the period of the inversive congruential pseudorandom numbers, given by (26.1) and (26.2) with $q = p$, is at least $p^{1/2+\varepsilon(p)}$. This means that this period length is within the nontriviality region of Theorem 26.1.

It is also known [172] that under the Extended Riemann Hypothesis $t_p \ge p^{1+o(1)}$ for almost all p.

Some analogues of the above results are also known for the multiplicative order t_m modulo an integer m. An analogue of Artin's conjecture modulo arbitrary integer numbers has been studied in [337, 338]. It has also been shown in [315] that, for almost all m, $t_m \ge m^{1/2+o(1)}$. Furthermore, under the Extended Riemann Hypothesis $t_m \ge m^{1+o(1)}$ for almost all m, see [314]. In fact, this bound has recently been improved in [339] as

$$t_m \ge m/(\ln m)^{2 \log\log\log m}$$

for almost all m.

Finally, the distribution of multiplicative orders of elements of $\mathbb{Z}_m^*$ has been studied in [336, 354, 355]. These results, together with lower bounds on the Carmichael function, see [174, 192], show that typically the multiplicative order of a random element of $\mathbb{Z}_m^*$ is $m^{1+o(1)}$.

The arithmetic structure of the multiplicative order of 2 modulo primes and odd numbers has been studied in [173, 402, 466]. For example, it has been shown in [29, 466] that for almost all primes p and odd integers m the multiplicative order of 2 modulo p and m, respectively, is not smooth. The method certainly applies to the multiplicative order of any fixed integer $g \ge 2$. In particular, this means that the subgroup of $\mathbb{F}_p^*$ generated by 2 is almost always "safe" against the *Pohlig–Hellman attack*.

Finally, it has been shown in [29] that for almost all integers m and for almost all primes p the values of $\varphi(m)$ and $\varphi(p-1)$ are not smooth. Accordingly, this shows that almost all primes p are "strong". We recall, see [389], that a prime p is called *strong* if both $p-1$ and $p+1$ have a large prime divisor, and if $p-1$ has a prime divisor r such that $r-1$ has a large prime divisor.

Chapter 6

Lattices and the Hidden Number Problem

It is well known that lattices provide a very powerful technique for cryptology, see [421, 422] for very detailed surveys of most important achievements in both cryptanalysis and design of new cryptosystems. However traditionally only heuristic results have obtained with their help. Here we outline some tools which, combined with other number-theoretic techniques, allow us to give rigorous proofs to some of these results, see also a brief survey in [525].

Let $\{\mathbf{b}_1, \dots, \mathbf{b}_s\}$ be a set of linearly independent vectors in $\mathbb{R}^s$. The set of vectors

$$L = \{\mathbf{z} \mid \mathbf{z} = \sum_{i=1}^{s} c_i \mathbf{b}_i, \quad c_1, \dots, c_s \in \mathbb{Z}\}$$

is called an s-dimensional full rank lattice. The set $\{\mathbf{b}_1, \dots, \mathbf{b}_s\}$ is called a *basis* of L.

It has been remarked in Section 2.1 of [392], then in Section 2.4 of [421] and then in Section 2.4 of [422] that the following statement holds which is somewhat stronger than that usually used in the literature.

Lemma 6.1. *There exists a deterministic polynomial time algorithm which, given an s-dimensional full rank lattice L and a vector $\mathbf{r} = (r_1, \dots, r_s) \in \mathbb{R}^s$, finds a lattice vector $\mathbf{v} = (v_1, \dots, v_s)$ satisfying the inequality*

$$\left(\sum_{i=1}^{s} (v_i - r_i)^2\right)^{1/2} \leq \exp\left(O\left(\frac{s \log^2 \log s}{\log s}\right)\right) \times \min\left\{\left(\sum_{i=1}^{s} (z_i - r_i)^2\right)^{1/2}, \quad \mathbf{z} = (z_1, \dots, z_s) \in L\right\}.$$

Proof. The statement is a combination of the Schnorr modification [495] of the lattice basis reduction algorithm of Lenstra, Lenstra and Lovász [329] with a result of Kannan [291] about reduction of the approximate closest vector problem to the approximate shortest vector problem. □

The lattice reduction algorithm, in particular a slightly weaker form of the above statement, has been one of the main tools in solving the *hidden number problem* in [64, 65]. In fact, to simplify the computation here we also use Lemma 6.1 with the same weaker constant as in [64, 65]. Namely, we ignore the logarithmic factors in the exponent, accordingly, our results are weaker by logarithmic factors than they could be.

On the other hand, because the final algorithms in all our applications of Lemma 6.1 are probabilistic one can also use its probabilistic analogue of [6] which gives a slightly better constant and thus would produce another logarithmic refinement of our results.

Let l be a prime number. In the most general settings the hidden number problem is the problem of finding an unknown integer $\alpha \in [0, l-1]$ from the information on k most significant bits of $\lfloor \alpha t_i \rfloor_l$, $i = 1, \ldots, d$, where t_i are chosen uniformly and independently at random from the elements of a certain sequence $\mathcal{T}$ and k and d are some integers.

In [64] an algorithm has been designed to solve this problem for $\mathcal{T} = \mathbb{F}_l^*$ and k and d of order $\log^{1/2} l$. Here we show that their approach works for much more general sets $\mathcal{T}$ which are uniformly distributed modulo l. This generalisation, combined with bounds of exponential sums (and Lemma 4.1 of course) leads to several improvements and extensions of the results of [64]. In fact in almost all cryptographic applications mentioned in [64] the multipliers t_i are selected from certain sets $\mathcal{T}$ which are different from $\mathbb{F}_l^*$ (this explains why proofs of several statements from [64] are not quite correct).

For integers $t_1, \ldots, t_d$, selected in the interval $[0, l-1]$, we denote by $L(t_1, \ldots, t_d)$ the following full rank $d+1$-dimensional lattice generated by the rows of the following $(d+1) \times (d+1)$-matrix

$$\begin{pmatrix} l & 0 & \ldots & 0 & 0 \\ 0 & l & \ldots & 0 & 0 \\ \vdots & \vdots & \ddots & \vdots & \vdots \\ 0 & 0 & \ldots & l & 0 \\ t_1 & t_2 & \ldots & t_d & 1/2^{k+1} \end{pmatrix}. \tag{6.1}$$

The following result is an extension of Theorem 5 of [64], which also generalises several other similar statements which have appeared in [166, 222, 223, 414, 417, 418].

We say that a finite sequence $\mathcal{T}$ of integers is Δ*-homogeneously distributed modulo an integer* $m \geq 2$ if for any integer a with $\gcd(a,m) = 1$, the discrepancy $\mathcal{D}_a(\mathcal{T})$ of the sequence of fractional parts

$$\left\{\frac{at}{m}\right\}, \qquad t \in \mathcal{T},$$

satisfies $\mathcal{D}_a(\mathcal{T}) \leq \Delta$.

We define the *modular norm* of an integer γ modulo an integer $m \geq 1$ as

$$\|\gamma\|_m = \min_{b \in \mathbb{Z}} |\gamma - bm| = \min\left\{\lfloor\gamma\rfloor_m, m - \lfloor\gamma\rfloor_m\right\}.$$

Lemma 6.2. *For a prime* l, *we define*

$$d = 4\left\lceil \log^{1/2} l \right\rceil \qquad \textit{and} \qquad \eta = 0.5\log^{1/2} l + 3.$$

Let α *be a fixed integer in the interval* $[0, l-1]$. *Assume that* $t_1, \ldots, t_l$ *are chosen uniformly and independently at random from the elements of a finite* $2^{-\eta}$*-homogeneously distributed modulo* l *sequence* $\mathcal{T}$ *of integer numbers. Then with probability* $P \geq 1 - 1/l$ *for any vector* $\mathbf{s} = (s_1, \ldots, s_d, 0)$ *with*

$$\left(\sum_{i=1}^{d} \left(\lfloor\alpha t_i\rfloor_l - s_i\right)^2\right)^{1/2} \leq l2^{-\eta},$$

all vectors $\mathbf{v} = (v_1, \ldots, v_d, v_{d+1}) \in L(t_1, \ldots, t_d)$ *satisfying*

$$\left(\sum_{i=1}^{d} \left(v_i - s_i\right)^2\right)^{1/2} \leq l2^{-\eta},$$

are such that

$$v_i \equiv \beta t_i \pmod{l}, \quad i = 1, \ldots, l, \qquad v_{d+1} = \beta/2^{k+1}$$

with some $\beta \equiv \alpha \pmod{l}$.

Proof. For any γ such that $\gamma \not\equiv 0 \pmod{l}$ the probability $P(\gamma)$ of

$$\|\gamma t\|_l > l2^{-\eta+1}$$

for an integer t chosen uniformly at random from the elements of a Δ-homogeneously distributed modulo l sequence is

$$P(\gamma) \geq 1 - 2^{-\eta+2} - \Delta.$$

Thus for the $2^{-\eta}$-homogeneously distributed modulo l sequence $\mathcal{T}$ we have

$$P(\gamma) \geq 1 - \frac{5}{2^\eta}.$$

Therefore, for any $\beta \not\equiv \alpha \pmod{l}$,

$$\begin{aligned}\Pr\left[\exists i \in [1,d] \mid \|\beta t_i - \alpha t_i\|_l > l2^{-\eta+1}\right] &= 1 - (1 - P(\beta - \alpha))^d \\ &\geq 1 - \left(\frac{5}{2^\eta}\right)^d,\end{aligned}$$

where the probability is taken over integers $t_1, \ldots, t_d$ chosen uniformly and independently at random from the elements of $\mathcal{T}$.

Since for $\beta \not\equiv \alpha \pmod{l}$ there are only $l-1$ possible values for $\lfloor \beta \rfloor_l$, we obtain

$$\begin{aligned}&\Pr\left[\forall \beta \not\equiv \alpha \pmod{l},\ \exists i \in [1,d] \mid \|\beta t_i - \alpha t_i\|_l > l2^{-\eta+1}\right] \\ &\qquad \geq 1 - (l-1)\left(\frac{5}{2^\eta}\right)^d > 1 - 1/l\end{aligned}$$

because

$$d(\eta - \log 5) > 2\left\lceil \log^{1/2} l \right\rceil \log^{1/2} l + 2\left\lceil \log^{1/2} l \right\rceil (3 - \log 5) > 2\log l + 1.3\log^{1/2} l.$$

The rest of the proof is identical to the proof of Theorem 5 of [64], we outline it for the sake of completeness.

Let us fix some integers $t_1, \ldots, t_d$ with

$$\min_{\beta \not\equiv \alpha \pmod{l}} \max_{i \in [1,d]} \|\beta t_i - \alpha t_i\|_l > l2^{-\eta+1}. \tag{6.2}$$

Let $\mathbf{v}$ be a lattice point satisfying

$$\left(\sum_{i=1}^{d} (v_i - s_i)^2\right)^{1/2} \leq l2^{-\eta}.$$

Clearly, since $\mathbf{v} \in L(t_1, \ldots, t_d)$, there are integers $\beta, z_1, \ldots, z_d$ such that

$$\mathbf{v} = (\beta t_1 - z_1 l, \ldots, \beta t_d - z_d l, \beta/2^{k+1}).$$

If $\beta \equiv \alpha \pmod{l}$, then we are done.

Now suppose that $\beta \not\equiv \alpha \pmod{l}$. In this case we have

$$\begin{aligned}\left(\sum_{i=1}^{d} (v_i - s_i)^2\right)^{1/2} &\geq \min_{i \in [1,d]} \|\beta t_i - s_i\|_l \\ &\geq \min_{i \in [1,d]} \left(\|\beta t_i - \alpha t_i\|_l - \|s_i - \alpha t_i\|_l\right) \\ &> l2^{-\eta+1} - l2^{-\eta} = l2^{-\eta}\end{aligned}$$

that contradicts our assumption. As we have seen, the condition (6.2) holds with probability exceeding $1 - 2^{-\log^{1/2} l}$ and the result follows. □

We recall the definition $\mathrm{MSB}_{k,p}(t)$ given by (1.1).

Using Lemma 6.2 in the same way as Theorem 5 of [64] is used in the proof of Theorem 1 of that paper, we obtain

Lemma 6.3. *For a prime l, we define*

$$k = \left\lceil \log^{1/2} l + \log\log l \right\rceil, \qquad d = 4\left\lceil \log^{1/2} l \right\rceil, \qquad \eta = 0.5\log^{1/2} l + 3.$$

Let $\mathcal{T}$ be a $2^{-\eta}$-homogeneously distributed modulo l sequence of integer numbers. There exists a deterministic polynomial time algorithm $\mathcal{A}$ such that for any fixed integer $\alpha \in [0, l-1]$ given $2d$ integers

$$t_i \qquad \text{and} \qquad s_i = \mathrm{MSB}_{k,l}(\alpha t_i), \qquad i = 1, \ldots, d,$$

its output satisfies

$$\Pr_{t_1,\ldots,t_d \in \mathcal{T}} \left[\mathcal{A}(t_1, \ldots, t_d; s_1, \ldots, s_d) = \alpha\right] \geq 1 - \frac{1}{l}$$

if $t_1, \ldots, t_d$ are chosen uniformly and independently at random from the elements of $\mathcal{T}$.

Proof. We follow the same arguments as in the proof Theorem 1 of [64] which we briefly outline here for the sake of completeness. We refer to the first d vectors in the defining matrix of $L(t_1, \ldots, t_d)$ as l-vectors.

Let us consider the vector $\mathbf{s} = (s_1, \ldots, s_d, 0)$. Multiplying the last row vector $(t_1, \ldots, t_d, 1/2^{k+1})$ of the matrix (6.1) by α and subtracting certain multiples of l-vectors, we obtain a lattice point

$$\mathbf{u}_\alpha = (u_1, \ldots, u_d, \alpha/2^{k+1}) \in L(t_1, \ldots, t_d)$$

such that

$$|u_i - s_i| < l2^{-k}, \qquad i = 1, \ldots, d+1.$$

Therefore,

$$\left(\sum_{i=1}^{d+1} (u_i - s_i)^2\right)^{1/2} \leq (d+1)^{1/2} l 2^{-k}.$$

Now we can use Lemma 6.1 (with a slightly rougher constant $2^{(d+1)/4}$) to find in polynomial time a lattice vector

$$\mathbf{v} = (v_1, \ldots, v_d, v_{d+1}) \in L(t_1, \ldots, t_d)$$

such that

$$\begin{aligned}
&\left(\sum_{i=1}^{d} (v_i - s_i)^2\right)^{1/2} \\
&\qquad \le 2^{(d+1)/4} \min\left\{ \left(\sum_{i=1}^{d+1} (z_i - s_i)^2\right)^{1/2}, \quad \mathbf{z} = (z_1, \ldots, z_d, z_{d+1}) \in L \right\} \\
&\qquad \le 2^{(d+1)/4} l(d+1)^{1/2} 2^{-k} \le l 2^{-\eta},
\end{aligned}$$

provided that l is sufficiently large. We also have

$$\left(\sum_{i=1}^{d} (u_i - s_i)^2\right)^{1/2} \le l d^{1/2} 2^{-k} \le l 2^{-\eta}.$$

Therefore,

$$\left(\sum_{i=1}^{d} (u_i - v_i)^2\right)^{1/2} \le l 2^{-\eta}.$$

Applying Lemma 6.2, we see that $\mathbf{v} = \mathbf{u}_\alpha$ with probability at least $1 - 1/l$, and therefore, α can be recovered in polynomial time. □

As in [222, 223] we remark that using the full strength of Lemma 6.1 leads to a version of Lemma 6.3 with some

$$k = O\left((\log l)^{1/2} (\log\log l)^{-1/2} \log\log\log l\right).$$

In turn, it implies stronger versions of Theorems 14.1, 14.2, 14.3 and 18.2 (with a slightly larger number of approximations d). The result of [6] implies further refinements, reducing this value to

$$k = O\left((\log l)^{1/2} (\log\log l)^{-1/2} (\log\log\log l)^{1/2}\right).$$

Exactly the same arguments work modulo composite numbers as well, although in this case the output is not α but only some integer β with $\gcd(\alpha - \beta, m) > 1$. In a special situation, corresponding to the RSA encryption, this is enough to either recover the message, or to factor the RSA modulus (and then recover the message anyway), see Chapter 18.

Lemma 6.4. *For an prime m, we define*

$$k = \left\lceil \log^{1/2} m + \log\log m \right\rceil, \qquad d = 4\left\lceil \log^{1/2} m \right\rceil, \qquad \eta = 0.5\log^{1/2} m + 3.$$

Let $\mathcal{T}$ be a $2^{-\eta}$-homogeneously distributed modulo m sequence of integer numbers. There exists a deterministic polynomial time algorithm $\mathcal{A}$ such that for any fixed integer $\alpha \in [0, m-1]$ given $2d$ integers

$$t_i \quad \text{and} \quad s_i = \mathrm{MSB}_{k,m}(\alpha t_i), \qquad i = 1, \ldots, d,$$

its output satisfies

$$\Pr_{t_1,\ldots,t_d \in \mathcal{T}} \left[\gcd\left(\mathcal{A}(t_1,\ldots,t_d; s_1,\ldots,s_d) - \alpha, m\right) > 1\right] \ge 1 - \frac{1}{m}$$

if $t_1, \ldots, t_d$ are chosen uniformly and independently at random from the elements of $\mathcal{T}$.

In fact, even if the sequence $\mathcal{T}$ is not known to be homogeneously distributed but at least admits a non-trivial upper bound for the number of its elements in an interval, one can still obtain some analogues of Lemma 6.3. In particular, several such generalisations of Lemma 6.3 have been obtained in [520, 521] (which are based on Lemmas 2.6).

Here we present a result of [342] which in many cases improves the result of [521] and also has an application to the XTR and LUC key exchange protocol, see Chapter 19.

Let $\omega_1, \ldots, \omega_m$ be a fixed basis of $\mathbb{F}_{p^m}$ over $\mathbb{F}_p$.

For d elements $t_1, \ldots, t_d \in \mathbb{F}_{p^m}$, we denote by $\mathcal{L}_m(t_1, \ldots, t_d)$ the full rank $d+m$-dimensional lattice generated by the rows of the following $(d+m)\times(d+m)$-matrix

$$\begin{pmatrix} p & \ldots & 0 & 0 & \ldots & 0 \\ 0 & \ldots & 0 & 0 & \ldots & 0 \\ \vdots & \ddots & \vdots & \vdots & \ddots & \vdots \\ 0 & \ldots & p & 0 & \ldots & 0 \\ \mathrm{Tr}_{\mathbb{F}_{p^m}/\mathbb{F}_p}(\omega_1 t_1) & \ldots & \mathrm{Tr}_{\mathbb{F}_{p^m}/\mathbb{F}_p}(\omega_1 t_d) & 1/2^{k+1} & \ldots & 0 \\ \vdots & & \vdots & \vdots & \ddots & \vdots \\ \mathrm{Tr}_{\mathbb{F}_{p^m}/\mathbb{F}_p}(\omega_m t_1) & \ldots & \mathrm{Tr}_{\mathbb{F}_{p^m}/\mathbb{F}_p}(\omega_m t_d) & 0 & \ldots & 1/2^{k+1} \end{pmatrix}. \qquad (6.3)$$

Lemma 6.5. *Let p be a prime number and let s be a divisor of m, $s|m$. Let $\mathcal{G}$ be a subgroup of group $\mathcal{N}_{m.s}(p)$ given by (3.5), of cardinality $|\mathcal{G}| \geq p^{(m-m/s)/2+\varepsilon}$ for some fixed $\varepsilon > 0$. Then for*

$$\eta = \left\lceil \log^{1/2} p \right\rceil \qquad \textit{and} \qquad d = \left\lceil \frac{m+1}{\eta - 3} \log p \right\rceil$$

and sufficiently large p the following statement holds. Let

$$\alpha = \sum_{j=1}^{m} a_j \omega_j \in \mathbb{F}_{p^m}, \qquad a_1, \dots, a_m \in \mathbb{F}_p,$$

be a fixed element of $\mathbb{F}_{p^m}$. Assume that $t_1, \dots, t_d \in \mathcal{G}$ are chosen uniformly and independently at random. Then with probability $P \geq 1 - p^{-1}$ for any vector $\mathbf{s} = (s_1, \dots, s_d, 0, \dots, 0)$ with

$$\left(\sum_{i=1}^{d} \left(\mathrm{Tr}_{\mathbb{F}_{p^m}/\mathbb{F}_p}(\alpha t_i) - s_i \right)^2 \right)^{1/2} \leq 2^{-\eta} p,$$

all vectors

$$\mathbf{v} = (v_1, \dots, v_d, v_{d+1}, \dots, v_{d+m}) \in \mathcal{L}_m(t_1, \dots, t_d)$$

satisfying

$$\left(\sum_{i=1}^{d} (v_i - s_i)^2 \right)^{1/2} \leq 2^{-\eta} p,$$

are such that

$$v_i \equiv \sum_{j=1}^{m} b_j \mathrm{Tr}_{\mathbb{F}_{p^m}/\mathbb{F}_p}(\omega_j t_1) \pmod p, \qquad i = 1, \dots, d,$$

and

$$v_{d+j} = \frac{b_j}{2^{k+1}}, \qquad j = 1, \dots, m$$

with some integers $b_j \equiv a_j \pmod p$, $j = 1, \dots, m$.

Proof. We see from Lemma 4.2 that for any $\beta \in \mathbb{F}_{p^m}$ with $\beta \neq \alpha$ the probability $P(\beta)$ that

$$\left\| \mathrm{Tr}_{\mathbb{F}_{p^m}/\mathbb{F}_p}(\alpha t) - \mathrm{Tr}_{\mathbb{F}_{p^m}/\mathbb{F}_p}(\beta t) \right\|_p \leq 2^{-\eta+1} p$$

for $t \in \mathcal{G}$ selected uniformly at random is

$$P(\beta) \leq 2^{-\eta+2} + O(p^{(m-m/s)/2} |\mathcal{G}|^{-1} \log p) \leq 2^{-\eta+2} + O(p^{-\varepsilon} \log p) \leq 2^{-\eta+3},$$

provided that p is large enough.

Therefore, choosing $d = \lceil (m+1)/(\eta - 3) \rceil$, for any $\beta \in \mathbb{F}_{p^m}$ we obtain

$$\Pr\left[\forall i \in [1,d] \mid \left\|\operatorname{Tr}_{\mathbb{F}_{p^m}/\mathbb{F}_p}(\alpha t_i) - \operatorname{Tr}_{\mathbb{F}_{p^m}/\mathbb{F}_p}(\beta t_i)\right\|_p \le 2^{-\eta+1}p\right] = P(\beta)^d \le p^{-m-1},$$

where the probability is taken over $t_1, \ldots, t_d \in \mathcal{G}$ chosen uniformly and independently at random.

From here, we derive

$$\Pr\Big[\forall \beta \in \mathbb{F}_{p^m} \backslash \{\alpha\},\ \forall i \in [1,d] \mid \left\|\operatorname{Tr}_{\mathbb{F}_{p^m}/\mathbb{F}_p}(\alpha t_i) - \operatorname{Tr}_{\mathbb{F}_{p^m}/\mathbb{F}_p}(\beta t_i)\right\|_p \le 2^{-\eta+1}p\Big] \le p^{-1}.$$

The rest of the proof is identical to the proof of Theorem 5 of [64]. Indeed, we fix some $t_1, \ldots, t_d \in \mathcal{G}$ with

$$\min_{\beta \in \mathbb{F}_{p^m} \backslash \{\alpha\}} \max_{i \in [1,d]} \left\|\operatorname{Tr}_{\mathbb{F}_{p^m}/\mathbb{F}_p}(\alpha t_i) - \operatorname{Tr}_{\mathbb{F}_{p^m}/\mathbb{F}_p}(\beta t_i)\right\|_p > 2^{-\eta+1}p. \tag{6.4}$$

Let $\mathbf{v} \in \mathcal{L}_m(t_1, \ldots, t_d)$ be a lattice point satisfying

$$\left(\sum_{i=1}^{d} (v_i - s_i)^2\right)^{1/2} \le 2^{-\eta}p.$$

Since $\mathbf{v} \in \mathcal{L}_m(t_1, \ldots, t_d)$, there are integers $b_1, \ldots, b_m, z_1, \ldots, z_d$ such that

$$\mathbf{v} = \left(\sum_{j=1}^{m} b_j \operatorname{Tr}_{\mathbb{F}_{p^m}/\mathbb{F}_p}(\omega_j t_1) - z_1 p, \ldots, \sum_{j=1}^{m} b_j \operatorname{Tr}_{\mathbb{F}_{p^m}/\mathbb{F}_p}(\omega_j t_d) - z_d p, \frac{b_1}{2^{k+1}}, \ldots, \frac{b_m}{2^{k+1}}\right).$$

If $b_j \equiv a_j \pmod p$, $j = 1, \ldots, m$, then we are done.

Now suppose that $b_j \not\equiv a_j \pmod p$ for some $j = 1, \ldots, m$. Put

$$\beta = \sum_{j=1}^{m} b_j \omega_j.$$

In this case, using (6.4). we obtain

$$\begin{aligned}
\left(\sum_{i=1}^{d}(v_i - s_i)^2\right)^{1/2} &\geq \max_{i\in[1,d]} \left\| \sum_{j=1}^{m} b_j \operatorname{Tr}_{\mathbb{F}_{p^m}/\mathbb{F}_p}(\omega_j t_i) - s_i \right\|_p \\
&\geq \max_{i\in[1,d]} \left(\left\| \operatorname{Tr}_{\mathbb{F}_{p^m}/\mathbb{F}_p}(\alpha t_i) - \sum_{j=1}^{m} b_j \operatorname{Tr}_{\mathbb{F}_{p^m}/\mathbb{F}_p}(\omega_j t_i) \right\|_p \right. \\
&\qquad \left. - \left\| s_i - \operatorname{Tr}_{\mathbb{F}_{p^m}/\mathbb{F}_p}(\alpha t_i) \right\|_p \right) \\
&\geq \max_{i\in[1,d]} \left(\left\| \operatorname{Tr}_{\mathbb{F}_{p^m}/\mathbb{F}_p}(\alpha t_i) - \operatorname{Tr}_{\mathbb{F}_{p^m}/\mathbb{F}_p}(\beta t_i) \right\|_p \right. \\
&\qquad \left. - \left\| s_i - \operatorname{Tr}_{\mathbb{F}_{p^m}/\mathbb{F}_p}(\alpha t_i) \right\|_p \right) \\
&> 2^{-\eta+1}p - 2^{-\eta}p = 2^{-\eta}p
\end{aligned}$$

that contradicts our assumption. As we have seen, the condition (6.4) holds with probability exceeding $1 - p^{-1}$ and the result follows. □

Accordingly, using Lemma 6.5, we obtain the following result.

Lemma 6.6. *Let p be a prime number and let s be a divisor of m, $s|m$. Let $\mathcal{G}$ be a subgroup of group $\mathcal{N}_{m,s}(p)$ given by (3.5), of cardinality $|\mathcal{G}| \geq p^{(m-m/s)/2+\varepsilon}$ for some fixed $\varepsilon > 0$. Then for*

$$k = \left\lceil 2\log^{1/2} p \right\rceil, \qquad d = \left\lceil 4(m+1)\log^{1/2} p \right\rceil$$

and sufficiently large p, the following statement holds. There exists a deterministic polynomial time algorithm $\mathcal{A}$ such that for any $\alpha \in \mathbb{F}_{p^m}$ given $2d$ values $t_i \in \mathcal{G}$ and $s_i = \mathrm{MSB}_{k,p}\left(\operatorname{Tr}_{\mathbb{F}_{p^m}/\mathbb{F}_p}(\alpha t_i)\right)$, $i = 1, \ldots, d$, its output satisfies

$$\Pr_{t_1,\ldots,t_d \in \mathcal{G}} \left[\mathcal{A}(t_1, \ldots, t_d; s_1, \ldots, s_d) = \alpha\right] \geq 1 - p^{-1}$$

if $t_1, \ldots, t_d$ are chosen uniformly and independently at random from $\mathcal{G}$.

Proof. Write

$$\alpha = \sum_{j=1}^{m} a_j \omega_j \in \mathbb{F}_{p^m}, \qquad a_1, \ldots, a_m \in \mathbb{F}_p.$$

We consider the vector $\mathbf{s} = (s_1, \ldots, s_d, s_{d+1}, \ldots, s_{d+m})$ where

$$s_{d+j} = 0, \qquad j = 1, \ldots, m.$$

Taking a linear combination of the jth trace-vectors of the matrix (6.3) with the coefficients a_j, $j = 1, \dots, m$, and subtracting a certain linear combination of the p-vectors, we obtain a lattice point

$$\mathbf{u}_\alpha = (u_1, \dots, u_d, a_1/2^{k+1}, \dots, a_m/2^{k+1}) \in \mathcal{L}_m(t_1, \dots, t_d)$$

such that

$$|u_i - s_i| \le p2^{-k-1}, \qquad i = 1, \dots, d+m,$$

where $u_{d+j} = a_j/2^{k+1}$ for $j = 1, \dots, m$. Therefore,

$$\|\mathbf{u}_\alpha - \mathbf{s}\| \le (d+m)^{1/2} 2^{-k-1} p.$$

Let

$$\eta = \left\lceil \log^{1/2} p \right\rceil.$$

By Lemma 6.1 (used with a slightly rougher constant $2^{(d+m)/4}$) in polynomial time we can find a lattice vector

$$\mathbf{v} = (v_1, \dots, v_d, v_{d+1}, \dots, v_{d+m}) \in \mathcal{L}_m(t_1, \dots, t_d)$$

such that

$$\begin{aligned} \|\mathbf{v} - \mathbf{s}\| &\le 2^{o(d+m)} \min\{\|\mathbf{z} - \mathbf{s}\|, \quad \mathbf{z} \in \mathcal{L}_m(t_1, \dots, t_d)\} \\ &\le 2^{-k+o(d)} p \le 2^{-\eta-1} p, \end{aligned}$$

provided that p is large enough. We also have

$$\left(\sum_{i=1}^{d} (u_i - s_i)^2 \right)^{1/2} \le d^{1/2} 2^{-k-1} p \le 2^{-\eta-1} p.$$

Therefore,

$$\left(\sum_{i=1}^{d} (u_i - v_i)^2 \right)^{1/2} \le 2^{-\eta} p.$$

Applying Lemma 6.5, we see that $\mathbf{v} = \mathbf{u}_\alpha$ with probability at least $1 - p^{-1}$, and therefore the components $a_1, \dots, a_m$ of α can be recovered from the last m components of $\mathbf{v} = \mathbf{u}_\alpha$. □

The bound is non-trivial for subgroups $\mathcal{G}$ of cardinality exceeding

$$|\mathcal{G}| \ge p^{(m-m/s)/2} \log p.$$

For arbitrary subgroups $\mathcal{G} \in \mathbb{F}_{p^m}^*$, without the condition $\mathcal{G} \subseteq \mathcal{N}_{m,s}(q)$, a similar result obtained in [342] is non-trivial only for subgroups $\mathcal{G}$ of cardinality about $p^{m/2+\varepsilon}$. We remark that this general bound is too high for applications to the

XTR and LUC key exchange protocols, thus the bound of Lemma 3.7 is really an essential tool for these applications. The results of [526] can be used to bring the threshold $p^{m/2+\varepsilon}$ down to $p^{(m-1)/2+\varepsilon}$ (under some additional conditions) but even this improved bound remains too high.

In [520] a similar method, combined with bounds of exponential sums and bounds on the number of zeros of sparse polynomials, given by Lemmas 3.1 and 2.6, has been used for the problem of finding an m-sparse polynomial

$$f(X) = \sum_{j=1}^{m} \alpha_j X^{e_j} \in \mathbb{F}_p[X]$$

from approximate values of $\lfloor f(t) \rfloor_l$ at polynomially many points $t \in \mathbb{F}_p$ selected uniformly at random. Similar statements can be obtained for finding f from several least significant bits.

The results of [520], which we present in Chapter 30, are based on certain analogues of Lemmas 6.2 and 6.5.

Let $k \geq 0$ and let $\mathbf{e} = (e_1, \ldots, e_m)$ be an m-dimensional vector of positive integers.

For $t_1, \ldots, t_d \in \mathbb{F}_p$, we denote by $\mathcal{L}_{k.\mathbf{e}}(t_1, \ldots, t_d)$ the full rank $d+m$-dimensional lattice generated by the rows of the $(d+m) \times (d+m)$-matrix

$$\begin{pmatrix} p & 0 & \ldots & 0 & 0 & \ldots & 0 \\ 0 & p & \ldots & 0 & 0 & \ldots & 0 \\ \vdots & \vdots & \ddots & \vdots & \vdots & & \vdots \\ 0 & 0 & \ldots & p & 0 & \ldots & 0 \\ t_1^{e_1} & t_2^{e_1} & \ldots & t_d^{e_1} & 1/2^{k+1} & \ldots & 0 \\ \vdots & \vdots & & \vdots & \vdots & \ddots & \vdots \\ t_1^{e_m} & t_2^{e_m} & \ldots & t_d^{e_m} & 0 & \ldots & 1/2^{k+1} \end{pmatrix}. \tag{6.5}$$

Lemma 6.7. *Let p be a sufficiently large n-bit prime and let $m \geq 1$ be an integer. We define*

$$d = 4\left\lceil (mn)^{1/2} \right\rceil \qquad \text{and} \qquad \eta = 0.5(mn)^{1/2} + 3.$$

Let

$$f(X) = \sum_{j=1}^{m} \alpha_j X^{e_j} \in \mathbb{F}_p[X]$$

be a fixed polynomial over $\mathbb{F}_p$ of degree $\deg f \leq 2^{-2\eta} p^{1/2}$. Assume that $t_1, \ldots,$ $t_d \in \mathbb{F}_p$ are chosen uniformly and independently at random. Then with probability $P \geq 1 - p^{-1}$ for any vector $\mathbf{s} = (s_1, \ldots, s_d, 0, \ldots, 0)$ with

$$\left(\sum_{i=1}^{d} \left(\lfloor f(t_i) \rfloor_p - s_i \right)^2 \right)^{1/2} \leq 2^{-\eta} p,$$

all vectors

$$\mathbf{v} = (v_1, \dots, v_d, v_{d+1}, \dots, v_{d+m}) \in \mathcal{L}_{k,\mathbf{e}}(t_1, \dots, t_d)$$

satisfying

$$\left(\sum_{i=1}^{d} (v_i - s_i)^2\right)^{1/2} \le 2^{-\eta} p,$$

are of the form

$$\mathbf{v} = \left(\left\lfloor \sum_{j=1}^{m} \beta_j t_1^{e_j} \right\rfloor_p, \dots, \left\lfloor \sum_{j=1}^{m} \beta_j t_d^{e_j} \right\rfloor_p, \right.$$
$$\left. \beta_1/2^{k+1}, \dots, \beta_m/2^{k+1}\right)$$

with some integers $\beta_j \equiv \alpha_j \pmod p$, $j = 1, \dots, m$.

Proof. Let $\mathcal{P}_f$ denote the set of $p^m - 1$ polynomials

$$g(X) = \sum_{j=1}^{m} \beta_j X^{e_j} \in \mathbb{F}_p[X]$$

with $g \ne f$.
We see from Lemma 3.1 and Lemma 4.1 (applied with $N = 1$, $L = p - 1$) that for any polynomial $g \in \mathcal{P}_f$ the probability $P(g)$ that

$$\|g(t) - f(t)\|_p \le 2^{-\eta+1} p$$

for $t \in \mathbb{F}_p$ selected uniformly at random is

$$P(g) = 2^{-\eta+2} + O\left(\deg f \cdot p^{-1/2} \log p\right) = 2^{-\eta+2} + O\left(2^{-2\eta} \log p\right) \le \frac{5}{2^\eta},$$

provided that p is large enough.
Therefore, for any $g \in \mathcal{P}_f$,

$$\Pr\left[\exists i \in [1,d] \mid \|g(t_i) - f(t_i)\|_p > 2^{-\eta+1} p\right] = 1 - P(g)^d \ge 1 - \left(\frac{5}{2^\eta}\right)^d,$$

where the probability is taken over $t_1, \dots, t_d \in \mathbb{F}_p$ chosen uniformly and independently at random.
Since $|\mathcal{P}_f| = p^m - 1$, we obtain

$$\Pr\left[\forall g \in \mathcal{P}_f,\ \exists i \in [1,d] \mid \|g(t_i) - f(t_i)\|_p > 2^{-\eta+1} p\right]$$
$$\ge 1 - (p^m - 1)\left(\frac{5}{2^\eta}\right)^d > 1 - 2^{-\eta}$$

because

$$\begin{aligned} d(\eta - \log 5) &> 4(mn)^{1/2}\left(0.5(mn)^{1/2} + 3 - \log 5\right) \\ &> 2mn + \eta \geq 2m \log p + \eta, \end{aligned}$$

provided that p is large enough.

The rest of the proof is identical to the proof of Lemma 6.2. Indeed, we fix some integers $t_1, \ldots, t_d$ with

$$\min_{g \in \mathcal{P}_f} \max_{i \in [1,d]} \|g(t_i) - f(t_i)\|_p > 2^{-\eta+1}p. \tag{6.6}$$

Let $\mathbf{v} \in \mathcal{L}_{k,\mathbf{e}}(t_1, \ldots, t_d)$ be a lattice point satisfying

$$\left(\sum_{i=1}^{d} (v_i - s_i)^2\right)^{1/2} \leq 2^{-\eta}p.$$

Since $\mathbf{v} \in \mathcal{L}_{k,\mathbf{e}}(t_1, \ldots, t_d)$, there are some integers $\beta_1, \ldots, \beta_m, z_1, \ldots, z_d$ such that

$$\mathbf{v} = \left(\sum_{j=1}^{m} \beta_j t_1^{e_j} - z_1 p, \ldots, \sum_{j=1}^{m} \beta_j t_d^{e_j} - z_d p, \beta_1/2^{k+1}, \ldots, \beta_m/2^{k+1}\right).$$

If $\beta_j \equiv \alpha_j \pmod p$, $j = 1, \ldots, m$, then for all $i = 1, \ldots, d$ we have

$$\sum_{j=1}^{m} \beta_j t_i^{e_j} - z_i p = \left\lfloor \sum_{j=1}^{m} \beta_j t_i^{e_j} \right\rfloor_p = \lfloor f(t_i) \rfloor_p,$$

since otherwise there is $i \in [1,d]$ such that $|v_i - s_i| > 2^{-\eta}p$.

Now suppose that $\beta_j \not\equiv \alpha_j \pmod p$ for some $j \in [1,m]$. In this case we have

$$\begin{aligned} \left(\sum_{i=1}^{d} (v_i - s_i)^2\right)^{1/2} &\geq \min_{i \in [1,d]} \left\| \sum_{j=1}^{m} \beta_j t_i^{e_j} - s_i \right\|_p \\ &\geq \min_{i \in [1,d]} \left(\left\| f(t_i) - \sum_{j=1}^{m} \beta_j t_i^{e_j} \right\|_p - \|s_i - f(t_i)\|_p \right) \\ &> 2^{-\eta+1}p - 2^{-\eta}p = 2^{-\eta}p \end{aligned}$$

that contradicts our assumption. As we have seen, the condition (6.6) holds with probability exceeding $1 - p^{-1}$ and the result follows. □

One of the crucial ingredients of Lemma 6.7 is the Weil bound. On the other hand, in the case of "dense polynomials", that is, when $e_j = j$, $j = 1, \ldots, m$, the Weil bound is not necessary. Similar results can be obtained from the trivial estimate $P(g) \le \deg f \left(2^{-\eta+2} + 1/p\right)$. For polynomials f of large degree we can use Lemma 2.6 (in fact the inequality (2.5)) instead of Lemma 3.1 to estimate $P(g)$.

Lemma 6.8. *Let p be a sufficiently large n-bit prime and let $m \ge 1$ be an integer. Let*

$$f(X) = \sum_{j=1}^{m} \alpha_j X^{e_j} \in \mathbb{F}_p[X]$$

be a fixed polynomial over $\mathbb{F}_p$ with the exponents

$$1 \le e_1 < \ldots < e_m \le p - 2$$

satisfying

$$\max_{0 \le i < j \le m} \gcd(e_j - e_i, p - 1) \le p^{1-\rho},$$

for some $\rho > 0$, where $e_0 = 0$. Then for any $\varepsilon > 0$ and

$$\eta = (1 - \rho/m + \varepsilon)n + 6, \qquad d = \lceil (m+1)/\varepsilon \rceil$$

the following statement holds. Assume that $t_1, \ldots, t_d \in \mathbb{F}_p$ are chosen uniformly and independently at random. Suppose that $k \ge \eta$, then with probability $P \ge 1 - p^{-1}$ for any vector $\mathbf{s} = (s_1, \ldots, s_d, 0, \ldots, 0)$ with

$$\left(\sum_{i=1}^{d} \left(\lfloor f(t_i) \rfloor_p - s_i \right)^2 \right)^{1/2} \le 2^{-\eta} p,$$

all vectors

$$\mathbf{v} = (v_1, \ldots, v_d, v_{d+1}, \ldots, v_{d+m}), \in \mathcal{L}_{k,\mathbf{e},p}(t_1, \ldots, t_d)$$

satisfying

$$\left(\sum_{i=1}^{d} (v_i - s_i)^2 \right)^{1/2} \le 2^{-\eta} p,$$

are of the form

$$\mathbf{v} = \left(\left\lfloor \sum_{j=1}^{m} \beta_j t_1^{e_j} \right\rfloor_p, \ldots, \left\lfloor \sum_{j=1}^{m} \beta_j t_d^{e_j} \right\rfloor_p, \beta_1/2^{k+1}, \ldots, \beta_m/2^{k+1} \right)$$

with some integers $\beta_j \equiv \alpha_j \pmod{p}$, $j = 1, \ldots, m$.

Proof. In the notations of the proof of Lemma 6.7 we see from (2.5) that

$$\begin{aligned} P(g) &\le 3\left(2^{-\eta+2}p+1\right)p^{-\rho/m} \le 2^{-\eta+5}p^{1-\rho/m} \\ &\le 2^{-\eta-(1-\rho/m)n+6} = 2^{-\varepsilon n-1} \le p^{-\varepsilon} \end{aligned}$$

provided that p is large enough.

Therefore, for any $g \in \mathcal{P}_f$,

$$\Pr\left[\exists i \in [1,d] \mid \|g(t_i) - f(t_i)\|_p > 2^{-\eta+1}p\right] = 1 - P(g)^d \ge 1 - p^{-\varepsilon d},$$

where the probability is taken over $t_1, \dots, t_d \in \mathbb{F}_p$ chosen uniformly and independently at random.

Since $|\mathcal{P}_f| = p^m - 1$, we obtain

$$\begin{aligned} \Pr\left[\forall g \in \mathcal{P}_f,\ \exists i \in [1,d] \mid \|g(t_i) - f(t_i)\|_p > 2^{-\eta+1}p\right] \\ \ge 1 - (p^m - 1)p^{-\varepsilon d} > 1 - p^{-1} \end{aligned}$$

for $d = \lceil (m+1)/\varepsilon \rceil$.

The rest of the proof is quite analogous to the proof of Lemma 6.7. □

In the above results the value of the multiplier or the evaluation point t is known. Now we consider the situation where this is no longer the case. Namely, we consider the following modification of the hidden number problem: recover a number $\alpha \in \mathbb{F}_p$ such that for many random $t \in \mathbb{F}_p$ approximations to the values of both $\lfloor \alpha t \rfloor_p$ and t are known. In fact we consider this problem modulo an arbitrary integer m.

The paper [275] provides a probabilistic polynomial time algorithm for this problem when the absolute error of approximation is $O(m^{1/5-\varepsilon})$. On the other hand, it is shown in [275] that for a prime $m = p$ this problem cannot be solved in polynomial time if the absolute error of approximation is up to $p^{1/2+\varepsilon}$. This lower bound makes use of bounds of exponential sums and uniformity of distribution results. Thus this contrasts quite dramatically to the situation where t is known as in the original hidden number problem, in which case one can recover α from very crude approximations to $\lfloor \alpha t \rfloor_p$ of order $\Delta = p\exp\left(-c(\log p)^{1/2}\right)$ with some absolute constant $c > 0$.

We apply this result to study the bit security of "timed-release crypto" introduced by Rivest, Shamir and Wagner [477], see Theorem 15.1 and the bit security of the inversive Diffie–Hellman mapping, see Theorem 14.4. It has also been used in [275] to design a "correction" algorithm for "noisy" exponentiation black-boxes, see Theorem 30.6. That result is similar to those known for algebraic functions over finite fields [12, 13, 175, 217, 300]. Finally, it has also been noted in [275] that the result can be used for reconstructing "noisy" *Weil pairing* on elliptic curves.

As before our main tool is Lemma 6.1. We also need the following result which can be interpreted as a statement about short vectors in a certain two-dimensional lattice, and thus has some links with Lemma 6.1. However, as usual with two-dimensional lattices, continued fractions provide stronger statements and shorter proofs.

Lemma 6.9. *There exists a polynomial time algorithm which, for a given* $Q \in [1, m]$, *for any* $A \in \mathbb{Z}_m$, *finds* $\lambda \in \mathbb{Z}_m$ *such that*

$$0 < \|\lambda\|_m < Q \qquad \text{and} \qquad \|\lambda A\|_m \le m/Q.$$

Proof. Let P_i and Q_i denote respectively the numerator and denominator of the ith continued fraction convergents to A/m, $i \ge 1$. There exists j such that $Q_j < Q \le Q_{j+1}$. Then we have

$$\left|\frac{A}{m} - \frac{P_j}{Q_j}\right| \le \frac{1}{Q_j Q_{j+1}}.$$

Therefore $|AQ_j - mP_j| \le m/Q_{j+1}$. Selecting $\lambda = Q_j$, we obtain the desired statement. □

We are now able to formulate an algorithm for the hidden number problem with approximately given multipliers.

Lemma 6.10. *Let* $\varepsilon > 0$ *and* $1 \ge \rho > 1/2$ *be fixed real numbers and let* m *be a sufficiently large integer. Let* $\mathcal{T}$ *be a subset of* $\mathbb{Z}_m^*$ *of cardinality*

$$|\mathcal{T}| \ge 0.5m^\rho.$$

We define

$$\Delta = m^{(2\rho-1)/5-\varepsilon} \qquad \text{and} \qquad d = \lceil 4\varepsilon^{-1}/5 \rceil.$$

Assume that the smallest prime divisor p *of* m *satisfies the inequality*

$$p \ge (m/\Delta)^{1/2}.$$

There exists a deterministic polynomial time algorithm $\mathcal{A}$ *such that for any* $a \in \mathbb{Z}_m^*$, *given* $2d+1$ *integers* $A, T_i, S_i \in [0, m-1]$ *with*

$$|a - A| \le \Delta, \quad |t_i - T_i| \le \Delta, \quad |\lfloor at_i \rfloor_m - S_i| \le \Delta, \qquad i = 1, \ldots, d,$$

where $t_1, \ldots, t_d$ *are chosen uniformly and independently at random from* $\mathcal{T}$, *it outputs* a *with probability at least* $1 + O(m^{-1})$.

Proof. Using Lemma 6.9 we find a non-zero integer λ with

$$\|\lambda\|_m \le (m/\Delta)^{1/2} \qquad \text{and} \qquad \|\lambda A\|_m \le (m\Delta)^{1/2}.$$

We define $V_i = \lfloor \lambda T_i \rfloor_m$, $U_i = \lfloor \lambda S_i \rfloor_m$, $i = 1, \dots, d$. Let $T_i - t_i = \delta_i$, $\lfloor \alpha t_i \rfloor_m - S_i = \Delta_i$, $i = 1, \dots, d$, and $\alpha - A = a$. Then for $i = 1, \dots, d$, we have

$$\begin{aligned} \alpha V_i - U_i &\equiv \alpha\lambda T_i - \lambda S_i \equiv \alpha\lambda(t_i + \delta_i) - \alpha\lambda t_i + \lambda\Delta_i \\ &\equiv \alpha\lambda\delta_i + \lambda\Delta_i \equiv A\lambda\delta_i + a\lambda\delta_i + \lambda\Delta_i \pmod m. \end{aligned}$$

Therefore

$$\begin{aligned} \|\alpha V_i - U_i\|_m &< \|A\lambda\|_m \|\delta_i\|_m + \|a\|_m \|\lambda\|_m \|\delta_i\|_m + \|\lambda\|_m \|\Delta_i\|_m \\ &\le \|A\lambda\|_m \Delta + \|\lambda\|_m \Delta^2 + \|\lambda\|_m \Delta \le 3m^{1/2}\Delta^{3/2}. \end{aligned}$$

We denote by $\mathcal{L}$ the $d+1$-dimensional lattice generated by the rows of the following $(d+1) \times (d+1)$-matrix

$$\begin{pmatrix} m & 0 & \dots & 0 & 0 \\ 0 & m & \dots & 0 & 0 \\ \vdots & \vdots & \ddots & \vdots & \vdots \\ 0 & 0 & \dots & m & 0 \\ V_1 & V_2 & \dots & V_d & 1/m \end{pmatrix}.$$

Then we conclude that this lattice contains a vector $\mathbf{v} = (v_1, \dots, v_d, v_{d+1})$ with $v_i \equiv \alpha V_i \pmod m$, $i = 1, \dots, d$, $0 \le v_{d+1} = \alpha/m < 1$ and such that

$$\sum_{i=1}^{d+1} (v_i - U_i)^2 \le 9dm\Delta^3 + 1 \le 10dm\Delta^3 \tag{6.7}$$

where $U_{d+1} = 0$. To find $\mathbf{v}$ and thus to recover a we use the algorithm of Lemma 6.1, getting a lattice vector $\mathbf{w} = (w_1, \dots, w_d, w_{d+1})$ with

$$\sum_{i=1}^{d+1} (w_i - U_i)^2 \le 10d2^d m\Delta^3. \tag{6.8}$$

We show that with probability $1 + O(m^{-1})$ we have $\mathbf{v} = \mathbf{w}$.

Let $\mathbf{w} = (w_1, \dots, w_d, w_{d+1})$ where $w_i \equiv \beta V_i \pmod m$, $i = 1, \dots, d$, and $w_{d+1} = \beta/m$ with some integer $\beta \in [0, m-1]$.

From (6.7) and (6.8) we derive

$$(w_i - v_i)^2 \le (v_i - U_i)^2 + (w_i - U_i)^2 \le 20d2^d m\Delta^3, \qquad i = 1, \dots, d+1.$$

Therefore for every $i = 1, \dots, d$ the inequality

$$\|(\alpha - \beta)\lambda T_i\|_m = \|(\alpha - \beta)V_i\|_m = |w_i - v_i| \le 5d^{1/2}2^{d/2}m^{1/2}\Delta^{3/2} \tag{6.9}$$

holds.

If $\alpha \equiv \beta \pmod{m}$, then $\mathbf{v} = \mathbf{w}$.

For each $c \in \mathbb{Z}_m$ with $\gcd(c, m) = f$ and every integer $h \in [0, m/2)$ there are at most

$$f\left(\left\lfloor \frac{2h+1}{f} \right\rfloor + 1\right) \le 2h + f + 1$$

values of $T \in \mathbb{Z}_m$ with $\|cT\|_m \le h$.

Because of the conditions of the theorem $0 < |\lambda| \le (m/\Delta)^{1/2} < p$ thus $\gcd(\lambda, m) = 1$. Assume that $\alpha \not\equiv b \pmod{m}$. Then

$$\gcd((\alpha - \beta)\lambda, m) \le (m\Delta)^{1/2}.$$

Hence the inequality

$$\|(\alpha - \beta)\lambda T\|_m \le 5d^{1/2}2^{d/2}m^{1/2}\Delta^{3/2}$$

holds for at most

$$10d^{1/2}2^{d/2}m^{1/2}\Delta^{3/2} + (m\Delta)^{1/2} + 1 \le 11d^{1/2}2^{d/2}m^{1/2}\Delta^{3/2}$$

values of $T \in \mathbb{Z}_m$. For each T there could be at most $2\Delta + 1 \le 3\Delta$ values of $t \in \mathbb{Z}_m^*$ with $|t - T| \le \Delta$.

Thus for each $i = 1, \ldots, d$ there are at most $33d^{1/2}2^{d/2}m^{1/2}\Delta^{5/2}$ values of $t_i \in \mathbb{Z}_m^*$ for which (6.9) holds. Hence the probability that (6.9) holds for every $i = 1, \ldots, d$ when the multipliers $t_1, \ldots, t_d$ are chosen uniformly and independently at random from $\mathbb{Z}_m^*$ is at most

$$\begin{aligned}\left(\frac{33d^{1/2}2^{d/2}m^{1/2}\Delta^{5/2}}{|T|}\right)^d &\le \left(\frac{33d^{1/2}2^{d/2}m^{1/2}\Delta^{5/2}}{0.5m^\rho}\right)^d \\ &\le \left(66d^{1/2}2^{d/2}\right)^d m^{-5d\varepsilon/2}\end{aligned}$$

provided that m is large enough. Thus the probability that this happens for at least one $b \in \mathbb{Z}_m$ is at most

$$\left(66d^{1/2}2^{d/2}\right)^d m^{1-5d\varepsilon/2} \le \left(66d^{1/2}2^{d/2}\right)^d m^{-1}$$

and the result follows. □

Recalling that the smallest prime divisor p of m satisfies the inequality

$$p \ge (m/\Delta)^{1/2} \ge m^{(3-\rho)/5+\varepsilon/2} \ge m^{2/5} > m^{1/3}$$

we see that m has at most two prime divisors. Therefore

$$|\mathbb{Z}_m^*| \ge m(1 - 1/p)^{O(1)} = m + O(m^{3/5}).$$

Hence for $\mathcal{T} = \mathbb{Z}_m^*$ one can apply Lemma 6.10 with $\rho = 1$ and thus with $\Delta = m^{1/5-\varepsilon}$.

It is easy to see that Lemma 6.10 applies to prime values of $m = p$ as well as to RSA moduli $m = pl$ which are products of two primes $p < l < 2p$.

The problem of recovering a hidden number α from several least significant bits of $\lfloor 1/(\alpha + t) \rfloor_p$, for polynomially many randomly selected $t \in \mathbb{F}_p$, has been studied in [57].

Several more modifications of the hidden number problem and their applications have been considered in [298]. Variants of the hidden number problem related to the Chinese Remainder Theorem can be found in [529, 530]. More precisely, the paper [529] considers the Chinese Remainder Theorem in a "hidden" algebraic number field; the paper [530] considers a "noisy" variant of the Chinese Remainder Theorem.

It has been shown in [199] that Lemma 6.6 yields certain bit security results for several new cryptographic schemes, relying on pairing on elliptic curves or abelian varieties, including Weil and Tate pairing. Its weaker analogue from [521] (which however applies to arbitrary subgroups), derived from Lemma 2.6, can be used as well.

A detailed classification and survey of various types of the hidden number problem, together with a brief summary of their applications, are given in [524].

Chapter 7

Complexity Theory

The relation between the CREW PRAM complexity and the sensitivity of a Boolean function is given by the following inequality which is essentially Theorem 4.7 of [453].

Lemma 7.1. *For the CREW PRAM complexity of any Boolean function B the inequality*

$$\mathit{CREW\ PRAM}(B) \geq 0.5 \log \sigma(B) + O(1)$$

holds.

We recall the following fundamental property of the Fourier coefficients of a Boolean function B of r variables, which is known as the *Parseval identity*:

$$\sum_{a \in \mathfrak{B}_r} \left| \widehat{B}(a) \right|^2 = 2^{2r}. \tag{7.1}$$

We make use of the following estimate which is due to [345], see also Theorem 11.7 of [364].

We recall that as before $\|a\|$ is the Hamming weight of $a \in \mathfrak{B}_r$.

Lemma 7.2. *Assume that a Boolean function B of r variables is computed by a circuit $C \in \mathrm{UBC}(d, S)$. Then for any $w \leq r$ the bound*

$$2^{-2r} \sum_{\substack{a \in \mathfrak{B}_r \\ \|a\| > w}} \left| \widehat{B}(a) \right|^2 \leq S 2^{-0.05 w^{1/d} + 1},$$

holds.

Some refinements of Lemma 7.2 have recently been given in [253], however they do not seem to improve our applications.

For the decision tree complexity an analogous result is known as well, see [40, 72, 147]). Namely we make use of the following statement.

Lemma 7.3. *For the decision tree complexity* $\mathrm{DT}(B)$ *of a Boolean function* B *of* r *variables the bound*

$$\mathrm{DT}(B) > 2^{-2r} \sum_{\substack{a \in \mathfrak{B}_r \\ \|a\| \geq w}} 2^{\|a\|} \left|\widehat{B}(a)\right|^2$$

holds.

In the next statement, which has been obtained in [40], we establish an upper bound on the Fourier coefficients of Boolean functions δ-approximated at level k.

Lemma 7.4. *Let a Boolean function* B *of* r *variables be* δ*-approximated at level* k*. Then,*

$$\left|\widehat{B}(a)\right| \leq \delta 2^r$$

for any $a \in \mathfrak{B}_r$ *such that* $1 \leq \|a\| \leq r - k$*.*

Proof. Let $a = (a_1, a_2, \ldots, a_r) \in \mathfrak{B}_r$ be a Boolean vector such that $0 < \|a\| = r - l \leq r - k$. Moreover, let $\mathcal{I} = \{i \mid a_i = 1\}$; thus $|\mathcal{I}| = r - l$.

For any vector $u \in \mathfrak{B}_{r-l}$, let $B_{a,u}$ denote the subfunction defined by the restriction $\rho_{a,u}$ that assigns to the variables x_i such that $i \in \mathcal{I}$, the $r - l$ values taken from the vector u, and leaves free the other l variables. Then, we have

$$\begin{aligned}\widehat{B}(a) &= \sum_{u \in \mathfrak{B}_r} (-1)^{B(u) + \langle au \rangle} = \sum_{u \in \mathfrak{B}_r} (-1)^{B(u) + \sum_{i \in \mathcal{I}} u_i} \\ &= \sum_{w \in \mathfrak{B}_{r-l}} (-1)^{\|w\|} \sum_{v \in \mathfrak{B}_l} (-1)^{B_{a,w}(v)} = \sum_{w \in \mathfrak{B}_{r-l}} (-1)^{\|u\|} \widehat{B}_{a,w}(0).\end{aligned}$$

For any $w \in \mathfrak{B}_{r-l}$, the subfunction $B_{a,w}$ depends on $l \geq k$ variables and, since B is δ-approximated at level k, we have

$$\left|\frac{1}{2^l} \widehat{B}_{a,w}(0) - \frac{1}{2^r} \widehat{B}(0)\right| \leq \delta.$$

Therefore,

$$\begin{aligned}\left|\widehat{B}(a)\right| &= \left|\frac{2^l}{2^r}\widehat{B}(0)\sum_{w\in\mathfrak{B}_{r-l}}(-1)^{\|w\|}+\sum_{w\in\mathfrak{B}_{r-l}}(-1)^{\|w\|}\left(\widehat{B}_{a,w}(0)-\frac{2^l}{2^r}\widehat{B}(0)\right)\right|\\ &= \left|\sum_{w\in\mathfrak{B}_{r-l}}(-1)^{\|w\|}\left(\widehat{B}_{a,w}(0)-\frac{2^l}{2^r}\widehat{B}(0)\right)\right|\\ &= 2^l\left|\sum_{w\in\mathfrak{B}_{r-l}}(-1)^{\|w\|}\left(\frac{1}{2^l}\widehat{B}_{a,w}(0)-\frac{1}{2^r}\widehat{B}(0)\right)\right|\le 2^l\sum_{u\in\mathfrak{B}_{r-l}}\delta,\end{aligned}$$

and the result immediately follows. □

We also need some relations between average sensitivity of a Boolean function and its real and real approximate degrees, $\Delta(B)$ and $\delta(B)$.

These relations are by Corollary 2.5 and by Lemma 3.8 of [447] which we summarize in the following statement.

Lemma 7.5. *Let B be a Boolean function. Then*

$$\Delta(B)\ge\sigma_{av}(B) \qquad \textit{and} \qquad \delta(B)\ge(\sigma_{av}(B)/6)^{1/2}.$$

The following result from [252], Theorem 4.2, establishes links between classes $\mathcal{P}^r_{k,s}$ and combinatorial complexity $CC(B)$.

Lemma 7.6. *For any function $B\in\mathcal{P}^r_{3,5}$ the bound*

$$CC(B)\ge\frac{7r-4}{6}$$

holds.

We also note that $B\in\mathcal{P}^r_{1,2}$ implies $CC(B)\ge r-1$ and that for each k there are infinitely many r and functions $B\in\mathcal{P}^r_{k,2^k}$ with $CC(B)\le 13(r+1)$; see Lemma 3.1 and Theorem 6.2 from [252]. Thus, although bounds on combinatorial complexity obtained this way are not trivial, there is no hope to get a non-linear bound.

We also need the following statement which provides a link the communication complexity $\mathrm{Comm}(B)$ and combinatorial discrepancy $\Gamma(B)$. A partial case of Lemma 2.2 from [15] gives a lower bound

$$\mathrm{Comm}(B)\ge\log\left(\frac{1}{\Gamma(B)}\right) \tag{7.2}$$

(which has actually been used in [509]). However, as it has been noticed in [299], using a certain result of [448], one can get a better value for the constant.

Lemma 7.7. *For any Boolean function $B(X,Y)$ depending on two arguments X and Y of r bits each the bound*

$$\mathrm{Comm}\,(B) \geq \frac{3}{2}\log\left(\frac{1}{\Gamma(B)}\right) + O(1)$$

holds.

Proof. Let us we define the matrix

$$M(B) = (B(u_1, \ldots, u_r, v_1, \ldots, v_r))_{x,y=0}^{2^r-1}, \tag{7.3}$$

whose rows are indexed by x and columns are indexed by y and which has the value of $B(u_1, \ldots, u_r, v_1, \ldots, v_r)$ at the position (x, y), where $x = u_1 \ldots u_r$ and $y = v_1 \ldots v_r$ are the bit representations of x and y. It is well known that

$$\mathrm{Comm}\,(B) \geq \log\,(\mathrm{rank}\,M(B)),$$

where $\mathrm{rank}\,M(B)$ is the rank of $M(B)$ over the field of real numbers, for example, see Lemma 1.28 of [316] (see also Theorem 2.2.2.7 of [278]). On the other hand, Theorem 3 of [448] asserts that

$$\Gamma(B)^{-1} = O\left((\mathrm{rank}\,M(B))^{3/2}\right)$$

and the result follows. □

Many of the above relations among various complexity characteristics of Boolean functions (as well as some other relations) are conveniently summarized in the survey [84].

Part II

Approximation and Complexity of the Discrete Logarithm

Chapter 8

Approximation of the Discrete Logarithm Modulo p

Here we show that polynomials and algebraic functions approximating the discrete logarithm modulo p on sufficiently large sets must be of sufficiently large degree, in fact exponentially large (in terms of $\log p$). Many of the results of this chapter can also be found in [133].

We start with a rather simple statement, which nevertheless demonstrates one of our basic ideas. Namely, we repeatedly exploit the fact that the discrete logarithm satisfies certain functional relations which cannot be satisfied by a polynomial of small degree.

Theorem 8.1. *Let $f(X) \in \mathbb{Z}[X]$ be a polynomial of degree $n = \deg f \le p-2$ and of sparsity $t = \operatorname{spr} f$ such that*

$$\operatorname{ind} x \equiv f(x) \pmod p, \qquad x \in S, \tag{8.1}$$

for a set $S \subseteq \{1, \dots, p-1\}$ of cardinality $|S| = p-1-s$. Then

$$n \ge p-2-2s, \qquad t \ge (p-1)/(2s+1) - 1.$$

Proof. Let R be the set of $x \in \{1, \dots, p-1\}$ for which both

$$\operatorname{ind} x \equiv f(x) \pmod p \quad \text{and} \quad \operatorname{ind} gx \equiv f(gx) \pmod p.$$

Then $|R| \ge p-1-2(p-1-|S|) = p-1-2s$. We have $\operatorname{ind} gx = 1 + \operatorname{ind} x$ if $x \not\equiv g^{p-2} \pmod p$. Hence

$$f(gx) \equiv \operatorname{ind} gx = 1 + \operatorname{ind} x \equiv 1 + f(x) \pmod p$$

for $x \in R$ with $x \not\equiv g^{p-2} \pmod p$. Therefore the polynomial $h(X) = f(gX) - f(X) - 1$ has at least $|R| - 1$ zeros modulo p and is not identical to zero modulo p (because $h(0) = -1$). Thus $n \geq \deg h \geq |R| - 1$.

Also, if f contains $t = \operatorname{spr} f$ monomials, then h contains $\operatorname{spr} h \leq t+1$ monomials. Applying Lemma 2.5 we see that $p-1-(|R|-1) \geq (p-1)/(t+1)$ and the desired result follows. □

In particular, if $s = o(p)$, then $\deg f \sim p$ and $\operatorname{spr} f \to \infty$.

Certainly, for any S one can satisfy (8.1) with a polynomial f of degree $\deg f \leq |S| - 1 = p - 2 - s$.

Theorem 8.1 is non-trivial if the set S is dense enough, $|S| > p/2$. The next result is applicable to quite sparse sets S beginning with $|S| > (2p)^{1/2}$.

Theorem 8.2. *Let $p \geq 3$ and let $f(X) \in \mathbb{Z}[X]$ be a polynomial of degree $n = \deg f$ such that*

$$\operatorname{ind} x \equiv f(x) \pmod p, \qquad x \in S,$$

for a set $S \subseteq \{1, \dots, p-1\}$. Then

$$n \geq \frac{|S|(|S|-1)}{2(p-2)}.$$

Proof. Let us consider the set

$$D = \{a \equiv yx^{-1} \pmod p, \ 2 \leq a \leq p-1, \ x, y \in S\}.$$

Trivially $|D| \leq p - 2$.

On the other hand, obviously there is $a \in D$ such that there are at least $|S|(|S|-1)/|D|$ representations $a \equiv yx^{-1} \pmod p$, $x, y \in S$. Select this a and let R be the set of $x \in \{1, \dots, p-1\}$ for which both

$$\operatorname{ind} x \equiv f(x) \pmod p \quad \text{and} \quad \operatorname{ind} ax \equiv f(ax) \pmod p.$$

We see that $|R| \geq |S|(|S|-1)/(p-2)$. Indeed for each representation $a \equiv yx^{-1} \pmod p$ we get a pair x and $y \equiv ax \pmod p$ of elements of S.

Also, we have $\operatorname{ind} ax = \operatorname{ind} a + \operatorname{ind} x$ or $\operatorname{ind} ax = \operatorname{ind} a + \operatorname{ind} x - p + 1$. Hence either

$$f(ax) \equiv \operatorname{ind} ax = \operatorname{ind} a + \operatorname{ind} x \equiv \operatorname{ind} a + f(x) \pmod p$$

or

$$f(ax) \equiv \operatorname{ind} ax = \operatorname{ind} a + \operatorname{ind} x - p + 1 \equiv 1 + \operatorname{ind} a + f(x) \pmod p$$

for $x \in R$. Therefore at least one of the polynomials $h_1(X) = f(aX) - f(X) - \operatorname{ind} a$ and $h_2(X) = f(aX) - f(X) - \operatorname{ind} a - 1$ has at least $|R|/2$ zeros modulo p. Because of our choice of D neither of these polynomials is identical to zero modulo p. Indeed,

$$h_1(0) \equiv -\operatorname{ind} a \not\equiv 0 \pmod p$$

since $a \neq 1$, and

$$h_2(0) \equiv -\operatorname{ind} a - 1 \not\equiv 0 \pmod p$$

since $0 \le \operatorname{ind} a \le p-2$. Thus $n \ge |R|/2$ and the desired result follows. □

Certainly, for any S one can satisfy (8.1) with a unique polynomial f of degree $\deg f \le |S|-1$. Now we show that for a randomly selected set S of size $o(p)$ the degree cannot be smaller. In particular, with probability $1-o(1)$ we have $\deg f = |S|-1$ for that polynomial.

Theorem 8.3. *Let S be a set of m random elements picked uniformly from $\{1,\dots,p-1\}$. Then the probability $P_k(p,m)$ that there exists a polynomial $f(X) \in \mathbb{Z}[X]$ of degree*

$$\deg f < m - k$$

and such that

$$\operatorname{ind} x \equiv f(x) \pmod p, \qquad x \in S,$$

satisfies the bound

$$P_k(p,m) \le \left(\frac{2m}{p-2}\right)^{k/2}.$$

Proof. We say that a set T is satisfied by a polynomial $f(X) \in \mathbb{Z}[X]$ if the condition of the theorem is fulfilled for this pair (T,f). Accordingly, we say that a set T is maximally satisfied by a polynomial $f(X) \in \mathbb{Z}[X]$ if it is satisfied by this polynomial but any superset of T is not.

Suppose there are N distinct sets $S_i \subseteq \{1,\dots,p-1\}$, $i=1,\dots,N$, that are maximally satisfied by polynomials f_i of degree at most $n = m-k-1$. In particular, polynomials f_i, $i=1,\dots,N$, are pairwise distinct.

Therefore, $|S_i \cap S_j| \le n$, $1 \le i < j \le N$, otherwise we would have $f_i = f_j$ being the unique polynomial on the intersection $S_i \cap S_j$, and hence on their union. Thus,

$$\sum_{i=1}^{N} \binom{|S_i|}{n+1} = \sum_{i=1}^{N} \sum_{\substack{T \subseteq S_i \\ |T|=n+1}} 1 \le \sum_{\substack{T \subseteq \{1,\dots,p-1\} \\ |T|=n+1}} 1 = \binom{p-1}{n+1}. \tag{8.2}$$

From Theorem 8.2 we see that $|S_i| \le (2n(p-2))^{1/2} + 1/2$.

For an $(n+1)$-element set $T \subseteq \{1,\dots,p-1\}$, denote by f_T the unique polynomial of degree at most n such that T is satisfied by this polynomial. Also, denote by R_T the set which is maximally satisfied by f_T. Each m-element set S is the union of an $(n+1)$-element set T and a set of k elements selected outside of T. Thus for each T there are precisely

$$\binom{p-n-2}{k}$$

such m-element sets. Each such set is satisfied by f_T if and only if $S \subseteq R_T$. Therefore,

$$\begin{aligned}
P_k(p,m) &= \sum_{|T|=n+1} \binom{p-1}{n+1}^{-1} \sum_{\substack{T\subseteq S\subseteq R_T \\ |S|=m}} \binom{p-n-2}{k}^{-1} \\
&= \binom{p-1}{n+1}^{-1} \binom{p-n-2}{k}^{-1} \sum_{i=1}^{N} \sum_{\substack{T\subseteq S_i \\ |T|=n+1}} \sum_{\substack{T\subseteq S\subseteq S_i \\ |S|=m}} 1 \\
&= \binom{p-1}{n+1}^{-1} \binom{p-n-2}{k}^{-1} \sum_{i=1}^{N} \binom{|S_i|}{n+1} \binom{|S_i|-n-1}{k}.
\end{aligned}$$

We remark that

$$\binom{u}{v}^{-1} \binom{w}{v} \le (w/u)^v \tag{8.3}$$

for any integers $u, v, w \ge 1$ with $w \le u$.

Therefore we have,

$$\begin{aligned}
\binom{p-n-2}{k}^{-1} \binom{|S_i|-n-1}{k} &\le \left(\frac{|S_i|-n-1}{p-n-2}\right)^k \\
&\le \left(\frac{|S_i|-1}{p-2}\right)^k \le \left(\frac{2n}{p-2}\right)^{k/2}.
\end{aligned}$$

Substituting this in the previous inequality and using (8.2), we derive the result. □

Selecting $k = 1$, we obtain that if $m = o(p)$, for almost all sets of size m the smallest degree of the polynomials which they satisfy is $m-1$.

Now we show that over intervals of length $H \ge p^{1/2+\varepsilon}$ polynomial approximation of the discrete logarithm requires polynomials of large degree.

Theorem 8.4. *Let $0 \le N < N+H \le p-1$ and let $f(X) \in \mathbb{Z}[X]$ be a polynomial of degree $n = \deg f \ge 1$ such that*

$$\operatorname{ind} x \equiv f(x) + \delta_x \pmod p, \qquad N+1 \le x \le N+H,$$

where the "error" vector $(\delta_x)_{x=N+1}^{N+H}$ *is of* L_1*-norm at most*

$$\sum_{x=N+1}^{N+H} |\delta_x| \leq \Delta H p$$

with $0 \leq \Delta < 0.5\pi^{-1}$. *Then*

$$n \geq \left(1 - 2\pi(\Delta + p^{-1})\right) \frac{H}{p^{1/2}\log p}.$$

Proof. For any real α we have

$$|\,\mathbf{e}(\alpha) - 1| = 2|\sin \pi\alpha| \leq 2\pi|\alpha|.$$

Therefore

$$\left|\sum_{x=N+1}^{N+H} \mathbf{e}_p\left(\operatorname{ind} x - f(x)\right)\right| = \left|\sum_{x=N+1}^{N+H} \mathbf{e}_p\left(\delta_x\right)\right| \geq H - 2\pi\Delta H.$$

We also have

$$0 \leq \frac{\operatorname{ind} x}{p-1} - \frac{\operatorname{ind} x}{p} = \frac{\operatorname{ind} x}{p(p-1)} < \frac{1}{p},$$

thus

$$|\,\mathbf{e}_p\left(\operatorname{ind} x\right) - \mathbf{e}_{p-1}\left(\operatorname{ind} x\right)| < \frac{2\pi}{p}.$$

Recalling that

$$\chi(x) = \mathbf{e}_{p-1}\left(\operatorname{ind} x\right)$$

is a non-trivial multiplicative character modulo p, from Lemma 3.5 we derive

$$\begin{aligned}\left|\sum_{x=N+1}^{N+H} \mathbf{e}_p\left(\operatorname{ind} x - f(x)\right)\right| &\leq \left|\sum_{x=N+1}^{N+H} \chi(x)\,\mathbf{e}_p\left(-f(x)\right)\right| + \frac{2\pi H}{p} \\ &\leq np^{1/2}\log p + \frac{2\pi H}{p}.\end{aligned}$$

Therefore $H - 2\pi\Delta H \leq np^{1/2}\log p + 2\pi H/p$ and the bound follows. □

In particular, if $\operatorname{ind} x \equiv f(x) \pmod p$ for a set $S \subseteq \{N+1, \ldots, N+H\}$ of cardinality $|S| = H - s$, then we can put $\Delta = s/H$. However, in this particular case Theorem 8.2 is stronger.

In the following theorem we consider the possibility of representing the discrete logarithm via algebraic functions.

Theorem 8.5. *Let* $F(X,Y) \in \mathbb{Z}[X,Y]$ *be a bivariate polynomial of the form*

$$F(X,Y) = \sum_{i=1}^{t} X^{n_i} f_i(Y),$$

where $0 \le n_1 < \ldots < n_t < p-1$ *and polynomials* $f_i(Y) \in \mathbb{Z}[Y]$, $i = 1, \ldots, t$, *are of degree at most* n *not all identical to zero modulo* p. *Assume that*

$$F(x, \operatorname{ind} x) \equiv 0 \pmod p, \qquad x \in S,$$

for a set $S \subseteq \{1, \ldots, p-1\}$ *of cardinality* $|S| = p - 1 - s$. *Then*

$$t(n+1) \ge (p-1)/(s+1).$$

Proof. From the condition of the theorem one sees that

$$F(g^y, y) \equiv 0 \pmod p, \qquad y \in R,$$

for a set $R \in \{0, \ldots, p-2\}$ of cardinality $|R| = |S| = p - 1 - s$. On the other hand, from (2.2) one finds that $u(y) = F(g^y, y)$ is a linear recurrence sequence of order at most $t(n+1)$. Its characteristic polynomial has the constant term

$$c_0 = g^{n_1} \ldots g^{n_t} \not\equiv 0 \pmod p.$$

Moreover, because $0 \le n_i < n_j < p-1$, then

$$g^{n_i} \not\equiv g^{n_j} \pmod p, \qquad 1 \le i < j \le t.$$

Taking into account that the polynomials $f_i(Y)$, $i = 1, \ldots, t$, are not identical to zero modulo p, we conclude that the sequence $(u(y))$ is not identical to zero modulo p either. It follows from Lemma 2.4 that

$$s \ge \left\lfloor \frac{p-1}{t(n+1)} \right\rfloor \ge \frac{p-1}{t(n+1)} - 1$$

and the desired estimate follows. □

In particular, let D_p denote the smallest degree

$$d = \max\{\deg_X F, \deg_Y F\}$$

of all non-zero modulo p polynomials $F(X,Y) \in \mathbb{Z}[X,Y]$ such that the congruence $F(x, \operatorname{ind} x) \equiv 0 \pmod p$ holds for all $x = 1, \ldots, p-1$. Thus, D_p is the degree of the discrete logarithm as an algebraic function over $\mathbb{F}_p$. For $(d+1)^2$ coefficients of such a polynomial we obtain a system of $p-1$ homogeneous equations which has a non-zero solution whenever $(d+1)^2 \ge p$. Hence, together with Theorem 8.5 (setting $t = d+1$, $n = d$, $s = 0$), we obtain a very tight bound:

$$p^{1/2} \ge D_p \ge (p-1)^{1/2} - 1.$$

Theorem 8.5 is non-trivial if the set S is of cardinality $|S| \sim p$. The next result is applicable to quite sparse sets S beginning with $|S| > 3^{1/2} p^{1/2}$.

Theorem 8.6. *Let $F(X,Y) \in \mathbb{Z}[X,Y]$ be a polynomial of total degree $n = \deg F$, non-zero modulo $p \geq 3$, such that*

$$F(x, \operatorname{ind} x) \equiv 0 \pmod p, \qquad x \in S,$$

for a set $S \subseteq \{1, \dots, p-1\}$. Then

$$n \geq \frac{|S|}{3^{1/2}p^{1/2}}.$$

Proof. In this proof it will be convenient to use the language of finite fields rather than congruences. Let us consider the complete factorisation of $F(X,Y)$ over the algebraic closure of $\mathbb{F}_p$ (thus all factors are absolutely irreducible polynomials). Let $\Psi(X,Y)$ be an irreducible factor of $F(X,Y)$, of total degree $d = \deg \Psi$, for which $\Psi(x, \operatorname{ind} x) = 0$ for at least $|S|d/n$ values of $x \in S$. Denote this set of x by T, $|T| \geq |S|d/n$.

As in the proof of Theorem 8.2 we select $a \neq 1$ such that there are at least $|T|(|T|-1)/(p-2)$ representations of $a = yx^{-1}$, $x, y \in T$. Let R be the set of $x \in \{1, \dots, p-1\}$ for which both

$$\Psi(x, \operatorname{ind} x) = 0 \quad \text{and} \quad \Psi(ax, \operatorname{ind} ax) = 0 \tag{8.4}$$

hold.

We see that

$$|R| \geq \frac{|S|d(|S|d-n)}{n^2(p-2)}.$$

We have $\operatorname{ind} ax = \operatorname{ind} a + \operatorname{ind} x$ or $\operatorname{ind} ax = \operatorname{ind} a + \operatorname{ind} x - p + 1$. Hence either

$$\Psi(ax, \operatorname{ind} x + \operatorname{ind} a) = 0 \qquad \text{or} \qquad \Psi(ax, \operatorname{ind} x + \operatorname{ind} a + 1) = 0$$

for $x \in R$. Therefore at least one of the polynomials $\Psi(aX, X + \operatorname{ind} a)$ and $\Psi(aX, X + \operatorname{ind} a + 1)$ has at least $|R|/2$ zeros. We see that $\operatorname{ind} a \neq 0$ since $a \neq 1$, and $\operatorname{ind} a \neq -1$ since $0 \leq \operatorname{ind} a \leq p-2$. Therefore, there is a $b \neq 0$ such that the system of equations

$$\Psi(X,Y) = \Psi(aX, Y+b) = 0$$

has at least $|R|/2$ solutions.

If the polynomials $\Psi(X,Y)$ and $\Psi(aX, Y+b)$ are relatively prime, then it follows from Bézout's theorem that this system has at most d^2 solutions and we obtain

$$d^2 \geq \frac{|S|d(|S|d-n)}{2n^2(p-2)}.$$

We may assume that $n \leq |S|/3$, otherwise the bound is trivial. Then

$$|S|d - n \geq 2|S|d/3,$$

so that

$$d^2 \geq \frac{S^2 d^2}{3n^2 p},$$

and the desired inequality follows.

If $\Psi(X,Y)$ and $\Psi(aX,Y+b)$ are not relatively prime, then recalling that $\Psi(X,Y)$ is absolutely irreducible (thus so is $\Psi(aX.Y+b)$) we see that $\Psi(aX,Y+b) = \mu\Psi(X,Y)$ for some constant $\mu \neq 0$. If

$$\Psi(X,Y) = \sum_{i=0}^{d} X^i f_i(Y),$$

then for each $i = 0, \dots, n$, $f_i(Y)$ divides $f_i(Y+b)$. That implies $f_i(Y) = \mu_i f_i(Y+b)$ for some constant $\mu_i \neq 0$. If $n < p$ (otherwise there is nothing to prove) then this is possible only if $f_i(Y)$ is a constant polynomial and $\mu_i = 1$. Thus $\Psi(X,Y) = \Psi(X)$ is a polynomial in one variable. Therefore, the system (8.4) has at most d solutions. Hence

$$d \geq \frac{|S|d(|S|d-n)}{2n^2(p-2)},$$

thus

$$n^2 \geq \frac{|S|(|S|d-n)}{2p}.$$

If $n > |S|/3$, then there is nothing to prove. Otherwise $|S|d - n \geq |S| - n \geq 2|S|/3$ and the desired result follows. □

It is obvious that for any set $S \subseteq \{1, \dots, p-1\}$ there is a polynomial $F(X,Y) \in \mathbb{Z}[X,Y]$ of degree at most $(2|S|)^{1/2} + 1$ which satisfies the condition of Theorem 8.6. Now we show that for almost all sets S of not too large size a lower bound of the same order holds.

Theorem 8.7. *Let p be sufficiently large, $0 < \varepsilon < \delta < 1$ and $m \leq p^{1-\delta}$. Let S be a set of m random elements picked uniformly from $\{1, \dots, p-1\}$. Then the probability $P_{\varepsilon,\delta}(p,m)$ that there exists a polynomial $F(X,Y) \in \mathbb{Z}[X,Y]$ of degree*

$$\deg F < \left\lfloor (\varepsilon m)^{1/2} \right\rfloor - 1$$

and such that

$$F(x, \operatorname{ind} x) \equiv 0 \pmod p, \qquad x \in S,$$

satisfies the bound

$$P_{\varepsilon,\delta}(p,m) \leq 2^m p^{-(\delta-\varepsilon)m/2}.$$

Proof. We say that a set T is satisfied by a polynomial $F(X,Y) \in \mathbb{Z}[X,Y]$ if the condition of the theorem is fulfilled for this pair (T,F). Accordingly, we say that T is maximally satisfied by a polynomial $F(X,Y) \in \mathbb{Z}[X,Y]$ if it is satisfied by this polynomial but any superset of T is not.

Suppose there are N distinct sets $S_i \subseteq \{1,\dots,p-1\}$, $i=1,\dots,N$, that are maximally satisfied by polynomials $F_i(X,Y) \in \mathbb{Z}[X,Y]$ of degree at most $n = \lfloor (\varepsilon m)^{1/2} \rfloor - 2$. In particular, polynomials F_i, $i=1,\dots,N$, are pairwise distinct modulo p, thus

$$N \le p^{(n+2)(n+1)/2}.$$

From Theorem 8.6 we derive $|S_i| \le n(3p)^{1/2}$. Therefore, using the inequality (8.3) we derive

$$\begin{aligned} P_{\varepsilon,\delta}(p,m) &= \binom{p-1}{m}^{-1} \sum_{i=1}^{N} \binom{|S_i|}{m} \le \sum_{i=1}^{N} \left(\frac{|S_i|}{p-1}\right)^m \\ &\le p^{(n+2)(n+1)/2} \left(\frac{n(3p)^{1/2}}{p-1}\right)^m \\ &\le 2^m n^m p^{(n+2)(n+1)/2 - m/2} \le 2^m m^{m/2} p^{(\varepsilon-1)m/2} \\ &\le 2^m p^{-(\delta-\varepsilon)m/2}, \end{aligned}$$

and the result follows. □

Finally we consider representations via linear recurrence sequences. The following result as well as the results of Theorems 8.9, 9.1, 11.4 and 31.3 have been obtained in [305]. In fact, the results of [305] apply to representations via linear recurrence sequences modulo arbitrary integers m and thus we present them here in this form as well.

Theorem 8.8. *Let $0 \le N < N+H \le p-1$ and let $(u(x))$ be an integer linear recurrence sequence of order n such that*

$$\operatorname{ind} x \equiv u(x) \pmod m, \qquad N+1 \le x \le N+H.$$

Then the bound

$$n \ge \min\left\{\frac{H}{5\log m}, \frac{m^{2/3}H^{2/3}}{8p^{2/3}\log m}\right\} - 2$$

holds.

Proof. We put

$$L = \min\left\{\left\lceil \frac{H}{5} \right\rceil, \left\lceil \frac{m^{2/3}H^{2/3}}{8p^{2/3}} \right\rceil\right\} - 1 \tag{8.5}$$

and denote $\mathcal{E} = \{-1,1\}^L$.

For any $x = N+1, \dots, N+H-L$ we have

$$\frac{1}{2^L} \sum_{(\varepsilon_1,\dots,\varepsilon_L)\in\mathcal{E}} \left(\sum_{l=1}^{L} \varepsilon_l \operatorname{ind}(x+l)\right)^2 = \sum_{l_1=1}^{L}\sum_{l_2=1}^{L} \operatorname{ind}(x+l_1)\operatorname{ind}(x+l_2)\frac{1}{2^L}\sum_{(\varepsilon_1,\dots,\varepsilon_L)\in\mathcal{E}} \varepsilon_{l_1}\varepsilon_{l_2}.$$

Further,

$$\frac{1}{2^L}\sum_{(\varepsilon_1,\dots,\varepsilon_L)\in\mathcal{E}} \varepsilon_{l_1}\varepsilon_{l_2} = \begin{cases} 1, & \text{if } l_1 = l_2, \\ 0, & \text{if } l_1 \neq l_2, \end{cases} \qquad 1 \le l_1, l_2 \le L.$$

Therefore,

$$\frac{1}{2^L}\sum_{(\varepsilon_1,\dots,\varepsilon_L)\in\mathcal{E}} \left(\sum_{l=1}^{L} \varepsilon_l \operatorname{ind}(x+l)\right)^2 = \sum_{l=1}^{L} (\operatorname{ind}(x+l))^2 < Lp^2.$$

Taking the sum over $x = N+1, \dots, N+H-L$, we find

$$\frac{1}{2^L}\sum_{(\varepsilon_1,\dots,\varepsilon_L)\in\mathcal{E}} \sum_{x=N+1}^{N+H-L} \left(\sum_{l=1}^{L} \varepsilon_l \operatorname{ind}(x+l)\right)^2 < L(H-L)p^2.$$

Thus, there exists a set $\mathcal{F} \subset \mathcal{E}$ such that

$$|\mathcal{F}| = 2^{L-1} \tag{8.6}$$

and

$$\sum_{x=N+1}^{N+H-L} \left(\sum_{l=1}^{L} \varepsilon_l \operatorname{ind}(x+l)\right)^2 < 2L(H-L)p^2, \tag{8.7}$$

for all $(\varepsilon_1, \dots, \varepsilon_L) \in \mathcal{F}$.

Let us define the linear recurrence sequences $(w_i(x))$, $i = 1, \dots, n$, as in Lemma 2.3.

We also define n-dimensional vectors

$$W(x) = (w_1(x), \dots, w_n(x)), \qquad x = 1, 2, \dots,$$

and consider 2^{L-1} linear combinations

$$\sum_{l=1}^{L} \varepsilon_l W(l), \qquad (\varepsilon_1, \dots, \varepsilon_L) \in \mathcal{F}.$$

Assume that

$$2^{L-1} > m^n. \tag{8.8}$$

Then, by (8.6), there are two distinct vectors $(\xi_1, \dots, \xi_L), (\zeta_1, \dots, \zeta_L) \in \mathcal{F}$ such that the binary vectors $(\alpha_1, \dots, \alpha_L) \in \{0,1\}^L$ and $(\beta_1, \dots, \beta_L) \in \{0,1\}^L$ defined as $\alpha_l = (1+\xi_l)/2$, $\beta_l = (1+\zeta_l)/2$, $l = 1, \dots, L$, satisfy the congruence

$$\sum_{l=1}^{L} \alpha_l W(l) \equiv \sum_{l=1}^{L} \beta_l W(l) \pmod m. \tag{8.9}$$

Moreover, we also have the inequality

$$\begin{aligned}
&\sum_{x=N+1}^{N+H-L} \left(\sum_{l=1}^{L} \alpha_l \mathrm{ind}\,(x+l) - \sum_{l=1}^{L} \beta_l \mathrm{ind}\,(x+l) \right)^2 \\
&\quad = \frac{1}{4} \sum_{x=N+1}^{N+H-L} \left(\sum_{l=1}^{L} \xi_l \mathrm{ind}\,(x+l) - \sum_{l=1}^{L} \zeta_l \mathrm{ind}\,(x+l) \right)^2 \\
&\quad \le \frac{1}{4} \sum_{x=N+1}^{N+H-L} \left(2\left(\sum_{l=1}^{L} \xi_l \mathrm{ind}\,(x+l) \right)^2 + 2\left(\sum_{l=1}^{L} \zeta_l \mathrm{ind}\,(x+l) \right)^2 \right) \\
&\quad = \frac{1}{2} \sum_{x=N+1}^{N+H-L} \left(\sum_{l=1}^{L} \xi_l \mathrm{ind}\,(x+l) \right)^2 + \frac{1}{2} \sum_{x=N+1}^{N+H-L} \left(\sum_{l=1}^{L} \zeta_l \mathrm{ind}\,(x+l) \right)^2,
\end{aligned}$$

and by (8.7) we conclude

$$\sum_{x=N+1}^{N+H-L} \left(\sum_{l=1}^{L} \alpha_l \mathrm{ind}\,(x+l) - \sum_{l=1}^{L} \beta_l \mathrm{ind}\,(x+l) \right)^2 < 2L(H-L)p^2. \tag{8.10}$$

Combining Lemma 2.3 and (8.9), we derive that

$$\sum_{l=1}^{L} \alpha_l u(x+l) - \sum_{l=1}^{L} \beta_l u(x+l) \equiv 0 \pmod m.$$

Hence, for $x \in [N+1, N+H-L]$ we have

$$\sum_{l=1}^{L} \alpha_l \mathrm{ind}\,(x+l) - \sum_{l=1}^{L} \beta_l \mathrm{ind}\,(x+l) = K_x m$$

with some integer K_x. This is equivalent to the congruence

$$\prod_{l=1}^{L} (x+l)^{\alpha_l} \equiv g^{K_x m} \prod_{l=1}^{L} (x+l)^{\beta_l} \pmod p.$$

Note that by (8.10)

$$\sum_{x=N+1}^{N+H-L} K_x^2 m^2 < 2L(H-L)p^2.$$

Let $\mathcal{X} = \{x \in [N+1, N+H-L] \mid |K_x| < 2pL^{1/2}/m\}$. We have

$$|\mathcal{X}| > (H-L)/2.$$

Remarking that K_x for $x \in \mathcal{X}$ can take at most $4pL^{1/2}/m + 1$ distinct values, we see that for some $A \not\equiv 0 \pmod p$ the congruence

$$\prod_{l=1}^{L}(x+l)^{\alpha_l} \equiv A \prod_{l=1}^{L}(x+l)^{\beta_l} \pmod p.$$

has at least

$$\frac{|\mathcal{X}|}{4pL^{1/2}/m+1} > \frac{H-L}{8pL^{1/2}/m+2}$$

solutions.

However, the polynomials on the left- and right-hand sides of this congruence are distinct because they have different sets of roots. On the other hand, taking into account that they are of degree at most L, and recalling (8.5), we conclude that they cannot coincide at

$$\frac{H-L}{8pL^{1/2}/m+2} > L$$

points. The obtained contradiction shows that the assumption (8.8) is false and we obtain the desired result. □

We remark that the choice of L given by (8.5) is not optimal and the constants can be improved by selecting L in a better (but more complicated) way. For example in the most interesting case $m = p$ selecting

$$L = \left\lfloor \frac{\left(H - H^{2/3}\right)^{2/3}}{4} \right\rfloor,$$

we obtain

$$n \geq \frac{H^{2/3}}{5 \log p}$$

provided that $H \geq 600$. This also follows from Theorem 11.4.

Recalling that a polynomial $f(X) \in \mathbb{Z}[X]$ of degree $n = \deg f$ is also a linear recurrence sequence of order $n+1$ we obtain that the congruence

$$\operatorname{ind} x \equiv f(x) \pmod p, \qquad N+1 \leq x \leq N+H,$$

implies

$$n \geq \frac{H^{2/3}}{5 \log p} - 1,$$

provided that $H \geq 600$.

We also see that for large values of m, namely for $m \geq 2pH^{1/2}$, Theorem 8.8 yields the bound

$$n \geq \frac{H}{5 \log m} - 2.$$

In particular, for $2pH^{1/2} \leq m \leq p^2$ we get

$$n \geq \frac{H}{10 \log p} - 2 \tag{8.11}$$

which is close to the best possible (up to a multiple of $\log p$). The following result shows that the estimate (8.11) also holds for $m > p^2$.

Theorem 8.9. *Let $0 \leq N < N + H \leq p - 1$, $m > p^2$, and let $(u(x))$ be an integer linear recurrence sequence of order n such that*

$$\operatorname{ind} x \equiv u(x) \pmod{m}, \qquad N + 1 \leq x \leq N + H.$$

Then

$$n \geq \frac{H}{3 \log p} - 2.$$

Proof. Put $L = \lceil 2H/3 \rceil - 1$ and denote by $\mathcal{F}$ the set of $|\mathcal{F}| \geq 2^{L-1}$ binary vectors of length L and of Hamming weight at most $L/2$. Assume that

$$2^{L-1} > p^{2n} \tag{8.12}$$

and consider the n-dimensional vectors

$$V(\gamma) = (v_1(\gamma), \dots, v_n(\gamma)), \qquad \gamma = (\gamma_1, \dots, \gamma_L) \in \mathcal{F},$$

where

$$v_j(\gamma) = \sum_{l=1}^{L} \gamma_l \operatorname{ind}(N + j + l), \qquad j = 1, \dots, n.$$

We remark that the assumption (8.12) implies $n \leq H - L$. Thus

$$N + j + l \leq N + H, \qquad j = 1, \dots, n, \ l = 1, \dots, L.$$

Also, by (8.12), there are two distinct vectors $\alpha = (\alpha_1, \dots, \alpha_L) \in \mathcal{F}$ and $\beta = (\beta_1, \dots, \beta_L) \in \mathcal{F}$ such that $V(\alpha) = V(\beta)$, namely,

$$\sum_{l=1}^{L} \alpha_l \operatorname{ind}(x + l) = \sum_{l=1}^{L} \beta_l \operatorname{ind}(x + l), \qquad x = N + 1, \dots, N + n.$$

and we have the congruences

$$\sum_{l=1}^{L} \alpha_l u(x+l) \equiv \sum_{l=1}^{L} \beta_l u(x+l) \pmod m$$

for $x = N+1, \dots, N+n$. By the recurrence relation and induction on x, the last congruence holds for all $x > N$. Therefore,

$$\sum_{l=1}^{L} \alpha_l \mathrm{ind}\,(x+l) \equiv \sum_{l=1}^{L} \beta_l \mathrm{ind}\,(x+l) \pmod m, \qquad x = N+1, \dots, N+H-L.$$

Let us remark that

$$\left| \sum_{l=1}^{L} \alpha_l \mathrm{ind}\,(x+l) - \sum_{l=1}^{L} \beta_l \mathrm{ind}\,(x+l) \right| < pL < m.$$

Thus, the last congruence is actually the equality

$$\sum_{l=1}^{L} \alpha_l \mathrm{ind}\,(x+l) = \sum_{l=1}^{L} \beta_l \mathrm{ind}\,(x+l), \qquad x = N+1, \dots, N+H-L.$$

This equality implies the congruence

$$\prod_{l=1}^{L} (x+l)^{\alpha_l} \equiv \prod_{l=1}^{L} (x+l)^{\beta_l} \pmod p.$$

However, the polynomials on the left- and right-hand sides of this congruence are distinct (because they have different roots) and are of degree at most $L/2$. Thus they cannot coincide at $(H-L) > L/2$ points. The obtained contradiction shows that the assumption (8.12) is false and the result follows. □

Some analogies of Theorem 8.1 have been obtained in [324, 445, 561]. In particular, Theorem 8.1, as well as several other results which are based on Lemma 2.5, can be extended to the discrete logarithm in the group generated by an arbitrary element of $\mathbb{F}_q$, which is not necessary a primitive root, see [324].

Chapter 9

Approximation of the Discrete Logarithm Modulo $p-1$

In this chapter we consider various approximations and representations of the discrete logarithm modulo a divisor d of $p-1$. Certainly the case of $d=2$ is of special interest because it corresponds to the representation of the rightmost bit of $\operatorname{ind} x$.

Moreover, instead of polynomials we consider a much wider class of representations via linear recurrence sequences. We use the method of proof of Theorem 8.8 which in fact simplifies and produces stronger results when applied to congruences modulo $p-1$. In fact, we consider congruences modulo a divisor d of $p-1$.

Theorem 9.1. *Let $0 \le N < N+H \le p-1$ and let $(u(x))$ be an integer linear recurrence sequence of order n such that*

$$\operatorname{ind} x \equiv u(x) \pmod{d}, \qquad N+1 \le x \le N+H,$$

where d divides $p-1$. Put $r=(p-1)/d$. Then

$$n \ge \frac{2H}{(r+2)\log d} - 2.$$

Proof. Let us define the linear recurrence sequences $(w_i(x))$, $i=1,\dots,n$, as in Lemma 2.3.

We consider n-dimensional vectors

$$W(x) = (w_1(x),\dots,w_n(x)), \qquad x = 1,2,\dots.$$

Put $L = \lceil 2H/(r+2) \rceil - 1$ and denote by $\mathcal{F}$ the set of $|\mathcal{F}| \ge 2^{L-1}$ binary vectors of length L and of Hamming weight at most $L/2$.

Consider the linear combinations

$$\sum_{l=1}^{L} \gamma_l W(l), \qquad (\gamma_1, \dots, \gamma_L) \in \mathcal{F}.$$

Assume that

$$2^{L-1} > d^n. \tag{9.1}$$

Then there are two linear combinations which coincide modulo d. Therefore we have the identity

$$\sum_{l=1}^{L} \alpha_l W(l) \equiv \sum_{l=1}^{L} \beta_l W(l) \pmod d,$$

with two distinct binary vectors $(\alpha_1, \dots, \alpha_L) \in \mathcal{F}$ and $(\beta_1, \dots, \beta_L) \in \mathcal{F}$. Thus from Lemma 2.3 we derive that

$$\sum_{l=1}^{L} \alpha_l u(x+l) \equiv \sum_{l=1}^{L} \beta_l u(x+l) \pmod d, \qquad x = 1, 2, \dots .$$

Therefore for $x \in [N+1, N+H-L]$ we have

$$\sum_{l=1}^{L} \alpha_l \operatorname{ind}(x+l) \equiv \sum_{l=1}^{L} \beta_l \operatorname{ind}(x+l) \pmod d$$

hence, for some integers K_x,

$$\sum_{l=1}^{L} \alpha_l \operatorname{ind}(x+l) = \sum_{l=1}^{L} \beta_l \operatorname{ind}(x+l) + K_x d$$

which is equivalent to the congruence

$$\prod_{l=1}^{L} (x+l)^{\alpha_l} \equiv g^{K_x d} \prod_{l=1}^{L} (x+l)^{\beta_l} \pmod p.$$

Since

$$\left(g^{K_x d}\right)^r \equiv g^{K_x (p-1)} \equiv 1 \pmod p,$$

we see that $g^{K_x d}$ can take on only r distinct values. Therefore for some $A \not\equiv 0 \pmod p$ the congruence

$$\prod_{l=1}^{L} (x+l)^{\alpha_l} \equiv A \prod_{l=1}^{L} (x+l)^{\beta_l} \pmod p$$

has at least $(H-L)/r$ solutions. However, the polynomials on the left- and right-hand sides of this congruence are distinct (for example they have different roots) and are of degree at most $L/2$, and thus they cannot coincide at $(H-L)/r > L/2$ points. The obtained contradiction shows that the assumption (9.1) is false and the result follows. □

In particular, Theorem 9.1, implies the lower bound

$$L(H) = \Omega(H \log^{-1} p)$$

on the linear complexity profile of the discrete logarithm modulo $p-1$.

The result of Theorem 9.1 is non-trivial only if d is large enough. Now we obtain a similar result which is non-trivial for any divisor d of $p-1$.

Theorem 9.2. *Let $0 \le N < N+H \le p-1$ and let $d > 1$ be a divisor of $p-1$. Let $(u(x))$ be an integer linear recurrence sequence of order n such that*

$$\operatorname{ind} x \equiv u(x) \pmod{d}, \qquad x \in S,$$

for a set $S \subseteq \{N+1, \dots, N+H\}$ of cardinality $|S| = H - s$. Then

$$n \ge \frac{H}{2s + 2 + p^{1/2} \log p} - 1.$$

Proof. We see that for at least $H - n - (n+1)(H - |S|) \ge H - (n+1)(s+1)$ values of $x = N+1, \dots, N+H$,

$$\operatorname{ind}(x+i) \equiv u(x+i) \pmod{d}, \qquad i = 0, \dots, n.$$

Put $c_n = -1$. From (1.2) we see that

$$\sum_{i=0}^{n} c_i \operatorname{ind}(x+i) \equiv 0 \pmod{d} \tag{9.2}$$

for at least $H-(n+1)(s+1)$ values of $x = N+1, \dots, N+H$. The congruence (9.2) is equivalent to the statement that the product $x^{c_0}(x+1)^{c_1} \dots (x+n)^{c_n}$ is a dth power residue modulo p. Thus, for a non-trivial character χ of $\mathbb{F}_p^*$ of order d, we have

$$\left| \sum_{x=N+1}^{N+H} \chi\left(x^{c_0}(x+1)^{c_1} \dots (x+n)^{c_n}\right) \right| \ge H - 2(n+1)(s+1).$$

On the other hand, because $c_n = -1$, Lemma 3.4 can be applied. Therefore,

$$H - 2(n+1)(s+1) \le (n+1)p^{1/2} \log p$$

and the result follows. □

In particular, Theorem 9.2 implies that if $H \ge \max\{sp^{\delta}, p^{1/2+\delta} \log p\}$ with some fixed $\delta > 0$, then the order of the sequence must be exponentially large, $n \gg p^{\delta}$.

It is interesting to note that the lower bound does not depend on the divisor d. In particular, selecting $d = 2$, we derive that even the rightmost bit of $\operatorname{ind} x$ cannot be given by a linear recurrence sequence of small order.

Using $s = 0$ in Theorem 9.2, one immediately obtains a lower bound $L(H) = \Omega(Hp^{-1/2}\log^{-1} p)$ on the linear complexity profile of the discrete logarithm modulo a divisor d of $p-1$.

We note that for $d = 2$ the linear complexity L of the discrete logarithm (considered as an infinite period sequence of period p, where $\operatorname{ind} 0 = p-1$) has been evaluated precisely in [158] as follows:

$$L = \begin{cases} (p-1)/2, & \text{if } p \equiv 1 \pmod 8, \\ p, & \text{if } p \equiv 3 \pmod 8, \\ p-1, & \text{if } p \equiv 5 \pmod 8, \\ (p+1)/2, & \text{if } p \equiv 7 \pmod 8, \end{cases}$$

see also [140, 154, 156, 157, 381].

Also, assume that we are given a piecewise representation of the discrete logarithm via linear recurrence sequences, thus we are given $m+1$ integers $0 = N_1 < N_2 < \ldots < N_{m+1} = p-1$, m divisors $d_1, \ldots, d_m$ of $p-1$ and m linear recurrence sequences $(u_1(x)), \ldots, (u_m(x))$, of orders $n_1, \ldots, n_m$ respectively, such that

$$\operatorname{ind} x \equiv u_i(x) \pmod{d_i}, \qquad x \in [N_i+1, N_{i+1}], \quad i = 1, \ldots, m.$$

Then

$$\sum_{i=1}^{m} n_i \geq \frac{1}{p^{1/2}\log p}\sum_{i=1}^{m}(N_{i+1} - N_i) - m = (p-1)p^{-1/2}\log^{-1} p - m.$$

On the other hand, obviously

$$\sum_{i=1}^{m} n_i \geq m,$$

thus we obtain

$$\sum_{i=1}^{m} n_i \geq 0.5(p-1)^{1/2}\log^{-1} p.$$

Obviously one can apply the result above to the special case of polynomials.

Finally, we obtain a lower bound on the length of a non-linear recurrence relation which the rightmost bit of the discrete logarithm (therefore the discrete logarithm itself) may satisfy. Moreover, we allow the coefficients to be polynomials in x rather than constants.

Theorem 9.3. *Let* $0 \leq N < N + H \leq p-1$ *and*

$$\operatorname{ind}(x+m_n) \equiv F(x, \operatorname{ind}(x+m_0), \ldots, \operatorname{ind}(x+m_{n-1})) \pmod 2,$$

for all elements $x \in S$ *of a set* $S \subseteq \{N+1, \ldots, N+H\}$ *of cardinality* $|S| = H - s$ *and a polynomial* $F(X_0, X_1, \ldots, X_n) \in \mathbb{Z}[X_0, X_1, \ldots, X_n]$ *and* $n+1$ *pairwise distinct modulo* p *integers* $m_0, \ldots, m_n$. *Then*

$$n \geq \log\left(\frac{H}{s + p^{1/2}}\right) + O(\log\log p).$$

Proof. From the condition of the theorem we see that there exists no $x \in S$ such that simultaneously

$$x \equiv \operatorname{ind}(x+m_0) \equiv \ldots \equiv \operatorname{ind}(x+m_{n-1}) \equiv 0 \pmod 2$$

and

$$\operatorname{ind}(x+m_n) \not\equiv F_0 \pmod 2,$$

where $F_0 = F(0,\ldots,0)$. Therefore for any even $x \in S$ we have

$$\left(\chi(x+m_n) - (-1)^{F_0}\right) \prod_{i=0}^{n-1} (\chi(x+m_i)+1) = 0,$$

where $\chi(z)$ is the quadratic character modulo p. Therefore

$$\sum_{\substack{x \in S \\ x \equiv 0 \pmod 2}} \left(\chi(x+m_n) - (-1)^{F_0}\right) \prod_{i=0}^{n-1} (\chi(x+m_i)+1) = 0. \tag{9.3}$$

After simple evaluation one sees that the left-hand side contains one "main" term with absolute value at least $H/2 - s - 1$ and $2^{n+1}-1$ terms of the form

$$\pm \sum_{\substack{x \in S \\ x \equiv 0 \pmod 2}} \chi\left((x+j_1)\ldots(x+j_k)\right),$$

where $0 \le j_1 < \ldots < j_k \le p-1$, $k \le n+1$.

Applying Lemma 3.4, we see that the absolute value of each such term does not exceed

$$\begin{aligned}
&\left| \sum_{\substack{x \in S \\ x \equiv 0 \pmod 2}} \chi\left((x+j_1)\ldots(x+j_k)\right) \right| \\
&\qquad \le \left| \sum_{\substack{x=N+1 \\ x \equiv 0 \pmod 2}}^{N+H} \chi\left((x+j_1)\ldots(x+j_k)\right) \right| + s \\
&\qquad = \left| \sum_{(N+1)/2 \le x \le (N+H)/2} \chi\left((2x+j_1)\ldots(2x+j_k)\right) \right| + s \\
&\qquad \le s + kp^{1/2}\log p.
\end{aligned}$$

Thus from (9.3) we derive

$$\begin{aligned}
H/2 - s - 1 &\le \sum_{k=1}^{n+1} \binom{n+1}{k} (s + kp^{1/2}\log p) \\
&\le 2^{n+1}\left(s + (n+1)p^{1/2}\log p\right),
\end{aligned}$$

which implies the desired result. □

The case when the polynomial F does not depend on the first coordinate, $s = 0$, and $m_i = i$, $i = 0, \dots, n$, corresponds to the *non-linear complexity* of the discrete logarithm on the interval $N + 1 \le x \le N + H$. Thus if $H \ge p^{1/2+\varepsilon}$, then it is $\Omega(\log p)$.

The result can be extended to any divisor d of $p - 1$ without any difficulty.

Chapter 10

Approximation of the Discrete Logarithm by Boolean Functions

Here we consider the bitwise approximation of the discrete logarithm given the bit representation of the argument. Moreover, we concentrate on the rightmost bit of $\operatorname{ind} x$. This question is essentially equivalent to deciding quadratic residuacity of x.

In [200] (see also [204]) the identity

$$x^{(q-1)/2} = \begin{cases} 1, & \text{if } x \text{ is a quadratic residue in } \mathbb{F}_q, \\ -1, & \text{if } x \text{ is a quadratic non-residue in } \mathbb{F}_q, \end{cases}$$

has been used to obtain the lower bound $\Omega(\log q)$ on the depth of arithmetic circuits over $\mathbb{F}_q$ deciding whether $x \in \mathbb{F}_q^*$ is a quadratic residue (the most important thing is that the degree $(q-1)/2$ is large). Here we consider Boolean circuits. It should be noted that our bound $\Omega(\log\log p)$ (which we prove for prime fields $\mathbb{F}_p$ only) on their depth is weaker. This actually agrees with the expectation that for this particular question Boolean circuits are exponentially more powerful than arithmetic ones; see [204] for a discussion of this phenomenon and a survey of relevant results with concrete examples of problems for which Boolean circuits are more powerful.

Also, as we have mentioned, the function ψ_p given by (13.1) can be computed by a polynomial size Boolean circuit but probably cannot be computed by a polynomial size arithmetic circuit; see [542].

We represent each Boolean function $B(U_1, \dots, U_r)$ as a multilinear polynomial of degree n over $\mathbb{F}_2$ of the form

$$B(U_1, \dots, U_r) = \sum_{k=0}^{n} \sum_{1 \le i_1 < \dots < i_k \le r} A_{i_1 \dots i_k} U_{i_1} \dots U_{i_k}, \qquad (10.1)$$

where

$$A_{i_1\dots i_k} \in \mathbb{F}_2, \qquad 1 \le i_1 < \dots < i_k \le r.$$

We consider Boolean functions producing the rightmost bit of $\operatorname{ind} x$ from the bit representation of x. We also assume that all numbers contain the same number r of bits (adding several leading zeros if necessary), where $r = \lfloor \log p \rfloor$. Thus each such function is defined on a portion $1 \le x \le 2^r - 1 \le p - 1$ of the complete residue system modulo p (and we do not impose any restrictions on the values of such functions at $x = 0$).

Theorem 10.1. *Let a Boolean function $B(U_1, \dots, U_r)$ of $r = \lfloor \log p \rfloor$ variables be such that for any x, $1 \le x \le 2^r - 1$,*

$$B(u_1, \dots, u_r) = \begin{cases} 0, & \text{if } x \text{ is a quadratic residue modulo } p, \\ 1, & \text{if } x \text{ is a quadratic non-residue modulo } p, \end{cases}$$

where $x = u_1 \dots u_r$ is the bit representation of x. Then the bound

$$\operatorname{spr} B \ge 2^{-3/2} p^{1/4} \log^{-1/2} p - 1$$

holds.

Proof. Put $t = \operatorname{spr} B$ and define k by the inequalities

$$2^k > t + 1 \ge 2^{k-1}.$$

For each $m = 1, \dots, 2^k - 1$ we consider the function

$$B_m(V_1, \dots, V_{r-k}) = B(V_1, \dots, V_{r-k}, e_1, \dots, e_k),$$

where $m = e_1 \dots e_k$ is the bit representation of m. Obviously the total number of distinct monomials in $V_1, \dots, V_{r-k}$ occurring in all these functions does not exceed t. Therefore, because of the choice of k, one can find a non-trivial linear combination

$$\sum_{m=1}^{2^k-1} c_m B_m(V_1, \dots, V_{r-k}), \qquad c_1, \dots, c_{2^k-1} \in \mathbb{F}_2,$$

which vanishes identically.

Let $\chi(z)$ be the quadratic character modulo p. From the condition of the theorem, we see that

$$\chi(x) = (-1)^{B(u_1, \dots, u_r)}.$$

Therefore, for $0 \le y \le 2^{r-k} - 1$ we have

$$\prod_{m=1}^{2^k-1} \chi(2^k y + m)^{c_m} = (-1)^{\sum_{m=1}^{2^k-1} c_m B_m(v_1, \dots, v_{r-k})} = 1,$$

where $y = v_1 \dots v_{r-k}$ is the bit representation of y.

Combining this result with Lemma 3.4, we obtain

$$2^{r-k} = \sum_{y=0}^{2^{r-k}-1} \chi\left(\prod_{m=1}^{2^k-1} (2^k y + m)^{c_m}\right) \le 2^k p^{1/2} \log p.$$

Hence,

$$2^{2k} \ge 2^r p^{-1/2} \log^{-1} p \ge 0.5 p^{1/2} \log^{-1} p.$$

And finally, we derive that $t + 1 \ge 2^{k-1} \ge 2^{-3/2} p^{1/4} \log^{-1/2} p$. □

It easy to see that the same result holds for the monomials of the form $(a_1 U_1 + b_1) \dots (a_n U_n + b_n)$ with $a_i, b_i = 0, 1$, $i = 1, \dots, n$ as well. In other words, one can consider not only positive literals U_i but also their negations $\neg U_i$, $i = 1, \dots, r$.

To estimate the degree of $n = \deg B$ from below we recall the asymptotic formula

$$\log \binom{N}{\lfloor \gamma N \rfloor} \sim H(\gamma) N,$$

where $H(\gamma)$ is the binary entropy function (1.4), which holds for any fixed γ, $0 < \gamma < 1$, and $N \to \infty$; see [357], Section 10.11. Therefore for $n \le r/2$

$$\sum_{i=0}^{n} \binom{r}{i} \le (n+1) \binom{r}{n} = 2^{rH(n/r) + o(r)}. \tag{10.2}$$

Then from the inequality

$$t \le \sum_{i=0}^{n} \binom{r}{i}$$

one easily derives that under the condition of Theorem 10.1

$$n \ge \vartheta \log p + o(\log p), \tag{10.3}$$

where $\vartheta = 0.041\dots$ is the root of the equation

$$H(\vartheta) = 1/4, \qquad 0 < \vartheta < 1/2.$$

Certainly the bound is of the correct order because obviously $n \le r \le \log p$.

In fact for primes $p \equiv \pm 3 \pmod 8$ a better estimate can be obtained. Indeed, suppose $p \equiv \pm 3 \pmod 8$. Then 2 is a quadratic non-residue modulo p. Hence the polynomial

$$F(X_1, \dots, X_{r-1}) = B(0, X_1, \dots, X_{r-1}) + B(X_1, \dots, X_{r-1}, 0)$$

takes the value 1 for all non-zero binary $(r-1)$-tuples $(x_1, \dots, x_{r-1})$, where, as before, $r = \lfloor \log p \rfloor$. On the other hand, $F(0, \dots, 0) = 0$. Therefore

$$F(X_1, \dots, X_{r-1}) = \prod_{i=1}^{r-1} (1 + X_i) + 1.$$

Thus

$$\deg B \geq \deg F = r - 1 \tag{10.4}$$

and

$$\operatorname{spr} B \geq \lceil 0.5 \operatorname{spr} F \rceil = \lceil (2^{r-1} - 1)/2 \rceil = 2^{r-2}.$$

Boolean functions giving the rightmost bit of $\operatorname{ind} x$ for all but at most s values of $x = 1, \dots, 2^r - 1$ can be considered as well. Indeed, one easily sees that for such functions the bound of Theorem 10.1 can be modified as

$$t \geq C \min\{p^{1/4} \log^{-1/2} p, ps^{-1}\},$$

where C is an absolute constant.

If $p \equiv \pm 3 \pmod 8$, then, as in the proof of the bound (10.4), we obtain that the polynomial

$$F(X_1, \dots, X_{r-1}) = B(0, X_1, \dots, X_{r-1}) + B(X_1, \dots, X_{r-1}, 0)$$

takes the value 1 for all but at most $2s + 1$ binary $(r-1)$-tuples. On the other hand, this polynomial is not constant because $F(0, \dots, 0) = 0$. Applying the bound of R. Smolensky, presented in [238], which claims that if a polynomial

$$H(Y_1, \dots, Y_m) \in \mathbb{F}_q[Y_1, \dots, Y_m]$$

of degree n has one zero over $\mathbb{F}_q$, then it has at least q^{m-n} zeros, we obtain

$$2s + 1 \geq 2^{r-1-\deg F}.$$

Therefore

$$\deg B \geq \deg F = r - 1 - \log(2s + 1).$$

Now we obtain a lower bound for the parallel complexity of computing the rightmost bit of $\operatorname{ind} x$ by Boolean circuits.

We deal with circuits which use addition and multiplication modulo 2 as their basic operations. Certainly circuits using the more common Boolean operations $\vee, \wedge, \neg$ can be simulated by such circuits with only a constant factor increase in the depth.

We consider *bounded fan-in Boolean circuits* of depth d of the class $\mathrm{BC}(d)$ defined in Chapter 1.

Theorem 10.2. *Assume that there is a circuit $C \in \mathrm{BC}(d)$ such that, given the bit representation $u_1 \dots u_r$ of x, it computes*

$$C(u_1, \dots, u_r) = \begin{cases} 0, & \text{if } x \text{ is a quadratic residue modulo } p, \\ 1, & \text{if } x \text{ is a quadratic non-residue modulo } p, \end{cases}$$

where $1 \le x \le 2^r - 1$, $r = \lfloor \log p \rfloor$. Then the bound

$$d \ge \log \log p + O(1)$$

holds.

Proof. We can assume that the kth level has no more than 2^{d-k} gates. Indeed the dth level may utilize no more than two gates of the $(d-1)$th level, those may utilize no more than four gates of the $(d-2)$th level and so on.

Now it is easy to see that the output values $C(u_1, \dots, u_r)$ coincide with values of some Boolean function of degree at most 2^d. Applying (10.3), we obtain the result. □

Now we show that the same method as used in the proof of Theorem 10.1 can be used in studying the average sensitivity of the Boolean functions deciding quadratic residuacity.

Theorem 10.3. *Let a Boolean function $B(U_1, \dots, U_r)$ of $r = \lfloor \log p \rfloor$ variables be such that for any x, $1 \le x \le 2^r - 1$,*

$$B(u_1, \dots, u_r) = \begin{cases} 0, & \text{if } x \text{ is a quadratic residue modulo } p, \\ 1, & \text{if } x \text{ is a quadratic non-residue modulo } p, \end{cases}$$

where $x = u_1 \dots u_r$ is the bit representation of x. Then the bound

$$\sigma_{av}(B) \ge 0.5r + o(r)$$

holds.

Proof. We put $m = \lfloor r^{1/2} \rfloor$, $k = 2m + 1$, $l = \lfloor r - r^{1/2} \rfloor$, $R = 2^r - k2^l$. One sees that for any fixed i, $0 \le i \le l$, and any $x = 0, \dots, R-1$, the vector $\left(B(x + j2^i)\right)_{j=1}^k$ is defined. Over the range of x, the vector takes on the value of each possible binary k-tuple $T = (t_1, \dots, t_k)$ with multiplicity

$$N(T) = 2^{-k} \sum_{x=0}^{R-1} \prod_{j=1}^{k} \left(\chi(x + j2^i)(-1)^{t_j} + 1\right).$$

After simple evaluation one finds that the sum on the left-hand side contains one "main" term $R2^{-k}$ and $2^k - 1$ terms of the form

$$\pm 2^{-k} \sum_{x=0}^{R-1} \chi\left((x + j_1 2^i) \dots (x + j_s 2^i)\right),$$

where $s \le k$ and $1 \le j_1 < \ldots < j_s \le k$. Applying Lemma 3.4, we see that each term does not exceed $2^{-k} s p^{1/2} \log p$ in absolute value. Thus,

$$\begin{aligned} N(T) &= R2^{-k} + O\left(2^{-k} \sum_{s=1}^{k} \binom{k}{s} s p^{1/2} \log p\right) \\ &= R2^{-k} + O\left(k p^{1/2} \log p\right) \\ &= R2^{-k} + O(mr2^{r/2}) = R2^{-k} + o(R2^{-k}). \end{aligned}$$

It follows from probabilistic arguments that for $2^k + o(2^k)$ binary k-tuples $T = (t_1, \ldots, t_k)$, both of the following statements are true:

- $t_{2j} \ne t_{2j+1}$ for $0.5m + o(m)$ values of $j = 1, \ldots, m$;
- $t_{2j} \ne t_{2j-1}$ for $0.5m + o(m)$ values of $j = 1, \ldots, m$.

This means that, whatever the $(i+1)$th bit of x happens to be, if the vector $\left(B(x + j2^i)\right)_{j=1}^{k}$ is such a k-tuple T, then among the m values $B(x + j2^{i+1})$, $j = 1, \ldots, m$, about half differ from their respective

$$B\left((x + j2^{i+1})^{(i)}\right) = B(x + j2^{i+1} \pm 2^i) = B\left(x + (2j \pm 1)2^i\right).$$

So,

$$\begin{aligned} &\sum_{i=0}^{l} \sum_{x=0}^{R-1} \sum_{\substack{j=1 \\ B(x+j2^{i+1}) \ne B\left((x+j2^{i+1})^{(i)}\right)}}^{m} 1 \\ &\qquad \ge (l+1)\left(R2^{-k} + o(R2^{-k})\right)\left(2^k + o(2^k)\right)\left(0.5m + o(m)\right) \\ &\qquad = 0.5Rlm + o(Rlm). \end{aligned}$$

For every i, $0 \le i \le l$, and every j, $1 \le j \le m$, we find

$$\left| \sum_{\substack{x=0 \\ B(x+j2^{i+1}) \ne B\left((x+j2^{i+1})^{(i)}\right)}}^{R-1} 1 - \sum_{\substack{x=0 \\ B(x) \ne B\left(x^{(i)}\right)}}^{2^r-1} 1 \right| \le m2^{l+1} = o(2^r).$$

Therefore

$$\sigma_{av}(B) = 2^{-r} \sum_{i=0}^{l} \sum_{\substack{x=0 \\ B(x) \ne B\left(x^{(i)}\right)}}^{2^r-1} 1 \ge 0.5l + o(l) = 0.5r + o(r),$$

and we are done. □

Certainly the bound is of the correct order because obviously

$$r \geq \sigma(B) \geq \sigma_{av}(B) \geq 0.5r + o(r).$$

Combining this result with Lemma 7.1, one immediately obtains the lower bound $0.5 \log\log p + O(1)$ on the CREW PRAM complexity of B. The bound is slightly weaker than that of Theorem 10.2 but it concerns a more powerful (but maybe less realistic) computational model.

The lower bound of Theorem 10.3 on the average sensitivity can be combined with Lemma 23 of [36] in order to obtain a quadratic lower bound on the formula size $s(B)$ of a Boolean function B on bits of x giving of the rightmost of $\operatorname{ind} x$.

Finally, we consider one more characteristic of a Boolean function which also has many applications to complexity lower bounds, namely the largest value of the Fourier coefficients [33, 34, 35, 36, 67, 79, 80, 211, 226, 364, 469, 481] given by (1.7).

We also note that the difference

$$N(B) = 2^{r-1} - \frac{1}{2} \max_{a \in \mathfrak{B}_r} \left| \widehat{B}(a) \right|$$

is called the *non-linearity* of the Boolean function B and has numerous applications to cryptography [97, 98, 99, 100, 101, 102, 103, 104, 108, 110, 111, 176, 181, 358, 359, 360, 361, 362, 363, 489, 490, 491, 502, 538, 546, 547, 548, 563, 564, 566, 567, 568, 569, 570]. In particular, it is well known that Boolean functions with low non-linearity cannot be used in cryptographic applications.

Theorem 10.4. *Let a Boolean function $B(U_1, \ldots, U_r)$ of $r = \lfloor \log p \rfloor$ variables be such that for any x, $1 \leq x \leq 2^r - 1$,*

$$B(u_1, \ldots, u_r) = \begin{cases} 0, & \text{if } x \text{ is a quadratic residue modulo } p, \\ 1, & \text{if } x \text{ is a quadratic non-residue modulo } p, \end{cases}$$

where $x = u_1 \ldots u_r$ is the bit representation of x. Then the bound

$$\max_{a \in \mathfrak{B}_r} \left| \widehat{B}(a) \right| \leq 2^{5/4} p^{7/8} \log^{1/4} p + 1$$

holds.

Proof. One sees that for any $a \in \mathfrak{B}_r$

$$\widehat{B}(a) = \sum_{x=0}^{2^r - 1} \chi(x)(-1)^{\langle x, a \rangle} + (-1)^{B(0, \ldots, 0)},$$

where χ is the quadratic character modulo p, $\langle x, a \rangle = u \cdot a$ and $u = (u_1, \ldots, u_r)$ is the bit representation of x.

Let

$$S(a) = \sum_{x=0}^{2^r-1} \chi(x)(-1)^{\langle x,a\rangle}.$$

Then

$$|\widehat{B}(a)| \le |S(a)| + 1.$$

We put

$$k = \left\lceil 0.5 \log \left(2p^{3/2} \log p \right) \right\rceil, \qquad N = 2^k, \qquad M = 2^{r-k}.$$

Then, defining $\chi(0) = 0$, we obtain

$$S(a) = \sum_{y=0}^{N-1} \sum_{z=0}^{M-1} \chi(y + Nz)(-1)^{\langle y,b\rangle + \langle z,c\rangle},$$

where $b = (a_1, \dots, a_k)$, $c = (a_{k+1}, \dots, a_r)$ and $\langle y, b\rangle$, $\langle z, c\rangle$ stand for the dot product of the bit representations of y and z and vectors b and c, respectively. Therefore

$$|S(a)| \le \sum_{y=0}^{N-1} \left| \sum_{z=0}^{M-1} \chi(y + Nz)(-1)^{\langle z,c\rangle} \right|.$$

From here we derive

$$\begin{aligned}
|S(a)|^2 &\le \left(\sum_{y=0}^{N-1} \left| \sum_{z=0}^{M-1} \chi(y + Nz)(-1)^{\langle z,c\rangle} \right| \right)^2 \\
&\le N \sum_{y=0}^{N-1} \left| \sum_{z=0}^{M-1} \chi(y + Nz)(-1)^{\langle z,c\rangle} \right|^2 \\
&= N \sum_{y=0}^{N-1} \sum_{z_1,z_2=0}^{M-1} \chi(y + Nz_1)\chi(y + Nz_2)(-1)^{\langle z_1,c\rangle + \langle z_1,c\rangle} \\
&\le N \sum_{z_1,z_2=0}^{M-1} \left| \sum_{y=0}^{N-1} \chi\left((y + Nz_1)(y + Nz_2)\right) \right|.
\end{aligned}$$

For $M(M-1)$ pairs (z_1, z_2) with $z_1 \ne z_2$ we can apply Lemma 3.4 to the inner sum,

$$\left| \sum_{y=0}^{N-1} \chi\left((y + Nz_1)(y + Nz_2)\right) \right| \le 2p^{1/2} \log p.$$

For M pairs (z_1, z_2) with $z_1 = z_2$ we use the trivial identity

$$\sum_{y=0}^{N-1} \chi\left((y + Nz_1)(y + Nz_2)\right) = N.$$

Therefore

$$|S(a)|^2 \le N(2M^2p^{1/2}\log p + MN) \le 2p^{5/2}N^{-1}\log p + pN.$$

Noting that

$$2p^{3/2}N^{-1}\log p \le N \le 2\left(2p^{3/2}\log p\right)^{1/2},$$

we obtain

$$|S(a)|^2 \le 4p\left(2p^{3/2}\log p\right)^{1/2} = 2^{5/2}p^{7/4}\log^{1/2} p$$

and the desired estimate follows. □

Now we show how to apply the bound of Theorem 10.4 to study the unbounded fan-in circuits of the class $\mathrm{UBC}(d,S)$ computing the discrete logarithm.

Theorem 10.5. *Assume that there is a circuit $C \in \mathrm{UBC}(d,S)$ such that, given the bit representation $u_1 \ldots u_r$ of x, it computes*

$$C(u_1,\ldots,u_r) = \begin{cases} 0, & \text{if } x \text{ is a quadratic residue modulo } p,\\ 1, & \text{if } x \text{ is a quadratic non-residue modulo } p,\end{cases}$$

where $1 \le x \le 2^r - 1$, $r = \lfloor \log p \rfloor$. Then the bound

$$d\log\log S \ge (1+o(1))\log r$$

holds.

Proof. Let $B(U_1,\ldots,U_r)$ be a Boolean function such that

$$B(u_1,\ldots,u_r) = \begin{cases} 0, & \text{if } x \text{ is a quadratic residue modulo } p,\\ 1, & \text{if } x \text{ is a quadratic non-residue modulo } p,\end{cases}$$

where $x = u_1 \ldots u_r$ is the bit representation of x.

As before, we remark that the equation

$$H(\vartheta) = 1/4, \qquad 0 < \vartheta < 1/2,$$

with the binary entropy function (1.4) $H(\gamma)$, has the unique root $\vartheta = 0.041\ldots$. We put $w = \lceil 0.04r \rceil$. Then from Theorem 10.4 and the bound (10.2) we see that

$$\begin{aligned} 2^{-2r}\sum_{\substack{\mathfrak{a}\in\mathfrak{B}_r\\ \|\mathfrak{a}\|\le w}} \left|\widehat{B}(a)\right|^2 &= O\left(w\binom{r}{w}2^{-2r}p^{7/4}\log^{1/2} p\right)\\ &= O\left(2^{-r/4+H(w/r)r+o(r)}\right) = o(1). \end{aligned}$$

Finally, from the Parseval identity (7.1), we derive

$$2^{-2r} \sum_{\substack{a \in \mathfrak{B}_r \\ \|a\| > w}} \left|\widehat{B}(a)\right|^2 = 1 + o(1)$$

and from Lemma 7.2 the desired estimate follows. □

In particular, if d is sublogarithmic in the input size, $d \le \log^{\alpha} r$, with some constant $\alpha < 1$, then the size is superpolynomial

$$S \ge 2^{2^{(1+o(1))\log^{1-\alpha} r}}.$$

As it has been mentioned in Introduction, the rightmost bit of $\operatorname{ind} x$ for $1 \le x \le 2^r - 1$, like any other Boolean function, can be computed by an unbounded fan-in Boolean circuit of depth $d = 2$ and size $S \le 2^r$, that is, such that $d \log\log S \le 2\log r$.

An alternative proof of Theorem 10.5 can be obtained from Theorem 10.3 and some results of [33, 67].

Here we show that Boolean functions deciding quadratic residuacity belong to the classes $\mathcal{P}_{k,s}^r$ with sufficiently large k. Unfortunately, the best known way to use this information is Lemma 7.6.

Theorem 10.6. *Let a Boolean function $B(U_1, \dots, U_r)$ of $r = \lfloor \log p \rfloor$ variables be such that for any x, $1 \le x \le 2^r - 1$,*

$$B(u_1, \dots, u_r) = \begin{cases} 0, & \text{if } x \text{ is a quadratic residue modulo } p, \\ 1, & \text{if } x \text{ is a quadratic non-residue modulo } p, \end{cases}$$

where $x = u_1 \dots u_r$ is the bit representation of x. Then

$$B \in \mathcal{P}_{k.2^k}^r,$$

for $k = 1, \dots, \lfloor 0.25 \log p - 3 \rfloor$.

Proof. Let us fix k integers $1 \le i_1 < \dots < i_k \le r$. We consider the set $\mathcal{U}$ of r-bit integers, $x = u_1 \dots u_r$, $0 \le x \le 2^r - 1$, such that the digits at the positions i_j, $j = 1, \dots, k$, are fixed in an arbitrary way. We also put $i_0 = 0$ and $i_{k+1} = r + 1$. Obviously,

$$\mathcal{U} = \left\{ A + \sum_{j=0}^{k} x_j 2^{i_j} \mid 0 \le x_j \le 2^{i_{j+1} - i_j - 1} - 1, \ j = 0, \dots, k \right\},$$

where A depends on the fixed digits of x.

Let us put

$$h_j = \lceil 2^{i_{j+1}-i_j-2} \rceil, \qquad j = 0, \dots, k,$$

and define

$$D = A + \sum_{j=0}^{k} h_j 2^{i_j}, \quad \mathcal{V} = \left\{ \sum_{j=0}^{k} x_j 2^{i_j} \mid 0 \le x_j \le h_j - 1, \ j = 0, \dots, k \right\}.$$

We see that $D + v - w \in \mathcal{U}$ for all $v, w \in \mathcal{V}$.

Also, we see that for distinct selections of the digits of x at the positions $i_1, \dots, i_k$ we get distinct values of D.

So to prove the theorem it is enough to show that for any $0 \le D_1 < D_2 < p$ and the quadratic character χ of $\mathbb{F}_p$ we have

$$\left| \sum_{v,w \in \mathcal{V}} \chi\left((D_1 + v - w)(D_2 + v - w)\right) \right| < |\mathcal{V}|^2 - 2|\mathcal{V}|.$$

We remark that it is not enough to show that the left-hand side is just smaller than $|\mathcal{V}|^2$ because of the possible zero values for which the function D is defined in an arbitrary way.

Now, from Lemma 3.1 we have

$$\begin{aligned} &\left| \sum_{v,w \in \mathcal{V}} \chi\left((D_1 + v - w)(D_2 + v - w)\right) \right| \\ &\quad = \frac{1}{p} \left| \sum_{z=0}^{p-1} \chi\left((D_1 + z)(D_2 + z)\right) \sum_{c=0}^{p-1} \sum_{v,w \in \mathcal{V}} \mathbf{e}_p(c(z - v + w)) \right| \\ &\quad \le \frac{1}{p} \sum_{c=0}^{p-1} \left| \sum_{z=0}^{p-1} \chi\left((D_1 + z)(D_2 + z)\right) \mathbf{e}_p(cz) \right| \left| \sum_{v \in \mathcal{V}} \mathbf{e}_p(cv) \right|^2 \\ &\quad \le \frac{3p^{1/2}}{p} \sum_{c=0}^{p-1} \left| \sum_{v \in \mathcal{V}} \mathbf{e}_p(cv) \right|^2 = 3p^{1/2} |\mathcal{V}|. \end{aligned}$$

Taking into account that

$$|\mathcal{V}| \ge \prod_{j=0}^{k} 2^{i_{j+1}-i_j-2} = 2^{r-2k-2} \ge p2^{-2k-3},$$

we obtain that

$$3p^{1/2} |\mathcal{V}| < |\mathcal{V}|^2 - 2|\mathcal{V}|$$

for any $k \le 0.25 \log p - 3$. □

Applying Lemma 7.6 one immediately obtains that

$$CC(B) \geq \frac{7r-4}{6} \tag{10.5}$$

for the Boolean function of Theorem 10.6.

The next computational model we consider is decision trees (see Chapter 1), obtaining an exponential lower bound on their size.

Theorem 10.7. *Let a Boolean function $B(U_1, \ldots, U_r)$ of $r = \lfloor \log p \rfloor$ variables be such that for any x, $1 \leq x \leq 2^r - 1$,*

$$B(u_1, \ldots, u_r) = \begin{cases} 0, & \text{if } x \text{ is a quadratic residue modulo } p, \\ 1, & \text{if } x \text{ is a quadratic non-residue modulo } p, \end{cases}$$

where $x = u_1 \ldots u_r$ is the bit representation of x. Then the bound

$$\mathrm{DT}\,(B) \geq 2^{r/8+o(r)}$$

holds.

Proof. From Lemma 2.2 (taken with S empty) of [289] we obtain

$$\mathrm{DT}\,(B) \geq 2^{-r} \sum_{a \in \mathfrak{B}_r} \left| \widehat{B}(a) \right|.$$

On the other hand, from the Parseval identity (7.1) and the bound of Theorem 10.4 we see that

$$2^{2r} = \sum_{a \in \mathfrak{B}_r} \left| \widehat{B}(a) \right|^2 \leq 2^{5/4} p^{7/8} \log^{1/4} p \sum_{a \in \mathfrak{B}_r} \left| \widehat{B}(a) \right|.$$

Therefore

$$\sum_{a \in \mathfrak{B}_r} \left| \widehat{B}(a) \right| \geq 2^{9r/8+o(r)}$$

and the desired estimate follows. □

Several other relations between decision trees and Fourier coefficients are known, for example, see [40, 72, 147]. Combining them with Theorem 10.4, one can derive several other new results.

It is known (see Theorem 2.1 of [289]) that

$$\mathrm{DT}\,(B) \leq \exp\left(O\left(\log r \log^2 W(B)\right)\right),$$

where

$$W(B) = \mathrm{DNF}\,(B) + \mathrm{DNF}\,(\neg B)$$

is the *weight* $W(B)$ of B which is defined in [289] as the sum of the number of monomials DNF (B) in the minimal disjunctive normal form for B and the number of monomials DNF $(\neg B)$ in the minimal disjunctive normal form of its negation $\neg B$.

Thus Theorem 10.7 yields

$$W(B) \geq \exp\left(c(r/\log r)^{1/2}\right) \tag{10.6}$$

for some absolute constant $c > 0$.

Chapter 11

Approximation of the Discrete Logarithm by Real Polynomials

Here we consider some questions about approximation of the discrete logarithm by real and even complex polynomials. Unfortunately our results are weaker than those in our previous settings.

We start with exact representations via real polynomials of a given additive complexity.

Theorem 11.1. *Let $p \geq 3$ and let $f(X) \in \mathbb{R}[X]$ be a polynomial such that*

$$\operatorname{ind} x = f(x), \qquad x \in S,$$

for a set $S \subseteq \{1, \ldots, p-1\}$ of cardinality $|S| = p - 1 - s$. Then for the additive complexity $\mathrm{C}_{\pm}(f)$ of f the bound

$$\mathrm{C}_{\pm}(f) \geq \left(\frac{1}{20} \log\left(\frac{p-1}{4} - s\right)\right)^{1/2} - 1$$

holds.

Proof. Let R be the set of x, $1 \leq x \leq (p-1)/2$, for which both

$$\operatorname{ind} x = f(x) \quad \text{and} \quad \operatorname{ind} 2x = f(2x).$$

Then $|R| \geq (p-1) - 2(p-1-|S|) = (p-1)/2 - 2s$. We have either

$$\operatorname{ind} 2x = \operatorname{ind} 2 + \operatorname{ind} x$$

or

$$\operatorname{ind} 2x = \operatorname{ind} 2 + \operatorname{ind} x - p + 1.$$

Hence for any $x \in R$

$$f(2x) = \operatorname{ind} 2x = \operatorname{ind} 2 + \operatorname{ind} x = f(x) + \operatorname{ind} 2$$

or

$$f(2x) = \operatorname{ind} 2x = \operatorname{ind} 2 + \operatorname{ind} x - p + 1 = f(x) + \operatorname{ind} 2 - p + 1.$$

We remark that $0 < \operatorname{ind} 2 < p - 1$. Therefore for some integer $A \neq 0$ the polynomial $h(X) = f(2X) - f(X) - A$ vanishes for at least $|R|/2$ values of $x = 1, \dots, (p-1)/2$. On the other hand, $h(0) = A \neq 0$, thus $h(X)$ is a non-zero polynomial.

From Lemma 2.7 we see that

$$\mathrm{C}_{\pm}(h) \geq \left(\frac{1}{5}\log(|R|/2)\right)^{1/2} \geq \left(\frac{1}{5}\log\left(\frac{p-1}{4} - s\right)\right)^{1/2}.$$

Noting that $\mathrm{C}_{\pm}(h) \leq 2\mathrm{C}_{\pm}(f) + 2$, we obtain the desired result. □

Using Lemma 2.8 instead of Lemma 2.7 we derive the following statement on the additive complexity of exact representations via polynomials with integer coefficients.

Theorem 11.2. *Let $p \geq 3$ and let $f(X) \in \mathbb{Z}[X]$ be a polynomial such that*

$$\operatorname{ind} x = f(x), \qquad x \in S,$$

for a set $S \subseteq \{1, \dots, p-1\}$ of cardinality $|S| = p - 1 - s \geq 3$. Then for the additive complexity $\mathrm{C}_{\pm}(f)$ of f the bound

$$\mathrm{C}_{\pm}(f) \geq C \frac{\log(p - 4s - 1)}{\log\log(p - 4s - 1)}$$

holds, where $C > 0$ is an absolute constant.

The estimates of Theorem 11.1 and Theorem 11.2 provide the same lower bounds on sequential arithmetic RAM complexity with the division-free operation set $\{+, -, \times\}$ (for the constant sets $\mathbb{R}$ and $\mathbb{Z}$, respectively).

In the following result we estimate the $M_f(r)$-invariant (defined in Chapter 1) for functions f giving the rightmost bit of the discrete logarithm which provides a lower bound on the computational tree complexity and on the sequential arithmetic RAM complexity of the discrete logarithm modulo a prime p with the operation set $\{+, -, \times, /, \lfloor\cdot\rfloor\}$ consisting of the four arithmetic operations (over the reals) and the truncation function [365, 366].

It is shown in [366] (see Theorems 3.2 and 4.1 therein) that both the computational tree complexity and the sequential arithmetic RAM complexity of computation of $f(x)$ on r-bit inputs with operation set $\{+, -, \times, /, \lfloor\cdot\rfloor\}$ are at least $\Omega(\log^{1/2} \log M_f(r))$, which matches the known upper bounds for many important functions on r-bit integers; see [81, 365, 366, 380].

Theorem 11.3. *Let $f(x)$ be a function over the reals such that for any x, $1 \le x \le 2^r - 1$,*

$$f(x) = \begin{cases} 0, & \text{if } x \text{ is a quadratic residue modulo } p, \\ 1, & \text{if } x \text{ is a quadratic non-residue modulo } p, \end{cases}$$

where $x = u_1 \dots u_r$ is the bit representation of x, $r = \lfloor \log p \rfloor$. Then for any $\varepsilon > 0$ there exists a constant $C(\varepsilon) > 0$ such that the bound

$$M_f(r) \ge C(\varepsilon) p^{1-1/4e^{1/2}+\varepsilon}$$

holds.

Proof. From the definition of $M_f(r)$ we see that $f(x_1) = f(x_2)$ for all r-bit integers $1 \le x_1 < x_2 \le 2^r - 1$ divisible by $M_f(r)$. Let $N = \lfloor 2^r / M_f(r) \rfloor$. Then all the numbers $M_f(r), 2M_f(r), \dots, NM_f(r)$ are simultaneously either quadratic residues or quadratic non-residues. Therefore the same is true for numbers $1, \dots, N$ and because 1 is a quadratic residue we see that all those numbers are quadratic residues. From the Burgess bound of Lemma 3.10 which implies that the smallest quadratic non-residue is $O(p^{1/4e^{1/2}+\varepsilon})$, see [85], we derive that

$$N = O(p^{1/4e^{1/2}+\varepsilon})$$

for any fixed $\varepsilon > 0$ Therefore

$$M_f(r) \ge 2^{(1-1/4e^{1/2}+\varepsilon)r+O(1)}$$

for p large enough, and the result follows. □

Combining Theorem 11.3 with Theorems 3.2 and 4.1 of [366], we obtain the lower bound $\Omega(\log^{1/2} \log p)$ on the complexity of computing the rightmost bit of the discrete logarithm in both of the above mentioned computational models.

The method used in the proof of Theorem 8.8 also applies to studying the relation $\operatorname{ind} x = \lfloor u(x) \rfloor_m$. Obviously for $m \ge p - 1$ this equation is equivalent to the case of the congruence $\operatorname{ind} x \equiv u(x) \pmod m$ already studied in Theorem 8.8. Thus only the values $m < p - 1$ are of interest.

Theorem 11.4. *Let $0 \le N < N + H \le p - 1$ and let $(u(x))$ be an integer linear recurrence sequence of order n such that for an integer m,*

$$\operatorname{ind} x = \lfloor u(x) \rfloor_m, \qquad N + 1 \le x \le N + H.$$

Then

$$n \ge \frac{H^{2/3}}{5 \log m}$$

provided that $H \ge 600$.

Proof. Put

$$L = \left\lfloor \frac{\left(H - H^{2/3}\right)^{2/3}}{4} \right\rfloor .$$

We use the method from the proof of Theorem 8.8 but since $\operatorname{ind} x < m$ we obtain

$$\sum_{x=N+1}^{N+H-L} \left(\sum_{l=1}^{L} \varepsilon_l \operatorname{ind}(x+l) \right)^2 < 2L(H-L)m^2,$$

for all $(\varepsilon_1, \ldots, \varepsilon_L) \in \mathcal{F}$, instead of the inequality (8.7).

Therefore choosing $\mathcal{X} = \{x \in [N+1, N+H-L] : |K_x| < 2L^{1/2}\}$ we obtain the contradiction to $2^{L-1} > m^n$. For $H \geq 600$ we also have that $L - 1 \geq H^{2/3}/5$. □

Now we turn to various approximations by real polynomials.

We extend the function $\lfloor \alpha \rfloor_m$ to real $\alpha \in \mathbb{R}$ by defining its "residue" modulo an integer m as

$$\langle \alpha \rangle_m = \alpha - m \lfloor \alpha/m \rfloor .$$

We remark that for rational numbers the integer part function has also been extended in a different way in Chapter 1, thus although for a rational α both $\lfloor \alpha \rfloor_m$ and $\langle \alpha \rangle_m$ are defined, they do not coincide.

Theorem 11.5. *Let $0 \leq N < N+H \leq p-1$, $\delta \geq 0.5$ and let $f(X) \in \mathbb{R}(X)$ be such that*

$$|\operatorname{ind} x - \langle f(x) \rangle_{p-1}| < \delta, \qquad x \in S,$$

for a set $S \subseteq \{N+1, \ldots, N+H\}$ of cardinality $|S| = H - s$. Then the bound

$$\deg f \geq \min\{0.5 \log(H/\delta),\ H/2(s+1)\} - 2$$

holds.

Proof. Let $n = \deg f$. Using the recurrence equation (2.3) as in the proof of Theorem 9.2, we derive that for at least $H - (n+2)(s+1)$ values of $x = N+1, \ldots, N+H$,

$$\sum_{i=0}^{n+1} e_i \operatorname{ind}(x+i) \equiv \Delta_x \pmod{p-1},$$

where

$$e_i = (-1)^{n+1-i} \binom{n+1}{i}, \qquad i = 0, \ldots, n+1,$$

and

$$|\Delta_x| \leq 2^{n+1}\delta.$$

Therefore, the rational function

$$\prod_{i=0}^{n+1}(X+i)^{e_i}$$

of degree at most 2^n takes at most $2(2^{n+1}\delta+1)$ distinct values modulo p for at least $H-(n+2)(s+1)$ values of $x=N+1,\dots,N+H$. Therefore

$$H-(n+2)(s+1)\le 2^{n+1}(2^{n+1}\delta+1)\le 2^{2n+3}\delta$$

and the result follows. □

We also consider computing the rightmost bit of the discrete logarithm by real polynomials.

Theorem 11.6. *Let $0\le N<N+H\le p-1$. Assume that a polynomial $f(X)\in \mathbb{R}[X]$ is such that $f(x)\ge 0$ if x is a quadratic residue modulo p and $f(x)<0$ otherwise, for every element $x\in S$ from some set $S\subseteq\{N+1,\dots,N+H\}$ of cardinality $|S|\ge H-s$. Then for any $\varepsilon>0$ the bound*

$$\deg f\ge\begin{cases} H/2-2s-1-p^{1/2}\log p, & \text{for any } H,\\ C(\varepsilon)\min\{H,\,H^2p^{-1/2}\}-2s-1, & \text{if } p^{1/4+\varepsilon}\le H\le p^{1/2+\varepsilon},\\ (p-1)/2-2s-1, & \text{if } N=0, H=p-1,\end{cases}$$

holds, where $C(\varepsilon)>0$ depends only on ε.

Proof. Let $\chi(z)$ denote the quadratic character modulo p. Denote by T the set of $x\in\{N+1,\dots,N+H\}$ such that $\chi(x)\ne\chi(x+1)$. We see that the number of $x\in T$ such that either $x\notin S$ or $x+1\notin S$ is at most $2s+1$. For others we have $f(x)f(x+1)<0$. Therefore, f changes its sign at least $|T|-2s-1$ times, so it has at least that many zeros, thus

$$\deg f\ge |T|-2s-1. \tag{11.1}$$

On the other hand, we have

$$\begin{aligned}|T| &= -\sum_{x\in T}\chi(x(x+1)) = -\sum_{x=N+1}^{N+H}\chi(x(x+1))+\sum_{x\notin T}\chi(x(x+1))\\ &= -\sum_{x=N+1}^{N+H}\chi(x(x+1))+H-|T|.\end{aligned}$$

Therefore

$$|T|=\frac{1}{2}H-\frac{1}{2}\sum_{x=N+1}^{N+H}\chi(x(x+1)). \tag{11.2}$$

From (11.1) and (11.2) and Lemma 3.4 we derive the first estimate.

For $p^{1/4+\varepsilon} \le H \le p^{1/2}$ we use directly the bound

$$\sum_{x=N+1}^{N+H} \chi\left(x(x+1)\right) \le H - c(\varepsilon)H^2p^{-1/2}$$

of [407].

For $p^{1/2} \le H \le p^{1/2+\varepsilon}$ we split the sum into $\lfloor H/p^{1/2} \rfloor$ sums of length $\lfloor p^{1/2} \rfloor$ and perhaps one sum of smaller length (which we estimate trivially). Thus for such an H we have.

$$\begin{aligned} \sum_{x=N+1}^{N+H} \chi\left(x(x+1)\right) &\le H - c(1/4)\left\lfloor H/p^{1/2} \right\rfloor \left\lfloor p^{1/2} \right\rfloor^2 p^{-1/2} \\ &\le H - \frac{c(1/4)}{8} H. \end{aligned}$$

Combining with (11.1) and (11.2) we obtain the second estimate.

If $H = 0$, $N = p$, then the value of the character sum is precisely known:

$$\sum_{x=1}^{p-1} \chi\left(x(x+1)\right) = -1,$$

(see, for example, Exercise 10.a.α in Chapter 6 of [553]), and the third estimate follows. □

Now we consider real multivariate polynomials.

First of all, we mention that Theorem 10.3 and Lemma 7.5 imply that for any real polynomial

$$f(X_1, \dots, X_r) \in \mathbb{R}[X_1, \dots, X_r]$$

such that for any x, $1 \le x \le 2^r - 1$,

$$f(u_1, \dots, u_r) = \begin{cases} 0, & \text{if } x \text{ is a quadratic residue modulo } p, \\ 1, & \text{if } x \text{ is a quadratic non-residue modulo } p, \end{cases}$$

where $x = u_1 \dots u_r$ is the bit representation of x, $r = \lfloor \log p \rfloor$, the bound

$$\deg f \ge 0.5r + o(r) \tag{11.3}$$

holds.

Also, Theorem 10.3 and Lemma 7.5 imply that for any real polynomial

$$f(X_1, \dots, X_r) \in \mathbb{R}[X_1, \dots, X_r]$$

approximating the Boolean function of Theorems 10.1 and 10.3 with an absolute error at most $1/3$, that is, a polynomial f such that $|f(u_1, \dots, u_r) - 1| \le 1/3$ if x is a not quadratic residue modulo p and $|f(u_1, \dots, u_r)| \le 1/3$ otherwise, where $x = u_1 \dots u_r$ is the bit representation of x, $1 \le x \le 2^r - 1$, $r = \lfloor \log p \rfloor$, the bound

$$\deg f \ge 12^{-1/2} r^{1/2} + o(r^{1/2}) \tag{11.4}$$

holds.

Here we show that the bound on the Fourier coefficients of Theorem 10.4 leads to a better estimate (11.7).

Moreover, instead of approximation of a Boolean function B, one can consider a more general *threshold representation* of that function via real multivariate polynomials f taking non-negative values on inputs for which the function B equals 1 and negative values on the other inputs [79, 80, 226, 312, 481]. If the largest value of such a polynomial on $\{0,1\}$-inputs is bounded by a constant then, by using linear transformations $af + b$ with some real a and b, this question can be reduced to the previous question on approximation with a fixed absolute error $\delta < 1/2$. However, if there are no restrictions on the size of the polynomial, then we obtain a wider class of representations. Moreover, the degree of such polynomials can be essentially smaller, see the remark after Theorem 11.7 below.

Furthermore, in the case of real polynomials, the Boolean values 0 and 1 can be interpreted as two arbitrary real values α_0 and α_1, not necessarily $\alpha_0 = 0$, $\alpha_1 = 1$. It is easy to see that the degree of the corresponding polynomials does not depend on the particular choice of α_0, α_1 because they are equivalent under a linear transformation of variables [312]. However it is shown in [312] that the number of monomials, $\operatorname{spr} f$, depends on the choice of α_0 and α_1. In fact, there are examples of Boolean functions demonstrating that for $(\alpha_0, \alpha_1) = (0, 1)$ and $(\alpha_0, \alpha_1) = (1, -1)$ the gap between the numbers of monomials of the corresponding polynomials for these two representations can be exponentially large [312].

In the next statement we estimate the degree and number of monomials in the most general situation. The result about the number of monomials seems to be new even for the aforementioned case of approximating polynomials with a fixed absolute error $\delta < 1/2$.

Theorem 11.7. *Let α_0, α_1 be two distinct real numbers, $r = \lfloor \log p \rfloor$, and let a polynomial*

$$f(X_1, \dots, X_r) \in \mathbb{R}[X_1, \dots, X_r]$$

be such that

$$f(\alpha_{u_1}, \dots, \alpha_{u_r}) \ge 0$$

if x is a quadratic residue modulo p and

$$f(\alpha_{u_1}, \dots, \alpha_{u_r}) < 0$$

otherwise, where $x = u_1 \dots u_r$ *is the bit representation of* x, $1 \le x \le 2^r - 1$. *Then* f *is of degree*

$$\deg f \ge \log r + o(\log r)$$

and contains at least

$$\operatorname{spr} f \ge 0.25r + o(r)$$

distinct monomials.

Proof. Assuming that p is large enough we put

$$m = \lfloor \log(r - 2\log^2 r) \rfloor - 1.$$

Let τ be the number of monomials $\mu_j(y)$, $j = 1, \dots, \tau$, in $y = (y_1, \dots, y_m)$ such that for every m-dimensional vector $y = (y_1, \dots, y_m) \in \{\alpha_0, \alpha_1\}^m$ we have a representation of the form

$$f(X_1, \dots, X_{r-m}, y) = \sum_{i=1}^{\tau} \mu_i(y) f_i(X_1, \dots, X_{r-m})$$

with some polynomials $f_j(Y_1, \dots, Y_{r-k}) \in \mathbb{R}[Y_1, \dots, Y_{r-k}]$.
Obviously

$$\tau \le \binom{\deg f + m}{\deg f} \quad \text{and} \quad \tau \le \operatorname{spr} f. \tag{11.5}$$

If $\tau < 2^m$, then there exist 2^m real coefficients c_y, $y \in \{\alpha_0, \alpha_1\}^m$, not all equal to zero and such that

$$\sum_{y \in \{\alpha_0, \alpha_1\}^m} c_y \mu_i(y) = 0, \qquad i = 1, \dots, \tau.$$

Therefore we have the identity

$$\sum_{y \in \{\alpha_0, \alpha_1\}^m} c_y f(X_1, \dots, X_{r-m}, y) = 0. \tag{11.6}$$

On the other hand, as in the proofs of Theorems 10.1 and 10.3, we see that for p large enough, any 2^m-dimensional pattern of signs occurs

$$2^{r-m} 2^{-2^m} + O(r 2^{r/2+m}) \ge 2^{r/2 + \log^2 r + O(\log r)} - 2^{r/2 + O(\log r)} \ge 1$$

times among the coordinates of the 2^m-dimensional vectors

$$\left(f(\alpha_{x_1}, \dots, \alpha_{x_{r-m}}, y)\right)_{y \in \{\alpha_0, \alpha_1\}^m}, \qquad x = 0, \dots, 2^{r-m} - 1,$$

where $x = x_1 \dots x_{r-m}$ is the bit representation of x.

In particular, there exists $x = x_1 \dots x_{r-m}$ such that

$$c_y f\left(\alpha_{x_1}, \dots, \alpha_{x_{r-m}}, y\right) > 0$$

for every $c_y \neq 0$, $y \in \{\alpha_0, \alpha_1\}^m$. Thus,

$$\sum_{y \in \{\alpha_0, \alpha_1\}^m} c_y f(\alpha_{x_1}, \dots, \alpha_{x_{r-m}}, y) > 0$$

which contradicts (11.6). Hence

$$\tau \geq 2^m \geq 2^{\log(r - \log^2 r) - 2} = 0.25r + O(\log^2 r).$$

From (11.5) we obtain

$$\deg f \geq m \quad \text{and} \quad \operatorname{spr} f \geq 2^m,$$

and the desired result follows. □

In particular, one sees that a linear threshold representation (corresponding to linear polynomials) of the rightmost bit of $\operatorname{ind} x$ is not possible.

The lower bound of Theorem 11.7 on the degree of f is much weaker than the bound (11.3) for real polynomials giving a precise representation, and than the bound (11.4) for approximating polynomials. In fact for the Boolean function

$$\mathrm{OR}_r(u_1, \dots, u_r) = \begin{cases} 0, & \text{if } u_1 = \dots = u_r = 0, \\ 1, & \text{otherwise,} \end{cases}$$

the gap is even bigger. It is shown in [447] that the precise representation requires a polynomial of degree r, any $1/3$-approximating polynomial must be of degree of order $r^{1/2}$, while it is obvious that the linear polynomial

$$f(X_1, \dots, X_r) = X_1 + \dots + X_r - 1/2$$

provides a threshold representation of this function.

The result of Theorem 11.7 can be extended to a wider class of functions. Using the same method, one can show that any exponential polynomial $\Phi(X_1, \dots, X_r)$ of the form

$$\Phi(X_1, \dots, X_r) = \sum_{j=1}^{k} \exp\left(L_j(X_1, \dots, X_r)\right) f_j(X_1, \dots, X_r)$$

with linear forms L_j and polynomials f_j, $j = 1, \dots, k$, with real coefficients and such that $\Phi(\alpha_{u_1}, \dots, \alpha_{u_r}) > 0$ if $x = u_1 \dots u_r$ is a quadratic residue modulo p and $\Phi(\alpha_{u_1}, \dots, \alpha_{u_r}) \leq 0$ otherwise, satisfies the inequality

$$\sum_{j=1}^{k} \operatorname{spr} f_i \geq 0.25r + o(r).$$

In fact, in the most general form the method is applicable to functions of the shape

$$\Phi(X_1, \dots, X_r) = \sum_{j=1}^{t} \prod_{i=1}^{r} g_{i,j}(X_i),$$

where $g_{i,j}(X)$ are arbitrary real-valued functions. One can prove that $t \geq 0.25r + r$ for any threshold representation of the rightmost bit of $\operatorname{ind} x$ by a function of this form.

As we have mentioned, the degree of polynomials providing a threshold representation is independent of the particular choice of α_0, α_1. Below we select $\alpha_0 = 1$, $\alpha_1 = -1$. This allows us to combine the estimate of the Fourier coefficients of Theorem 10.4 with some general results of [79, 80] and to get an exponential lower bound on the sparsity of the corresponding polynomials.

Theorem 11.8. *Let a multilinear polynomial*

$$f(X_1, \dots, X_r) \in \mathbb{R}[X_1, \dots, X_r]$$

be such that

$$f((-1)^{u_1}, \dots, (-1)^{u_r}) \geq 0$$

if x is a quadratic residue modulo p and

$$f((-1)^{u_1}, \dots, (-1)^{u_r}) < 0$$

otherwise, where $x = u_1 \dots u_r$ is the bit representation of x, $1 \leq x \leq 2^r - 1$, $r = \lfloor \log p \rfloor$. Then the bound

$$\operatorname{spr} f \geq 2^{r/8 + o(r)}$$

holds.

Proof. Let $w = \operatorname{spr} f$. From Theorem 5.1 of [79] or Theorem 3 of [80] we see that

$$w \geq \frac{2^r}{\max_{a \in \mathfrak{B}_r} \left| \widehat{B}(a) \right|},$$

where $B(U_1, \dots, U_r)$ is a Boolean function such that for any x, $1 \leq x \leq 2^r - 1$,

$$B(u_1, \dots, u_r) = \begin{cases} 0, & \text{if } x \text{ is a quadratic residue modulo } p, \\ 1, & \text{if } x \text{ is a quadratic non-residue modulo } p, \end{cases}$$

where $x = u_1 \dots u_r$ is the bit representation of x. Applying Theorem 10.4, we derive the desired estimate. □

Denote by $\vartheta = 0.017\ldots$ the unique root of the equation

$$H(\vartheta) = 1/8, \qquad 0 < \vartheta < 1/2,$$

where $H(\gamma)$ is the binary entropy function (1.4). Then the following refinement

$$\deg f \geq \vartheta r + o(r) \tag{11.7}$$

of the bound (11.4) holds for any polynomial f satisfying the conditions of Theorem 11.7. Indeed, if $n = \deg f$, then the sparsity of this polynomial does not exceed

$$\operatorname{spr} f \leq \sum_{i=0}^{n} \binom{r}{i} \leq (n+1)\binom{r}{n}$$

and from Theorem 11.8 the bound follows. We note that this bound is of correct order because, obviously, $\deg f \leq r$.

Finally, we show how to combine some results of [549] with our approach in order to obtain lower bounds for the size of formulas deciding quadratic residuacity over various bases of real-valued functions.

We recall that the *formula size* $s(\Phi)$ of a formula Φ is the total number of symbols contained in the formula Φ including variables and basis functions.

Theorem 11.9. *Let $\Phi(X_1, \ldots, X_r)$ be a formula in some basis containing finitely many multilinear polynomials over* $\mathbb{R}$ *and all constants from* $\mathbb{R}$. *Assume that*

$$\Phi(u_1, \ldots, u_r) \geq 0$$

if x is a quadratic residue modulo p and

$$\Phi(u_1, \ldots, u_r) < 0$$

otherwise, where $x = u_1 \ldots u_r$ is the bit representation of x, $1 \leq x \leq 2^r - 1$, $r = \lfloor \log p \rfloor$. Then for the size $s(\Phi)$ the bound

$$s(\Phi) \geq C \frac{r^2}{\log^2 r}$$

holds, where $C > 0$ is an absolute constant.

Proof. We put $l = \lfloor \log r \rfloor - 2$, $m = \lfloor r/l \rfloor - 1$ and define the sets

$$S_i = \{li + 1, \ldots, l(i+1)\}, \qquad i = 1, \ldots, m.$$

For each $i = 1, \ldots, m$ we denote by t_i the number of Boolean functions of l variables obtained from B over all possible specializations of variables with subscripts outside of S_i.

Finally, we define the numbers r_i by means of the equation

$$(2^l r_i)^{r_i} = t_i, \qquad i = 1, \dots, m.$$

Then Theorem 6.1 of [549] claims that

$$s(\Phi) \geq c \sum_{i=1}^{m} r_i.$$

where $c > 0$ is an absolute constant.

It is enough to show that, for our choice of parameters, every t_i takes its largest possible value

$$t_i = 2^{2^l}, \qquad i = 1, \dots, m. \tag{11.8}$$

Indeed, in this case

$$r_i \geq \frac{r}{2 \log r}, \qquad i = 1, \dots, m,$$

and the result follows.

We remark that each specialization of variables outside of S_i corresponds to a formula $\Phi_{i,\lambda,\eta}(w_1, \dots, w_l)$ satisfying the condition

$$\Phi_{i,\lambda,\eta}(w_1, \dots, w_l) \geq 0$$

if $2^{l(i+1)}\lambda + w2^{li} + \eta$ is a quadratic residue modulo p and

$$\Phi_{i,\lambda,\eta}(w_1, \dots, w_l) < 0$$

otherwise, where $w = w_1 \dots w_l$ is the bit representation of w, $0 \leq w \leq 2^l - 1$. The parameters λ and η, corresponding to particular specialization, take the following values: $\lambda = 0, \dots, 2^{r-l(i+1)} - 1$, $\eta = 1, \dots, 2^{li} - 1$.

To show that for each $i = 1, \dots, m$ all possible Boolean functions of l variables can be represented by the formula $\Phi_{i,\lambda,\eta}(w_1, \dots, w_l)$ for some choice of λ and η, we show that for any $L = 2^l$ numbers $\tau_w = \pm 1$, $w = 0, \dots, L-1$, we can satisfy

$$\chi\left(2^{l(i+1)}\lambda + w2^{li} + \eta\right) = \tau_w, \qquad w = 0, \dots, L-1,$$

for some $\lambda = 0, \dots, 2^{r-l(i+1)} - 1$, $\eta = 1, \dots, 2^{li} - 1$, where χ is the quadratic character modulo p.

We fix some $i = 1, \dots, m$ and put

$$H = 2^{r-l(i+1)-1}, \qquad I = 2^{li}, \qquad J = 2^{li-1}, \qquad K = 2^{l(i+1)}.$$

It is enough to show that for some $\lambda_1, \lambda_2 = 0, \dots, H-1$ and some $\eta_1, \eta_2 = 1, \dots, J-1$ we can satisfy

$$\chi\left(K(H + \lambda_1 - \lambda_2) + \eta_1 - \eta_2 + J + Iw\right) = \tau_w, \qquad w = 0, \dots, L-1.$$

Put $A = KH + J$. Then the number of tuples $(\lambda_1, \lambda_2, \eta_1, \eta_2)$ with $\lambda_1, \lambda_2 = 0, \ldots, H-1$, $\eta_1, \eta_2 = 1, \ldots, J-1$ for which the last condition is satisfied is

$$\begin{aligned} N &= \frac{1}{2^L} \left| \sum_{\lambda_1,\lambda_2=0}^{H-1} \sum_{\eta_1,\eta_2=1}^{J-1} \prod_{w=0}^{L-1} \left(\chi\left(K(\lambda_1 - \lambda_2) + \eta_1 - \eta_2 + Iw + A\right) + \tau_w\right) \right| \\ &= \frac{H^2(J-1)^2}{2^L} + \frac{1}{2^L} R, \end{aligned}$$

where R contains $2^L - 1$ sums of the form

$$\sigma = \pm \sum_{\lambda_1,\lambda_2=0}^{H-1} \sum_{\eta_1,\eta_2=1}^{J-1} \prod_{k=1}^{s} \chi\left(K(\lambda_1 - \lambda_2) + \eta_1 - \eta_2 + Iw_k + A\right)$$

for some $0 \le w_1 < \ldots < w_s \le L-1$. For each such sum we have

$$\begin{aligned} \pm\sigma &= \frac{1}{p} \sum_{x=0}^{p-1} \prod_{k=1}^{s} \chi\left(x + Iw_k + A\right) \\ &\qquad \times \sum_{a=0}^{p-1} \sum_{\lambda_1,\lambda_2=0}^{H-1} \sum_{\eta_1,\eta_2=1}^{J-1} \mathbf{e}_p\left(a\left(x - K(\lambda_1 - \lambda_2) - \eta_1 + \eta_2\right)\right) \\ &= \frac{1}{p} \sum_{a=0}^{p-1} \sum_{x=0}^{p-1} \prod_{k=1}^{s} \chi\left(x + Iw_k + A\right) \mathbf{e}_p(ax) \\ &\qquad \times \left| \sum_{\lambda=0}^{H-1} \sum_{\eta=1}^{J-1} \mathbf{e}_p\left(a(K\lambda + \eta)\right) \right|^2 . \end{aligned}$$

Applying the Weil bound given by Lemma 3.1 to the sum over x, we derive

$$\begin{aligned} |\sigma| &\le Lp^{-1/2} \sum_{a=0}^{p-1} \left| \sum_{\lambda=1}^{H-1} \sum_{\eta=1}^{J-1} \mathbf{e}_p\left(a(K\lambda + \eta)\right) \right|^2 \\ &= Lp^{-1/2} \sum_{\lambda_1,\lambda_2=0}^{H-1} \sum_{\eta_1,\eta_2=1}^{J-1} \sum_{a=0}^{p-1} \mathbf{e}_p\left(a\left(K\lambda_1 + \eta_1 - K\lambda_2 - \eta_2\right)\right) \\ &= Lp^{1/2} T, \end{aligned}$$

where T is the number of solutions of the congruence

$$\begin{aligned} K\lambda_1 + \eta_1 &\equiv K\lambda_2 + \eta_2 \pmod p, \\ \lambda_1, \lambda_2 = 0, \ldots, H-1, &\qquad \eta_1, \eta_2 = 1, \ldots, J-1. \end{aligned}$$

Because $K > J$ and $K(H-1)+J-1 < p$, this is only possible for $\lambda_1 = \lambda_2$, $\eta_1 = \eta_2$. Therefore $T = H(J-1)$. Hence

$$N \geq \frac{H^2(J-1)^2}{2^L} - Lp^{1/2}H(J-1).$$

We see that $L \leq r/4$, thus

$$H(J-1)2^{-L} = 2^{r-l-L+O(1)} = 2^{3r/4+o(r)} > Lp^{1/2},$$

provided that p is large enough. Hence $N > 0$ and (11.8) follows. □

In the following theorem we add the sign function

$$\operatorname{sign}(x) = \begin{cases} 1, & \text{if } x > 0, \\ 0, & \text{if } x = 0, \\ -1, & \text{if } x < 0, \end{cases}$$

to the set of basis functions and estimate the size of formulas using this function as well.

Theorem 11.10. *Let $\Phi(X_1, \dots, X_r)$ be a formula in some basis containing finitely many multilinear polynomials over $\mathbb{R}$, all constants from $\mathbb{R}$ and the sign function* sign (x). *Assume that*

$$\Phi(u_1, \dots, u_r) \geq 0$$

if x is a quadratic residue modulo p and

$$\Phi(u_1, \dots, u_r) < 0$$

otherwise, where $x = u_1 \dots u_r$ is the bit representation of x, $1 \leq x \leq 2^r - 1$, $r = \lfloor \log p \rfloor$. Then for the size $s(\Phi)$ the bound

$$s(\Phi) \geq C \frac{r^{3/2}}{\log r}$$

holds, where $C > 0$ is an absolute constant.

Proof. By Theorem 6.5 of [549], in the notation of Theorem 11.9, we have

$$s(\Phi) \geq c \sum_{i=1}^{m} \log^{1/2} t_i,$$

where $c > 0$ is an absolute constant. Taking into account that (11.8) holds for $l = \lfloor \log r \rfloor - 2$, we obtain the desired estimate. □

It is interesting to remark that, using the result of [482], one can apply a similar technique to derive lower bounds for quantum formulas deciding whether a given integer is a quadratic residue modulo p.

Part III

Approximation and Complexity of the Diffie–Hellman Secret Key

Chapter 12

Polynomial Approximation and Arithmetic Complexity of the Diffie–Hellman Secret Key

Let g be a primitive root of a finite field $\mathbb{F}_q$ of q elements. One of the most popular public key cryptosystems, the Diffie–Hellman key exchange protocol, is based on the still unproved assumption that recovering the value of the Diffie–Hellman secret key

$$K(x, y) = g^{xy}$$

from the known values of g^x and g^y is essentially equivalent to the discrete logarithm problem and therefore is hard. Here we show that even computation of g^{x^2} from g^x cannot be realized by a polynomial of low degree.

We remark (although it is not essential for what follows) that, as it has been shown in Section 5.3 of [375], the general case can be reduced to this one via the identity

$$g^{(x+y)^2} g^{-x^2} g^{-y^2} = g^{2xy},$$

a deterministic polynomial time algorithm for finding square roots in cyclic groups with known generator (we recall that g is known) and the Pohlig–Hellman algorithm to select among the two square roots, see [16, 389, 541].

As in the case of the discrete logarithm, first of all we study polynomial approximations of different kinds. Many of the results of this chapter have been presented in [133].

First of all we remark that an analogue of the explicit representation given in [399]

$$f(x) \equiv -1 + \sum_{k=1}^{p-2} (g^{-k} - 1)^{-1} x^k \pmod p \tag{12.1}$$

for the polynomial $f(X) \in \mathbb{F}_q[X]$ of degree $\deg f \le q-2$ and such that

$$g^{x^2} = f(g^x), \qquad x = 0, \dots, q-2,$$

for a primitive root $g \in \mathbb{F}_q^*$, has recently been discovered in [384]. The explicit formulas for the coefficients depend on the residue class of q modulo 4 but in all cases the degree $\deg f$ is close to $q-2$. To be more precise, $\deg f = q-3$ if $q \equiv 1 \pmod 4$ and $\deg f = q-2$ otherwise. It also follows from those explicit formulas that sparsity $\operatorname{spr} f = (q-1)/2$ if q is odd and $\operatorname{spr} f = q-1$ if q is even. In fact, in this paper [384] the case of elements g of arbitrary multiplicative order has been considered.

Now we study partial representations.

Theorem 12.1. *Let $f(X) \in \mathbb{F}_q[X]$ be such that*

$$g^{x^2} = f(g^x), \qquad x \in S, \tag{12.2}$$

for a set $S \subseteq \{N+1, \dots, N+H\}$ of cardinality $|S| = H - s$ with $H \le q-1$. Then

$$\deg f \ge H - 2s - 3.$$

Proof. Let R be the set of $x \in \{N+1, \dots, N+H\}$ for which both $g^{x^2} = f(g^x)$ and $g^{(x+1)^2} = f(g^{x+1})$. We see that

$$|R| \ge H - 1 - 2(H - |S|) = H - 1 - 2s.$$

Now, for $u = g^x$ with $x \in R$ we have

$$f(gu) = f(g^{x+1}) = g^{x^2+2x+1} = g^{2x+1} f(g^x) = gu^2 f(u).$$

So the polynomial $h(X) = gX^2 f(X) - f(gX)$ has at least $|R|$ zeros in $\mathbb{F}_q$ and is obviously not identical to zero. Therefore $|R| \le \deg h = \deg f + 2$, thus $\deg f \ge |R| - 2 \ge H - 2s - 3$. □

Thus, as in the case of the discrete logarithm, if $s = o(H)$, then $\deg f \sim H$. Also, if $N = 0$, $H = q-1$, one can show that $\operatorname{spr} f \ge (q-1)/4s$ (to apply Lemma 2.5 one also needs the technical assumption $\deg f \le q-4$ which however can be eliminated by more careful examination).

Certainly, for any S one can satisfy (12.2) with a polynomial f of degree $\deg f \le |S| - 1 = H - s - 1$.

Theorem 12.1 is non-trivial if the set S is dense enough on some interval, $|S| > H/2$. The next result is applicable to arbitrary quite sparse sets S beginning with $|S| > 2H^{2/3}$.

Theorem 12.2. *Let* $f(X) \in \mathbb{F}_q[X]$ *be a polynomial of degree* $n = \deg f$ *such that*

$$g^{x^2} = f(g^x), \qquad x \in S,$$

for a set $S \subseteq \{N+1, \ldots, N+H\}$ *with* $H \le q-1$. *Then*

$$n \ge |S|^2/2H - 4H/|S| - 1.$$

Proof. Let us define $K = \lfloor 2H/|S| \rfloor$ and consider the $K+1$ shift-sets $S_i = S - i$, $i = 0, \ldots, K$. These all belong to the interval of length $H + K$, thus denoting $R_{i,j} = S_i \cap S_j$, from the inclusion–exclusion principle we obtain

$$(K+1)|S| - \sum_{0 \le i < j \le K} |R_{i,j}| = \sum_{i=0}^{K} |S_i| - \sum_{0 \le i < j \le K} |R_{i,j}| \le \left| \cup_{i=0}^{K} S_i \right| \le H + K.$$

Therefore, there is a pair $0 \le i < j \le K$ such that

$$|R_{0,j-i}| = |R_{i,j}| \ge \frac{2|S|}{K} - \frac{2(H+K)}{K(K+1)} \ge |S|/K - 1 \ge |S|^2/2H - 1.$$

For this pair we put $k = j - i$ and let $R = R_{0,k}$. Then for any $x \in R$ we have both

$$g^{x^2} = f(g^x) \quad \text{and} \quad g^{(x+k)^2} = f(g^{x+k}).$$

Therefore,

$$f(g^{x+k}) = g^{(x+k)^2} = g^{x^2} g^{2kx} g^{k^2} = g^{2kx} g^{k^2} f(g^x).$$

Thus the equation $f(g^k u) = g^{k^2} u^{2k} f(u)$ is satisfied for each $u = g^x$ with $x \in R$. On the other hand, it can be reduced to the form

$$g^{k^2} u^{2k} f(u) - f(g^k u) = 0$$

and therefore has at most $2k + n$ solutions (because $k > 0$ the polynomial on the left-hand side is not identical to zero). Hence $n \ge |R| - 2K$. □

Certainly, for any S one can satisfy (12.2) with a unique polynomial f of degree $\deg f \le |S| - 1$. Now we show that for a sufficiently small randomly selected set S that degree cannot be smaller. In particular, with probability $1 - o(1)$ we have $\deg f = |S| - 1$ for that polynomial.

Theorem 12.3. *Let* q *be sufficiently large and let* S *be a set of* m *random elements picked uniformly from* $\{0, \ldots, q-2\}$. *Then, for sufficiently large* q, *the probability* $P_k(q, m)$ *that there exists a polynomial* $f(X) \in \mathbb{F}_q[X]$ *of degree*

$$\deg f < m - k$$

and such that

$$g^{x^2} = f(g^x), \qquad x \in S,$$

satisfies the bound

$$P_k(q,m) \le \left(\frac{4m}{q-1}\right)^{k/2} + \begin{cases} 0, & \text{if } m-k \ge (4q)^{1/3}, \\ 3^{m+1}q^{-m/3}, & \text{if } m-k < (4q)^{1/3}. \end{cases}$$

Proof. We say that a set T is satisfied by a polynomial $f(X) \in \mathbb{F}_q[X]$ if the condition of the theorem is fulfilled for the pair (T,f), and that T is maximally satisfied by a polynomial $f(X) \in \mathbb{F}_q[X]$ if it is satisfied by this polynomial but any superset of T is not.

Suppose there are N distinct sets $S_i \subseteq \{0,\dots,q-2\}$, $i=1,\dots,N$, that are maximally satisfied by polynomials f_i of degree at most $n = m-k$. In particular, polynomials f_i, $i=1,\dots,N$, are pairwise distinct.

Therefore, $|S_i \cap S_j| \le n$, $1 \le i < j \le N$, otherwise we would have $f_i = f_j$ being the unique polynomial satisfying the intersection $S_i \cap S_j$. In particular,

$$\sum_{i=1}^{N} \sum_{\substack{T \subseteq S_i \\ |T|=n+1}} 1 \le \sum_{\substack{T \subseteq \{0,\dots,q-2\} \\ |T|=n+1}} 1 = \binom{q-1}{n+1}. \tag{12.3}$$

Also assume that only the first M sets are of size

$$|S_i| \ge 2n^{1/2}(q-1)^{1/2}.$$

First of all we remark that $M=0$ if $n \ge (4q)^{1/3}$. Indeed, from Theorem 12.2 (with $H = q-1$) we see that if $M \ne 0$, then

$$n \ge \frac{4n(q-1)}{2(q-1)} - \frac{4(q-1)}{2n^{1/2}(q-1)^{1/2}} - 1 = 2n - 2n^{-1/2}(q-1)^{1/2} - 1.$$

It is easy to verify that this inequality fails for $n \ge (4q)^{1/3}$.

Now we consider the case $n < (4q)^{1/3}$. Again from Theorem 12.2 we see that in this case $|S_i| \le (\alpha + o(1))\, q^{2/3}$, $i=1,\dots,N$, where $\alpha = 2.519\dots$ is the unique positive root of the equation

$$\alpha^2/2 - 4/\alpha = 4^{1/3}.$$

Hence

$$|S_i| \le 2.6q^{2/3}, \qquad i=1,\dots,N,$$

for q large enough. We also claim that

$$\sum_{i=1}^{M} |S_i| < 2q. \tag{12.4}$$

Indeed, assuming the inverse inequality, we select $L \le M$ with

$$2q \le \sigma = \sum_{i=1}^{L} |S_i| \le 2q + 2.6q^{2/3}.$$

We know that the number of S_i is at most

$$L \le \sum_{i=1}^{L} \frac{|S_i|}{2n^{1/2}(q-1)^{1/2}} \le \frac{2q + 2.6q^{2/3}}{2n^{1/2}(q-1)^{1/2}} = (1/2 + o(1))\, q^{1/2} n^{-1/2}.$$

By the inclusion–exclusion principle we know that

$$q \ge \sum_{i=1}^{L} |S_i| - \sum_{1 \le i < j \le L} |S_i \cap S_j| \ge \sigma - nL(L-1)/2 \ge (3/2 + o(1))\, q,$$

which is not possible for q large enough. Therefore the inequality (12.4) holds.

Now we estimate the sum

$$W = \sum_{i=1}^{M} \left(\frac{|S_i|}{q-1} \right)^{m+1}.$$

Obviously, $W = 0$ for $n \ge (4q)^{1/3}$. For $n < (4q)^{1/3}$, from the inequality (12.4) we derive

$$\begin{aligned} W &= \sum_{i=1}^{M} \left(\frac{|S_i|}{q-1} \right) \left(\frac{|S_i|}{q-1} \right)^{m} \le 2.6^m q^{-m/3} \sum_{i=1}^{M} \frac{|S_i|}{q-1} \\ &\le 3^{m+1} q^{-m/3} \end{aligned}$$

for q large enough.

For an $(n+1)$-element set $T \subseteq \{0, \ldots, q-2\}$ denote by f_T the unique polynomial of degree at most n such that T is satisfied by this polynomial. Also, denote by R_T the set which is maximally satisfied by f_T. Now we see that

$$\begin{aligned} P_k(q,m) &= \sum_{|T|=n+1} \binom{q-1}{n+1}^{-1} \sum_{\substack{T \subseteq S \subseteq R_T \\ |S|=m}} \binom{q-n-2}{k}^{-1} \\ &\le \binom{q-1}{n+1}^{-1} \binom{q-n-2}{k}^{-1} \sum_{i=1}^{N} \sum_{\substack{T \subseteq S_i \\ |T|=n+1}} \sum_{\substack{T \subseteq S \subseteq S_i \\ |S|=m}} 1 \\ &= P_1 + P_2, \end{aligned}$$

where P_1 is the part of the sum over $i = 1, \ldots, M$ and P_2 is the part over $i = M+1, \ldots, N$. Thus

$$\begin{aligned} P_1 &= \binom{q-1}{n+1}^{-1} \binom{q-n-2}{k}^{-1} \sum_{i=1}^{M} \sum_{\substack{T \subseteq S_i \\ |T|=n+1}} \sum_{\substack{T \subseteq S \subseteq S_i \\ |S|=m}} 1 \\ &= \binom{q-1}{n+1}^{-1} \binom{q-n-2}{k}^{-1} \sum_{i=1}^{M} \binom{|S_i|}{n+1} \binom{|S_i|-n-1}{k}. \end{aligned}$$

From the inequality (8.3) we derive

$$\binom{q-1}{n+1}^{-1} \binom{|S_i|}{n+1} \le \left(\frac{|S_i|}{q-1}\right)^{n+1}$$

and

$$\binom{q-n-2}{k}^{-1} \binom{|S_i|-n-1}{k} \le \left(\frac{|S_i|-n-1}{q-n-2}\right)^{k} \le \left(\frac{|S_i|}{q-1}\right)^{k}.$$

Therefore

$$P_1 \le W \le \begin{cases} 0, & \text{if } n \ge (4q)^{1/3}, \\ 3^{m+1} q^{-m/3}, & \text{if } n < (4q)^{1/3}. \end{cases} \tag{12.5}$$

For P_2 we obtain

$$\begin{aligned} P_2 &= \binom{q-1}{n+1}^{-1} \binom{q-n-2}{k}^{-1} \sum_{i=M+1}^{N} \sum_{\substack{T \subseteq S_i \\ |T|=n+1}} \sum_{\substack{T \subseteq S \subseteq S_i \\ |S|=m}} 1 \\ &= \binom{q-1}{n+1}^{-1} \binom{q-n-2}{k}^{-1} \sum_{i=M+1}^{N} \sum_{\substack{T \subseteq S_i \\ |T|=n+1}} \binom{|S_i|-n-1}{k} \\ &\le \binom{q-1}{n+1}^{-1} \sum_{i=M+1}^{N} \sum_{\substack{T \subseteq S_i \\ |T|=n+1}} \left(\frac{|S_i|}{q-1}\right)^{k} \\ &\le \binom{q-1}{n+1}^{-1} \left(\frac{2n^{1/2}(q-1)^{1/2}}{q-1}\right)^{k} \sum_{i=M+1}^{N} \sum_{\substack{T \subseteq S_i \\ |T|=n+1}} 1. \end{aligned}$$

From (12.3) and the previous inequality we derive

$$P_2 \le \left(\frac{2n^{1/2}(q-1)^{1/2}}{q-1}\right)^{k} = \left(\frac{4n}{q-1}\right)^{k/2} \le \left(\frac{4m}{q-1}\right)^{k/2}. \tag{12.6}$$

Combining (12.5) and (12.6), we obtain the result. □

We remark that the first term dominates if $k \le 2m/3$. Selecting $k = 1$, we obtain that if $m = o(q)$, for almost all sets of size m the smallest degree of the polynomials which they satisfy is $m-1$.

Now we consider representations by algebraic functions.

Theorem 12.4. *Let $F(U,V) \in \mathbb{F}_q[U,V]$ be a polynomial, not identically zero, with*

$$\deg_U F = n, \qquad \deg_V F = m.$$

Assume that

$$F(g^x, g^{x^2}) = 0, \qquad x \in S,$$

for a set $S \subseteq \{N+1,\dots,N+H\}$ of cardinality $|S| = H - s$ with $H \le q-1$. Then

$$(m+1)(n+s+1) + \frac{2}{3}m(m+1)(m+2) \ge H+1.$$

Proof. First of all we write down the polynomial $F(U,V)$ in the form

$$F(U,V) = \sum_{i=1}^{t} V^{k_i} f_i(U),$$

where $f_i(U) \in \mathbb{F}_q[U]$, $i = 1,\dots,t$, are non-zero polynomials, $0 \le k_1 < \dots < k_t = m$, $t \le m+1$.

We see that there is a set R of cardinality at least

$$|R| \ge H - t + 1 - t(H - |S|) = H - t(s+1) + 1 \ge H - (m+1)(s+1) + 1$$

such that for any $x \in R$

$$F(g^{x+j}, g^{(x+j)^2}) = 0, \qquad j = 0,\dots,t-1.$$

Therefore, for any $x \in R$, the homogeneous system of equations

$$\sum_{i=1}^{t} Z_i g^{2jk_ix + j^2k_i} f_i(g^{x+j}) = 0, \qquad j = 0,\dots,t-1,$$

has a non-zero solution $Z_i = g^{k_ix^2}$, $i = 1,\dots,t$. Hence its determinant equals zero. Thus we see that the polynomial

$$\Delta(U) = \det\left(g^{(j-1)^2k_i} U^{2k_i(j-1)} f_i(g^{j-1}U)\right)_{i,j=1}^{t}$$

has at least $|R|$ zeros $u = g^x$, $x \in R$. Now we show that $\Delta(U)$ is not identical to zero. Indeed, let π be a permutation of the set $\{0,\dots,t-1\}$. The term of Δ corresponding to the permutation π is of degree

$$\sum_{i=1}^{t} (\deg f_i + 2k_i\pi(i)) = \sum_{i=1}^{t} \deg f_i + 2\sum_{i=1}^{t} k_i\pi(i).$$

It is known that for any two sequences of non-negative numbers

$$0 \le a_1 < \ldots < a_t \qquad \text{and} \qquad 0 \le b_1 < \ldots < b_t$$

the sum $a_1 b_{\pi(1)} + \ldots + a_t b_{\pi(t)}$ attains its maximal value for the identity permutation and that value is strictly greater than values corresponding to other permutations. So $\Delta(U)$ is a non-zero polynomial of degree

$$\begin{aligned}\deg \Delta &= \sum_{i=1}^{t} \deg f_i + 2\sum_{i=1}^{t} k_i(i-1) \le (m+1)n + 2\sum_{i=1}^{m+1} i(i-1)\\ &= (m+1)n + \frac{2}{3}m(m+1)(m+2),\end{aligned}$$

and the result follows. □

Once again, one can see that Theorem 12.4 is quite precise. Indeed, it is easy to see that for any m and n with $(m+1)(n+1) > |S| = H - s$ there is a polynomial F satisfying the conditions of the theorem.

The following result is non-trivial for sparse sets with at least $H^{2/3+\varepsilon}$ elements.

Theorem 12.5. *Let $F(U,V) \in \mathbb{F}_q[U,V]$ be a polynomial of degree $n = \deg F$, not identically zero, such that*

$$F(g^x, g^{x^2}) = 0, \qquad x \in S,$$

for a set $S \subseteq \{N+1, \ldots, N+H\}$. Then there is an absolute effectively computable constant $C > 0$ such that the bound

$$n \ge C|S|^{3/2}/H$$

holds.

Proof. For a polynomial $G(U,V) \in \mathbb{F}_q[U,V]$ and integer k (not necessarily positive) let us introduce the shift transformation

$$\sigma_k\left(G(U,V)\right) = U^{-l} G(g^k U, g^{k^2} U^{2k} V),$$

where l is chosen in such a way that $\sigma_k(G)$ is a polynomial not divisible by U. One easily verifies that

$$\sigma_k\left(\sigma_m(G)\right) = \sigma_{k+m}(G)$$

and that

$$\sigma_k(G_1 G_2) = \sigma_k(G_1)\sigma_k(G_2).$$

In particular, if $\Psi(U,V)$ is an absolutely irreducible polynomial which is not a univariate polynomial (either in U or in V) then $\Phi = \sigma_k(\Psi)$ is absolutely irreducible as well. We also note that for an absolutely irreducible Ψ and for $k \neq 0$, we have $\sigma_k(\Psi) \neq c\Psi$ for any non-zero $c \in \mathbb{F}_q$. Indeed, assuming that

$$\Psi(U,V) = \sum_{i=0}^{v} V^i f_i(U),$$

for some integer l we would have $f_i(U) = cg^{ik^2} U^{2ik-l} f_i(g^k U)$, for each $i = 0, \ldots, v$. This is only possible if $i = -l/2k$. Thus there is only one non-zero polynomial among the polynomials $f_0(U), \ldots, f_v(U)$. Thus $\Psi(U,V) = V^h f(U)$, where $h \leq v$ and $f(U)$ is a non-zero polynomial of degree at most v, which is not possible because of our assumptions.

First of all we denote by $\varphi(U)$ and $\psi(V)$ two possible univariate factors of $F(U,V)$.

Let us consider the complete factorisation of the fraction

$$\frac{F(U,V)}{\varphi(U)\psi(V)}$$

over the algebraic closure of $\mathbb{F}_q$ (thus all factors are absolutely irreducible polynomials). Index the absolutely irreducible factors in this fraction as $\Psi_{ij}(U,V)$, that is,

$$F(U,V) = \varphi(U)\psi(V) \prod \Psi_{ij}(U,V),$$

in the following way. Two factors share the same first index if and only if one is essentially a shift of the other:

$$\Psi_{ij}(U,V) = c\sigma_k(\Psi_{im})$$

for some integer k and some non-zero $c \in \mathbb{F}_q$. It follows from the two aforementioned properties of the transformation σ_k that this breakup is legitimate.

Among each family Ψ_{ij} of factors sharing a first index i, assign the index $j = 0$ to that factor having minimal degree in U, and for the other members of the family let j denote the amount of shift, that is,

$$\Psi_{ij} = c\sigma_j(\Psi_{i0})$$

with some non-zero $c \in \mathbb{F}_q$.

Collect all factors $\Psi_{ij}(U,V)$ sharing the same second index j into a factor $F_j(U,V)$. So we have

$$F(U,V) = \varphi(U)\psi(V) \prod_{j \in J} F_j(U,V),$$

where J is the set of possible shifts among absolutely irreducible factors of F and for each $F_j(U,V)$, $j \in J$, we have that $\sigma_{-j}F_j$ is a factor of F_0.

For each $j \in J$ we define the set $T_j \subset S$ such that

$$F_j(g^x, g^{x^2}) = 0, \qquad x \in T_j.$$

As in the proof of Theorem 12.2, we select $1 \leq k_j \leq 2H/|T_j|$ for which both

$$F_j(g^x, g^{x^2}) = 0 \quad \text{and} \quad F_j(g^{(x+k_j)}, g^{(x+k_j)^2}) = 0 \tag{12.7}$$

hold for at least $|T_j|^2/2H - 1$ values of x. Then we see that the system of equations

$$F_j(U,V) = \sigma_{k_j}(F_j(U,V)) = 0$$

has at least $|T_j|^2/2H - 1$ solutions.

Let $F_j(U,V)$, $j \in J$, have degrees u_j and v_j in U and V, respectively. Then the U-degree of $\sigma_{k_j}F_j$ is at most $u_j + 2k_j v_j$ (its V-degree is still v_j). Now we claim that F_j is relatively prime to $\sigma_k(F_j)$ for any integer k and $j \in J$. Indeed, otherwise F_j would have two distinct absolutely irreducible factors Ψ and Φ satisfying $\Phi = c\sigma_k(\Psi)$ with some non-zero $c \in \mathbb{F}_q$, but then Φ is a divisor of F_{j+k} rather than of F_j. Therefore, from Bézout's theorem we derive the inequality

$$|T_j|^2/2H - 1 \leq u_j v_j + (u_j + 2k_j v_j)v_j = 2u_j v_j + 2k_j v_j^2. \tag{12.8}$$

Let J_1 be the set of $j \in J$ with $u_j \geq k_j v_j$ and J_2 be the set of $j \in J$ with $u_j < k_j v_j$.

For $j \in J_1$ we have

$$|T_j|^2/2H \leq 4u_j v_j + 1 \leq 5u_j v_j \leq 5(\deg F_j)^2.$$

Therefore

$$n \geq \sum_{j \in J_1} \deg F_j \geq (10H)^{-1/2} \sum_{j \in J_1} |T_j|. \tag{12.9}$$

Let us turn to J_2. We notice that

$$u_j \geq |j| v_j. \tag{12.10}$$

Indeed, assume that the polynomial $\Psi_{i0}(U,V)$ is an absolutely irreducible divisor of $F_0(U,V)$ such that $\Psi_{ij}(U,V)$ is a divisor of $F_j(U,V)$. Assume that

$$v = \deg_V \Psi_{i0} = \deg_V \Psi_{ij}, \quad w = \deg_U \Psi_{i0}(U,V), \quad u = \deg_U \Psi_{ij}(U,V).$$

We see that the coefficient of V^0 in $\Psi_{i0}(U,V)$ is a polynomial in U of some degree $0 \leq r \leq w$, and the coefficient of V^v is a polynomial in U of some degree $0 \leq s \leq w$.

The first polynomial is not zero because otherwise Ψ_{i0} would be divisible by V; the second one is not zero because the V-degree of $F_j(U,V)$ is v. Let l be the power of U in the definition of σ_j. We have

$$l \le \min\{r,\ s+2jv\}.$$

On the other hand,

$$u \ge \max\{r-l,\ s+2jv-l\}.$$

If $j>0$, then we see that

$$u \ge s+2jv-l \ge s+2jv-r \ge 2jv-r \ge 2jv-w.$$

If $j<0$, then

$$u \ge r-l \ge r-2jv-s \ge -2jv-s \ge -2jv-w.$$

From our selection of Ψ_{i0} we also see that $u \ge w$. Combining these inequalities, we derive $u \ge |j|v$ and (12.10) follows.

Then, for $j \in J_2$ we have

$$|T_j|^2/2H \le 4k_jv_j^2+1 \le 5k_jv_j^2 \le 10Hv_j^2/|T_j|.$$

Hence

$$v_j \ge 20^{-1/2}|T_j|^{3/2}H^{-1}, \qquad j \in J_2.$$

From this and (12.10) we derive

$$n \ge \sum_{j\in J_2} \deg F_j \ge \sum_{j\in J_2} u_j \ge \sum_{j\in J_2} |j|v_j \ge 20^{-1/2}H^{-1}\sum_{j\in J_2} |j||T_j|^{3/2}.$$

If $0 \in J_2$ we can include T_0 in the sum by

$$\deg F_0 \ge v_0 \ge 20^{-1/2}H^{-1}|T_0|^{3/2},$$

thus obtaining

$$n \ge 20^{-1/2}H^{-1}\sum_{j\in J_2} \max\{|j|,\ 1\}|T_j|^{3/2}.$$

We verify that

$$\sum_{j\in J_2}|T_j| \le \left(\sum_{j\in J_2}\max\{|j|,\ 1\}^{-2}\right)^{1/3}\left(\sum_{j\in J_2}\max\{|j|,\ 1\}|T_j|^{3/2}\right)^{2/3}$$

and

$$\sum_{j\in J_2}\max\{|j|,\ 1\}^{-2} < 1+2\sum_{j=1}^{\infty} j^{-2} = 1+2\frac{\pi^2}{6} < 5.$$

Therefore

$$n \geq (10H)^{-1} \left(\sum_{j \in J_2} |T_j| \right)^{3/2}. \tag{12.11}$$

The univariate factors φ and ψ are easier to treat. The set T_u of $x \in S$ for which $\varphi(g^x) = 0$ is of cardinality

$$|T_u| \leq \deg \varphi \leq n. \tag{12.12}$$

The set T_v of $x \in S$ for which $\psi(g^{x^2}) = 0$ satisfies the inequality

$$|T_v| = O(Hq^{-1/2} \deg \psi) = O(nHq^{-1/2}), \tag{12.13}$$

which follows from the general bound of [307] on the number of solutions of polynomial congruences over an incomplete residue system. Indeed in our case we have up to $\deg \psi$ congruences of the form

$$x^2 \equiv \operatorname{ind} v \pmod{q-1}$$

for each solution v of the equation $\psi(v) = 0$. Taking into account that

$$\max \left\{ |T_u|, |T_v|, \sum_{j \in J_1} |T_j|, \sum_{j \in J_2} |T_j| \right\} \geq |S|/4$$

from (12.9), (12.11), (12.12) and (12.13), we derive the result. □

The result of Theorem 12.5 has also been used in [247] to obtain a lower bound on the non-linear complexity profile of the quadratic exponential generator (26.10).

It is obvious that for any $S \subseteq \{0, \ldots, q-2\}$ there is a polynomial $F(U,V) \in \mathbb{F}_q[U,V]$ of degree at most $(2|S|)^{1/2}$ which satisfies the condition of Theorem 12.5. Now we show that for almost all sufficiently small sets S this bound is the best possible, to within a multiplicative constant.

Theorem 12.6. *Let q be sufficiently large, $0 < \varepsilon < 2\delta/3$, $\delta < 1$ and $m \leq q^{1-\delta}$. Let S be a set of m random elements picked uniformly from the set $\{0, \ldots, q-2\}$. Then the probability $P_{\varepsilon,\delta}(q,m)$ that there exists a polynomial $F(U,V) \in \mathbb{F}_q[U,V]$ of degree*

$$\deg F < \left\lfloor (\varepsilon m)^{1/2} \right\rfloor - 1$$

and such that

$$F(g^x, g^{x^2}) = 0, \qquad x \in S,$$

satisfies the bound

$$P_{\varepsilon,\delta}(q,m) \leq c^m q^{-(\delta/3 - \varepsilon/2)m},$$

where $c > 0$ is an absolute constant.

Proof. We say that a set T is satisfied by a polynomial $F(U,V) \in \mathbb{F}_q[U,V]$ if the condition of the theorem is fulfilled for the pair (T,F), and that T is maximally satisfied by a polynomial $F(U,V) \in \mathbb{F}_q[U,V]$ if it is satisfied by this polynomial but any superset of T is not.

Suppose there are N distinct sets $S_i \subseteq \{0,\dots,q-2\}$, $i = 1,\dots,N$, that are maximally satisfied by polynomials $F_i(U,V) \in \mathbb{F}_q[U,V]$ of degree at most $n = \lfloor (\varepsilon m)^{1/2} \rfloor - 2$. In particular, polynomials F_i, $i = 1,\dots,N$, are pairwise distinct, thus

$$N \le q^{(n+2)(n+1)/2}.$$

From Theorem 12.5 we derive $|S_i| = O\left((nq)^{2/3}\right)$. Therefore, using the inequality (8.3),

$$\begin{aligned}
P_{\varepsilon,\delta}(p,m) &= \binom{q-1}{m}^{-1} \sum_{i=1}^{N} \binom{|S_i|}{m} \le \sum_{i=1}^{N} \left(\frac{|S_i|}{q-1}\right)^m \\
&\le q^{(n+2)(n+1)/2}(cn^{2/3}q^{-1/3})^m \\
&\le c^m n^{2m/3} q^{(n+2)(n+1)/2-m/3} \\
&\le c^m m^{m/3} q^{(\varepsilon/2-1/3)m} \\
&\le c^m q^{-(\delta/3-\varepsilon/2)m}
\end{aligned}$$

with some constant $c > 0$. □

Now we obtain a lower bound on the probabilistic parallel complexity of breaking the Diffie–Hellman key exchange protocol. In fact the method above allows us to obtain a lower bound on the depth of circuits from quite a general class which answer a weaker question about verifying whether $\operatorname{ind} v = \operatorname{ind}^2 u$ for a given $u, v \in \mathbb{F}_q^*$. Moreover, it is enough to require that such a circuit works correctly on a very small portion of the input.

Now we consider *verification* of the property that two elements $u, v \in \mathbb{F}_q$ satisfy $u = g^x$, $v = g^{x^2}$ for some integer x which is apparently a harder problem than finding v from the given value of u.

We estimate the depth of circuits of the class $\mathrm{PAC}_q(d)$, defined in Chapter 1, which verify this property. Each gate of the starting level of such a circuit accepts either some constant or the value of one of the two variables $u, v \in \mathbb{F}_q$ for which we want to verify that $\operatorname{ind} v = \operatorname{ind}^2 u$.

Theorem 12.7. *Let $\gamma, \delta > 0$ be constants such that*

$$(1-\gamma)\delta > 2/3.$$

Assume that there is a circuit $C \in \mathrm{PAC}_q(d)$ such that given $u = g^x$, $v = g^{x^2}$, $u, v \in \mathbb{F}_q^*$ for every element x from some set $S \subseteq \{N+1,\dots,N+H\}$, $1 \le H \le q-1$, of cardinality $|S| \ge H^\delta$, it outputs **YES** *with probability*

$$\Pr\left[C(g^x, g^{x^2}) = \mathbf{YES}\right] \ge |S|^{-\gamma},$$

and for any $u, v \in \mathbb{F}_q^*$ *such that there is no* $x \in \{0, \dots, q-2\}$ *with* $u = g^x$, $v = g^{x^2}$ *it outputs* **NO** *with probability 1. Then*

$$d \geq C(\gamma, \delta) \log H,$$

where the constant $C(\gamma, \delta) > 0$ *depends only on* γ *and* δ.

Proof. We can assume that the kth level has no more than 5^{d-k} gates. Indeed the dth level may utilize no more than five gates of the $(d-1)$th level, which may utilize no more than 25 gates of the $(d-2)$th level and so on.

Let $\vartheta = (\vartheta_1, \dots, \vartheta_l)$ be the vector of random variables used by the circuit C. Obviously l does not exceed the total number of gates, that is, $l = O(5^d)$. For each fixed vector $\vartheta = (\vartheta_1, \dots, \vartheta_l)$ we obtain a deterministic circuit C_ϑ.

Let $R(\vartheta)$ be the set of $x \in S$ such that, given $u = g^x$ and $v = g^{x^2}$, C_ϑ works correctly, $C_\vartheta(u, v) =$ **YES**. Therefore, there is a $\vartheta_0 \in \mathfrak{B}_l$ for which C_{ϑ_0} produces the correct result for at least $|S|^{1-\gamma}$ values of $x \in S$.

We define the sequence of sets $R_1, \dots, R_d$ recursively. We put $R_1 = R(\vartheta_0)$. Assume that the set R_{i-1} has already been defined, $2 \leq i \leq d$. We define R_i as the set of $x \in R_{i-1}$ for which $\psi(g^x, g^{x^2}) \neq 0$ for all non-constant functions ψ defining branching on the ith level. Obviously, for $u \in R_{i-1}$ all such functions $\psi(V, U)$ of the ith level are rational functions in V and U of degree at most 2^{i-1}. It follows from Theorem 12.5 that the number of $x \in R_{i-1} \subseteq S$ for which such a function vanishes is at most $O(2^{2i/3} H^{2/3})$. Therefore

$$|R_{i-1}| \geq |R_i| + O(5^{d-i} 2^{2i/3} H^{2/3}).$$

Taking into account that because of our selection of ϑ_0, $|R_1| \geq |S|^{1-\gamma}$, we find that

$$|R_d| \geq |S|^{1-\gamma} + O(5^d H^{2/3}) \geq 0.5|S|^{1-\gamma}$$

if

$$d \leq \log_6(|S|^{1-\gamma} H^{-2/3}) \leq ((1-\gamma)\delta - 2/3) \log_6 H$$

and H is large enough (otherwise there is nothing to prove). So, there are at least $0.5|S|^{1-\gamma}$ values of $x \in S$ for which the circuit works without branching and outputs the correct answer. We note that the function ψ of the last level cannot be a constant, thus this is a non-constant rational function of degree at most 2^{d-1} having at least $|R_d|$ zeros. From Theorem 12.5 we see that

$$2^{d-1} \geq 0.3|R_d| H^{-2/3} \geq 0.3 H^{(1-\gamma)\delta - 2/3}$$

and the claim follows. □

Certainly the constant $C(\gamma, \delta)$ can be explicitly evaluated.

Also, we see that if $H \geq \exp((\log q)^{\varepsilon})$ for some $\varepsilon > 0$, then the depth cannot be polynomial in $\log\log q$.

Now we study some properties of polynomial transformations of g^{xy}.

It has been shown in [550], under certain natural conditions, that if we are given an oracle which for each pair (g^x, g^y) with some integers x and y returns the value of $F(g^{xy})$, than this oracle can be used to construct a polynomial time algorithm to compute the Diffie–Hellman secret key g^{xy}. We remark that polynomials F can be of very large degree (thus direct solving of the equation $F(g^{xy}) = A$ is not feasible) but contain a reasonably small number of monomials. The result has been motivated by applications to the proof of security of some new cryptosystems, see [69, 76, 332, 333, 334, 335, 550, 551].

Here we obtain a generalisation of Theorem 24 of [550] to the case of "unreliable" oracle which returns the result only for a certain very small fraction of inputs and returns an error message for other inputs.

We need a version of the bound (2.5) for exponential equations.

Let $\vartheta \in \mathbb{F}_q$ be of multiplicative order t. Then for $m \geq 2$ elements $a_1, \ldots, a_m \in \mathbb{F}_q^*$ and integers $e_1, \ldots, e_m$ the number of solutions of the equation

$$\sum_{i=1}^{m} a_i \vartheta^{e_i u} = 0, \qquad u \in [0, t-1].$$

does not exceed

$$W \leq 3t^{1-1/(m-1)} D^{1/(m-1)}, \tag{12.14}$$

where

$$D = \min_{1 \leq i \leq m} \max_{j \neq i} \gcd(e_j - e_i, t)$$

Indeed, we write $\vartheta = \gamma^{(q-1)/t}$ where γ is a primitive root of $\mathbb{F}_q$ and note that each solution $u \in [0, t-1]$ of the previous exponential equation gives rise to $(q-1)/t$ distinct solutions $z_j = \gamma^{u+tj}$, $j = 0, \ldots, (q-1)/t - 1$, of the equation

$$\sum_{i=1}^{r} a_i z^{\tau_i} = 0, \qquad z \in \mathbb{F}_q^*,$$

where $\tau_j = e_j(q-1)/t$. Remarking that

$$\gcd(\tau_j - \tau_i, q-1) = \frac{q-1}{t} \gcd(e_j - e_i, t),$$

from the bound (2.5) we obtain (12.14).

Let $g \in \mathbb{F}_q^*$ be an element of multiplicative order t.

As in [550] we consider an m-sparse polynomial

$$F(X) = \sum_{i=1}^{m} c_i X^{e_i} \in \mathbb{F}_q[X], \qquad (12.15)$$

where $c_1, \ldots, c_m \in \mathbb{F}_q^*$ and $e_1, \ldots, e_m$ are pairwise distinct modulo t.

We also fix some ε with $0 < \varepsilon \le 1$.

Assume that we are given an *oracle* $\widetilde{\mathcal{DH}}_{F,\varepsilon}$ such that for every $x \in [0, t-1]$, given the values of g^x and g^y, it evaluates and returns $F(g^{xy})$ for at least εt values of $y \in [0, t-1]$ and returns an error message for other values of $y \in [0, t-1]$.

The case $\varepsilon = 1$, that is, the case of a "error-free" oracle has been considered in [550].

We are ready to prove the main result. For simplicity we assume that t is a prime number, although analogues of our result hold for composite t as well. Nevertheless this case allows us to simplify some arguments and it is also one of the most practically important cases, see [69, 76, 332, 333, 334, 335, 550, 551].

Theorem 12.8. *Let t be prime, $m \ge 2$ and let an m-sparse polynomial F be given by (12.15). Assume that*

$$1 \ge \varepsilon \ge 6t^{-1/(m-1)}.$$

Given an oracle $\widetilde{\mathcal{DH}}_{F,\varepsilon}$, there exists a probabilistic algorithm which given g^x and g^y makes the expected number of at most $2m\varepsilon^{-1}$ calls of the oracle $\widetilde{\mathcal{DH}}_{F,\varepsilon}$, executes polynomial number $(m \log q)^{O(1)}$ arithmetic operations in $\mathbb{F}_q$ per each call and returns g^{xy} for all pairs $(x, y) \in [0, t-1]^2$.

Proof. If $x = 0$ the result is trivial. Let us consider a pair $(x, y) \in [0, t-1]^2$ with $x \ne 0$.

Let $\mathcal{U}$ be the set of $u \in [0, t-1]$ for which the oracle, given the values of g^x and g^{y+u} returns the value of $F\left(g^{x(y+u)}\right)$. By the conditions of the theorem $|\mathcal{U}| \ge \varepsilon t$. We also remark that if g^y is known, then for any $v \in [0, t-1]$ the value of g^{y+v} can easily be computed as well.

Put $\vartheta = g^x$.

Our algorithm consists of m steps which recursively build a certain sequence $u_1, \ldots, u_m \in [0, t-1]$ after which g^{xy} can be found by solving a nonsingular system of m linear equations in m variables over $\mathbb{F}_q$. At each step the expected number of oracle calls is at most $2\varepsilon^{-1}$.

At the first step we select a sequence of elements v uniformly and independently at random in the interval $[0, t-1]$ and for each of them feed g^x and g^{y+v} in the oracle $\widetilde{\mathcal{DH}}_{F,\varepsilon}$ until we find an element $u \in \mathcal{U}$ and thus find the value of $F\left(g^{x(y+u)}\right)$.

Let us call this element u_1. The expected number of oracle calls to find such an element is $\varepsilon^{-1} \leq 2\varepsilon^{-1}$.

At the kth step, $2 \leq k \leq m$, we assume that we have already found $k-1$ elements $u_1, \dots, u_{k-1} \in \mathcal{U}$ with

$$\det\left(\vartheta^{e_i u_j}\right)_{i,j=1}^{k-1} \neq 0. \tag{12.16}$$

We select elements v uniformly and independently at random in the interval $[0, t-1]$ until we find an element $u_k \in \mathcal{U}$ such that

$$\det\left(\vartheta^{e_i u_j}\right)_{i,j=1}^{k} \neq 0. \tag{12.17}$$

We remark that if the last determinant vanishes, then u_k satisfies an equation of the form

$$\Delta_1 \vartheta^{e_k u_k} + \ldots + \Delta_k \vartheta^{e_1 u_k} = 0$$

where, by the assumption (12.16), we have

$$\Delta_1 = \det\left(\vartheta^{e_i u_j}\right)_{i,j=1}^{k-1} \neq 0.$$

Applying the bound (12.14) we obtain that the number of elements $u_k \in \mathcal{U}$ which satisfy the condition (12.17) is at least

$$|\mathcal{U}| - 3t^{1-1/(k-1)} \geq |\mathcal{U}| - 3t^{1-1/(m-1)} \geq \frac{1}{2}|\mathcal{U}|.$$

Thus such an element $u_k \in \mathcal{U}$ can be found in the expected number of at most $2\varepsilon^{-1}$ oracle calls with g^x and g^{y+v} where elements v are selected uniformly and independently at random in the interval $[0, t-1]$. More precisely, if for the chosen value of v we have

$$\Delta_1 \vartheta^{e_k v} + \ldots + \Delta_k \vartheta^{e_1 v} = 0, \tag{12.18}$$

then we discard this value of v and chose another one. Otherwise, we call the oracle $\widetilde{\mathcal{DH}}_{F,\varepsilon}$ with g^x and g^{y+v} for a random $v \in [0, t-1]$. We repeat this procedure until it returns $F\left(g^{x(y+v)}\right)$ and denote the corresponding value by u_k. Because there are at least $0.5|\mathcal{U}| \geq 0.5\varepsilon t$ values of v for which (12.18) does not hold, the expected number of oracle calls is at most $2\varepsilon^{-1}$.

Therefore, after m steps and the expected number of at most $2m\varepsilon^{-1}$ oracle calls, we obtain m elements $u_1, \dots, u_m \in \mathcal{U}$ with corresponding values of $A_j = F\left(\vartheta^{y+u_j}\right)$ for each $j = 1, \dots, m$ and such that

$$\det\left(\vartheta^{e_i u_j}\right)_{i,j=1}^{m} \neq 0.$$

The rest of the proof follows essentially the same arguments as the proof of Theorem 24 of [550]. Indeed, we see from our construction that we have a nonsingular system of linear equations

$$\sum_{i=1}^{m} c_i \vartheta^{e_i u_j} \vartheta^{e_i y} = A_j, \qquad j = 1, \dots, m,$$

from which the vector $(c_1\vartheta^{e_1 y}, \ldots, c_m\vartheta^{e_m y})$ can be found and thus we obtain the values of $g^{e_1 xy}, \ldots, g^{e_m xy}$.

Because $m \geq 2$ and t is prime, at least one of $e_1, \ldots, e_m$ (which are pairwise distinct modulo t) is relatively prime to t. Say if $\gcd(e_1, t) = 1$ we define an integer $f_1 \in [1, t-1]$ from the congruence $f_1 e_1 \equiv 1 \pmod t$ and compute

$$g^{xy} = (g^{e_1 xy})^{f_1}.$$

Remarking that the expected number of oracle calls is $2m\varepsilon^{-1}$ and that the rest of the algorithm can be implemented in deterministic polynomial in $m \log q$ time, we obtain the desired result. □

Let $q = p^m$. Then the trace function

$$\mathrm{Tr}_{\mathbb{F}_{p^m}/\mathbb{F}_p}(X) = \sum_{i=0}^{m-1} X^{p^i}$$

provides a natural example of a polynomial of the form (12.15). This function as well as the function

$$L(X) = \sum_{\substack{i,j=0 \\ i \neq j}}^{m-1} X^{p^i + p^j}$$

have been studied in [76] (with $m = 6$). Our results imply a stronger version of Lemma 2.1 of [76] and thus give more security assurance to the cryptosystem proposed there. The same comment also applies to the XTR key exchange protocol, proposed and studied in [332, 333, 334, 335, 551], which is based on a more computationally efficient modification of the ideas of [76]. As well it applies to the LUC cryptosystem [47, 536].

It is easy to see that by making more oracle calls one can replace the oracle $\widetilde{\mathcal{DH}}_{F,\varepsilon}$ with a more natural and general oracle $\widetilde{\mathcal{DH}}^*_{F,\varepsilon}$ which returns $F(g^{xy})$ for at least εt^2 pairs $(x, y) \in [0, t-1]^2$. For $x \in [0, t-1]$ let M_x denote the number $y \in [0, t-1]$ for which the oracle $\widetilde{\mathcal{DH}}^*_{F,\varepsilon}$, given the values of g^x and g^y, returns $F(g^{xy})$. Thus,

$$\sum_{x=0}^{t} M_x \geq \varepsilon t^2.$$

Let L be the number of $x \in [0, t-1]$ for which $M_x \geq 0.5\varepsilon t$. Then

$$\sum_{x=0}^{t} M_x \leq 0.5\varepsilon t(t - L) + Lt = 0.5\varepsilon t^2 + (1 - 0.5\varepsilon)Lt.$$

Therefore

$$L \geq \frac{\varepsilon}{2(1 - 0.5\varepsilon)} t \geq 0.5\varepsilon t.$$

Now we select a random $u \in [0, t-1]$ and compute g^{x+u}. Using polynomially many random values of $v \in [0, t-1]$ with high probability we can test whether $M_{x+u} \geq 0.5\varepsilon t$. If this is not the case we select another value of u. After the expected number of $t/L \leq 2\varepsilon^{-1}$ random choices of u we find a value with $M_{x+u} \geq 0.5\varepsilon t$. Now we apply the same arguments as in the proof of Theorem 12.8 with g^{x+u} and g^y, recovering $g^{(x+u)y}$. Hence we can find $g^{xy} = g^{(x+u)y} \left(g^y\right)^{-u}$.

In fact we do not even need the oracle to return the error message. It is enough to assume that, when it does not compute $F\left(g^{xy}\right)$, it returns just a random element of $\mathbb{F}_q$. Then repeating each oracle call polynomially many times one can distinguish between correct outputs and random outputs with overwhelming probability.

The paper [167] provides analogues of Theorems 12.2 and 12.5 for the relations of the form $g^{xy} = f(g^x, g^y)$ and $F(g^x, g^y, g^{xy}) = 0$ for $(x, y) \in W$, where $W \subseteq [N+1, N+H]^2$. In particular, for integers N and H, $1 \leq H \leq q-1$, these results imply the upper bounds

$$12 \max\{H^{8/5},\ n^{1/2} H^{3/2}\} \qquad \text{and} \qquad 3nH^{8/5}$$

on the number of solutions $(x, y) \in [N+1, N+H]^2$ of the equations

$$g^{xy} = f(g^x, g^y) \qquad \text{and} \qquad F(g^x, g^y, g^{xy}) = 0,$$

respectively, with non-zero polynomials

$$f(U, V) \in \mathbb{F}_q[U, V] \qquad \text{and} \qquad F(U, V, Z) \in \mathbb{F}_q[U, V, Z]$$

of degree at most n. A different upper bound on the number of solutions of the equation $g^{xy} = f(g^x, g^y)$ has recently been given in [560]. For some special sets W it is stronger than that of [167]. Namely, the result of [560] applies only to "rectangular" sets $W = U \times V$ where at least one of the sets U and V is dense in the interval $[N+1, N+H]$, but instead the other set can be of arbitrary structure and cardinality.

Chapter 13

Boolean Complexity of the Diffie–Hellman Secret Key

The bounds obtained in Chapter 12 are related to the arithmetic model of computation. Intuitively, the arithmetic model seems more powerful than the Boolean model (and in some situations it is), but there is no proof that this is really the case for our particular situation. Moreover, as a dual question of computing powers in parallel shows in some cases (over finite fields of small characteristic), the Boolean model of computation is exponentially more powerful than the arithmetic model [200, 201].

Here we consider two special partial cases: fields $\mathbb{F}_{2^r}$ and prime fields $\mathbb{F}_p$.

We begin by deriving a lower bound on the degree of Boolean functions giving the binary coordinate vector of $g^{x^2} \in \mathbb{F}_{2^r}$ with respect to some fixed basis of $\mathbb{F}_{2^r}$ over $\mathbb{F}_2$ from the binary coordinate vector of $g^x \in \mathbb{F}_{2^r}$. As before, each Boolean function $B(U_1, \ldots, U_r)$ we consider as a multilinear polynomial over $\mathbb{F}_2$ of the form (10.1).

We also fix a basis $\omega_1, \ldots, \omega_r$ of $\mathbb{F}_{2^r}$ over $\mathbb{F}_2$.

Theorem 13.1. *Let Boolean functions* $B_i(U_1, \ldots, U_r)$, $i = 1, \ldots, r$, *be such that*

$$g^{x^2} = \omega_1 B_1(u_1, \ldots, u_r) + \ldots + \omega_r B_r(u_1, \ldots, u_r),$$

where

$$g^x = \omega_1 u_1 + \ldots + \omega_r u_r,$$

for at least $2^r - 1 - s$ *values of* $x \in [0, 2^r - 2]$. *Then*

$$\max_{1 \le i \le r} \deg B_i \ge r - \log(2s + 3).$$

Proof. We denote be $\vartheta_1, \ldots, \vartheta_r$ the dual basis to $\omega_1, \ldots, \omega_r$; see [343]. That is, $\mathrm{Tr}_{\mathbb{F}_{2^r}/\mathbb{F}_2}(\vartheta_i\omega_j) = 1$ if $i = j$ and $\mathrm{Tr}_{\mathbb{F}_{2^r}/\mathbb{F}_2}(\vartheta_i\omega_j) = 0$ if $i \neq j$, $1 \le i, j \le r$. Therefore,

$$u = \sum_{i=1}^{r} \omega_i \mathrm{Tr}_{\mathbb{F}_{2^r}/\mathbb{F}_2}(\vartheta_i u).$$

Thus from the condition of the theorem we obtain

$$g^{x^2} = \sum_{i=1}^{r} \omega_i B_i \left(\mathrm{Tr}_{\mathbb{F}_{2^r}/\mathbb{F}_2}(\vartheta_1 g^x), \ldots, \mathrm{Tr}_{\mathbb{F}_{2^r}/\mathbb{F}_2}(\vartheta_r g^x)\right)$$

for at least $2^r - 1 - s$ values of $x \in [0, 2^r - 2]$. Let us consider the polynomial

$$f(U) = \sum_{i=1}^{r} \omega_i B_i \left(\mathrm{Tr}_{\mathbb{F}_{2^r}/\mathbb{F}_2}(\vartheta_1 U), \ldots, \mathrm{Tr}_{\mathbb{F}_{2^r}/\mathbb{F}_2}(\vartheta_r U)\right).$$

We see that after the reduction of all exponents modulo $2^r - 1$ the new polynomial $h(U)$ contains exponents which are integer numbers with at most n non-zero bits, where

$$n = \max_{1 \le i \le r} \deg B_i.$$

Therefore, the largest such exponent is $\deg h \le 2^r - 2^{r-n}$.

If $n \ge r - 1$, then there is nothing to prove. So for the rest of the proof we may assume that $n < r - 1$.

As in the proof of Theorem 12.1, we obtain that the polynomial $\Psi(U) = gU^2h(U) - h(gU)$ has at least $2^r - 1 - 2s$ zeros over $\mathbb{F}_{2^r}$ and is of degree at most $\deg \Psi \le 2^r - 2^{r-n} + 2$. To finish the proof it is enough to show that $\Psi(U)$ is a non-zero polynomial.

First of all we note that $h(U)$ is a non-zero polynomial. Indeed, $h(g^x) = g^{x^2} \neq 0$, $x \in S$.

Let E denote the set of exponents of $h(U)$. Obviously E is a non-empty set. Hence, if $\Psi(U)$ is identical to zero, then E coincides with its shift by 2 taken modulo $2^r - 1$, that is, $E \equiv E + 2 \pmod{2^r - 1}$. However, this property means that $E \equiv E + 2m \pmod{2^r - 1}$ for any integer m, hence

$$E = \{0, 1, \ldots, 2^r - 2\}.$$

Therefore, $2^r - 2^{r-n} \ge \deg h = 2^r - 2$, which contradicts the assumption that $n < r - 1$. Hence $\Psi(U)$ is a non-zero polynomial and we obtain the desired result. □

An analogue of Theorems 10.2 and 12.7 can be obtained as well (for deterministic Boolean circuits). Using Theorem 13.1, one can show that for any $\alpha < 1$ there is a constant $C > 0$ such that any deterministic branching Boolean circuit computing the coordinates of g^{x^2} from the coordinates of $g^x \in \mathbb{F}_{2^r}^*$ for all but at most $2^{\alpha r}$ values of $x = 0, \dots, 2^r - 2$ must be of depth at least $C \log r$.

We remark that there are several distinct natural interpretations of the Boolean model of computation over $\mathbb{F}_{2^r}$. The model we use here (related to coordinates with respect to a fixed basis of $\mathbb{F}_{2^r}$ over $\mathbb{F}_2$) has also been studied in [71, 186, 542]. Several other important relations between various models of computation have been considered in these papers as well. In particular, a very interesting result is obtained in [542] which, informally, claims that the *Fermat quotient* function $\psi_p : \mathbb{F}_p \to \mathbb{F}_p$ given by

$$\psi_p(x) \equiv \frac{x^p - x}{p} \pmod p, \quad x = 0, 1, \dots, p-1, \tag{13.1}$$

(where we use integer arithmetic on the right-hand side) is the hardest function over $\mathbb{F}_p$ for the arithmetic circuit computational model in the following sense. While it is obvious that ψ_p has a small Boolean circuit (one can just compute $x^p - x \pmod{p^2}$), the fact that ψ_p has a polynomial size arithmetic circuit would imply that any other function over $\mathbb{F}_p$ also has such polynomial size arithmetic circuit.

We now demonstrate how a lower bound on the Boolean complexity of breaking the Diffie–Hellman key exchange protocol modulo a prime p can be derived from the general lower bounds of Theorems 29.1 and 29.8 of Chapter 29.

Theorem 13.2. *Let p be a prime and let $r = \lceil \log p \rceil$. Assume that a Boolean function $B(U_1, \dots, U_r, V_1, \dots, V_r)$ of $2r$ Boolean variables is such that for any integers $x, y \in [1, p-1]$ the value of $B(u_1, \dots, u_r, v_1, \dots, v_r)$ equals the second leftmost bit of the smallest non-negative residue $\lfloor g^t \rfloor_p$ of g^t modulo p, where $t = \operatorname{ind} x \operatorname{ind} y$ and $x = u_1 \dots u_r$ and $y = v_1 \dots v_r$ are the bit representations of x and y. Then there exists an absolute constant $c > 0$ such that the bound*

$$\sigma(B) \ge c r^{1/2}$$

holds.

Proof. Select $y \equiv g^3 \pmod p$. Then the function B gives the second leftmost bit of $\lfloor x^3 \rfloor_p$. Applying Theorem 29.1, we derive the result. □

Combining Theorem 13.2 and Lemma 7.1, we immediately see that the CREW PRAM complexity CREW PRAM (DH_p) of breaking the Diffie–Hellman key exchange protocol modulo a prime p satisfies the inequality

$$\text{CREW PRAM}(\mathrm{DH}_p) \ge 0.25 \log\log p + o(\log\log p).$$

Theorem 13.3. *Let p be a prime and let $r = \lfloor \log p \rfloor + 1$. Assume that a Boolean circuit $C \in \mathrm{UBC}(d, S)$ is such that $B(u_1, \ldots, u_r, v_1, \ldots, v_r)$ equals the rightmost bit of the smallest non-negative residue $\lfloor g^t \rfloor_p$ of g^t modulo p, where $t = \mathrm{ind}\, x\, \mathrm{ind}\, y$ and $x = u_1 \ldots u_r$ and $y = v_1 \ldots v_r$ are the bit representations of x and y. Then the bound*

$$d \log \log S \geq (1 + o(1)) \log r$$

holds.

Proof. Select $y \equiv g^3 \pmod p$. Then the circuit C gives the rightmost bit of $\lfloor x^3 \rfloor_p$. Applying Theorem 29.8, we derive the result. □

As before, we notice that if d is sublogarithmic in the input size, that is, $d \leq \log^{\alpha} r$ with some constant $\alpha < 1$, then the size is superpolynomial. More precisely,

$$S \geq 2^{2^{(1+o(1)) \log^{1-\alpha} r}}.$$

Certainly it is natural to ask about the communication complexity (see Chapter 1) of the Diffie–Hellman secret key

$$K_p(x, y) = \lfloor g^{xy} \rfloor_p \tag{13.2}$$

or, more precisely, about the communication complexity of some specific bit of $K_p(x, y)$. The communication complexity of the rightmost bit of $K_p(x, y)$ has been estimated in [509]. Here we obtain a similar result for the communication complexity of the leftmost bits of $K_p(x, y)$. More precisely let us define the Boolean function $B(U_1, \ldots, U_r, V_1, \ldots, V_r)$ such that

$$B(u_1, \ldots, u_r, v_1, \ldots, v_r) = \begin{cases} 0, & \text{if } K_p(x, y) \leq (p-1)/2, \\ 1, & \text{if } K_p(x, y) \geq (p+1)/2, \end{cases} \tag{13.3}$$

for every pair x, y, $0 \leq x, y \leq p - 2$, where $x = u_1 \ldots u_r$ and $y = v_1 \ldots v_r$ are the bit representations of x and y, $r = \lceil \log(p-1) \rceil$.

The following bound is slightly stronger than that of [509] where the inequality (7.2) has been used instead of Lemma 7.7. Here, as in [299], we apply Lemma 7.7 which improves the constant $1/24$ of [509] up to $1/16$.

Theorem 13.4. *For the communication complexity of the Boolean function B given by (13.3) the bound*

$$\mathrm{Comm}\,(B) \geq \frac{1}{16} r + o(r)$$

holds.

Proof. Fix two subsets $\mathcal{X}, \mathcal{Y} \subseteq \mathfrak{B}_r$. As before we denote by $N_\mu(\mathcal{X}, \mathcal{Y})$ the number of pairs $(x, y) \in \mathcal{X} \times \mathcal{Y}$ with $B(x, y) = \mu$. Applying Lemma 3.22 with $t = p - 1$ and Lemma 4.1 (applied with $N = 1,\ L = p - 1$) we obtain the bound

$$\left| N_\mu(\mathcal{X}, \mathcal{Y}) - \frac{|\mathcal{X}||\mathcal{Y}|}{2} \right| \ll |\mathcal{X}|^{1/2} |\mathcal{Y}|^{5/6} p^{5/8+\varepsilon/2} \ln p \ll p^{47/24+\varepsilon},$$

where $\mu = 0, 1$. In particular, for the combinatorial discrepancy we have

$$\Gamma(B) = O\left(p^{-1/24+\varepsilon}\right)$$

for any $\varepsilon > 0$. Now Lemma 7.7 yields the result. □

In the following result we use very similar arguments to derive an upper bound from [509] on the Fourier coefficients (1.7) of this function.

Theorem 13.5. *For the Boolean function B given by (13.3) and any $\varepsilon > 0$ the bound*

$$\max_{a \in \mathfrak{B}_{2r}} \left| \widehat{B}(a) \right| \ll p^{47/24+\varepsilon}$$

holds.

Proof. We fix some non-zero vectors $u, v \in \mathfrak{B}_r$ and denote by $\mathcal{X}_0$ and $\mathcal{Y}_0$ the sets of integers $x \in \mathfrak{B}_r$ and $y \in \mathfrak{B}_r$ for which $\langle ux \rangle = 0$ and $\langle vy \rangle = 0$, respectively. Similarly, we define the sets $\mathcal{X}_1$ and $\mathcal{Y}_1$ by the conditions $\langle ux \rangle = 1$ and $\langle vy \rangle = 1$, respectively. In particular $|\mathcal{X}_\eta| = |\mathcal{Y}_\mu| = 2^{r-1}$, $\eta, \mu = 0, 1$. Then we obtain,

$$\begin{aligned} \widehat{B}(u, v) &= \sum_{x \in \mathcal{X}_0} \sum_{y \in \mathcal{Y}_0} (-1)^{B(x,y)} + \sum_{x \in \mathcal{X}_1} \sum_{y \in \mathcal{Y}_1} (-1)^{B(x,y)} \\ &\qquad - \sum_{x \in \mathcal{X}_1} \sum_{y \in \mathcal{Y}_0} (-1)^{B(x,y)} - \sum_{x \in \mathcal{X}_0} \sum_{y \in \mathcal{Y}_1} (-1)^{B(x,y)}. \end{aligned}$$

It is easy to see that

$$\sum_{x \in \mathcal{X}_\eta} \sum_{y \in \mathcal{Y}_\mu} (-1)^{B(x,y)} = 2N_0(\mathcal{X}_\eta, \mathcal{Y}_\mu) - |\mathcal{X}_\eta||\mathcal{Y}_\mu|, \qquad \eta, \mu = 0, 1,$$

where, as before, $N_0(\mathcal{X}_\eta, \mathcal{Y}_\mu)$ is the number of pairs $(x, y) \in \mathcal{X}_\eta \times \mathcal{Y}_\mu$ with $B(x, y) = 0$. As in the proof of Theorem 13.4, from Lemmas 3.22 and 4.1 (applied with $N = 1,\ L = p - 1$) we derive that

$$N_0(\mathcal{X}_\eta, \mathcal{Y}_\mu) = \frac{1}{2}|\mathcal{X}_\eta||\mathcal{Y}_\mu| + O\left(p^{47/24+\varepsilon}\right) = 2^{2r-3} + O\left(p^{47/24+\varepsilon}\right),$$

for $\eta, \mu = 0, 1$, and we obtain the desired result for non-zero vectors u and v.

Now, if $u = \overline{0}$ is a zero vector and v is not, then defining $\mathcal{Y}_0$ and $\mathcal{Y}_1$ as before we obtain

$$\widehat{B}(\overline{0}, v) = \sum_{x \in \mathfrak{B}_r} \sum_{y \in \mathcal{Y}_0} (-1)^{B(x,y)} - \sum_{x \in \mathfrak{B}_r} \sum_{y \in \mathcal{Y}_1} (-1)^{B(x,y)}.$$

As before we derive

$$N_0(\mathfrak{B}_r, \mathcal{Y}_\mu) = 2^{2r-2} + O\left(p^{47/24+\varepsilon}\right), \qquad \mu = 0, 1,$$

which implies the desired estimate in this case. The same arguments apply if $v = \overline{0}$ is a zero vector and u is not.

Finally, if both $u = v = \overline{0}$ are zero vectors, then

$$\hat{B}(\overline{0}, \overline{0}) = \sum_{x \in \mathfrak{B}_r} \sum_{y \in \mathfrak{B}_r} (-1)^{B(x,y)} = \left(2N_0(\mathfrak{B}_r, \mathfrak{B}_r) - 2^{2r-1}\right)$$

and using the bound

$$N_0(\mathfrak{B}_r, \mathfrak{B}_r) = 2^{2r-1} + O(p^{47/24+\varepsilon}),$$

we conclude the proof. □

Some information about the distribution of the most significant bits of $K_p(x, y)$ can be derived from the upper bound of the exponential sum given by Lemma 3.21.

For a box

$$B = [k_1, k_1 + h_1 - 1] \times [k_2, k_2 + h_2 - 1] \times [k_3, k_3 + h_3 - 1],$$

where $0 \le k_i \le k_i + h_i - 1 \le p - 1$, of size $|B| = h_1 h_2 h_3$ we denote by $T(B)$ the number of triples (g^x, g^y, g^{xy}), $x, y = 1, \ldots, p-1$, whose smallest non-negative residues modulo p belong to the box B.

Theorem 13.6. *For any $\varepsilon > 0$ the bound*

$$\sup_B \left| T(B) - \frac{|B|}{p-1} \right| = O\left(p^{23/12} \log^3 p\right)$$

holds.

Proof. The stated result follows from Lemma 3.21 and Lemma 4.1 (applied with $N = 3$, $L = p - 1$). □

In particular, Theorem 13.6 implies that, in some sense, any portions $\alpha < 1/36$ of the most significant bits of $\lfloor g^x \rfloor_p$, $\lfloor g^y \rfloor_p$, $\lfloor g^{xy} \rfloor_p$ are independently and uniformly distributed. It is important to keep in mind that this is a very rough interpretation of Theorem 13.6 because in fact the rightmost bit which can be biased by the position of p between two consecutive powers of 2. It is, however, easy to see what kind of adjustments are necessary for a rigorous statement of this kind.

Moreover, bits on arbitrary positions can be studied as well although the results are weaker: instead of asymptotic estimates on their joint distribution, the paper [94] contains only a lower bound.

In a similar way Lemma 3.23 can be used to study the distribution of the triples $\left(\lfloor g^x \rfloor_p, \lfloor g^y \rfloor_p, \lfloor g^{xy} \rfloor_p\right)$ with "sparse" exponents $x, y \in \mathcal{W}_k$, where as before $\mathcal{W}_k$ denotes the set of r-bit integers which have precisely k non-zero bits in their bit representation and $r = \lfloor \log p \rfloor$ is the bit length of p.

Namely, for a box

$$B = [k_1, k_1 + h_1 - 1] \times [k_2, k_2 + h_2 - 1] \times [k_3, k_3 + h_3 - 1],$$

where $0 \le k_i \le k_i + h_i - 1 \le p - 1$, of size $|B| = h_1 h_2 h_3$ we denote by $Q_k(B)$ the number of triples (g^x, g^y, g^{xy}), $x, y \in \mathcal{W}_k$, whose smallest non-negative residues modulo p belong to the box B.

Let $\alpha_0 = 0.349\ldots$ be the root of the function $E(\alpha)$ given by (3.15) in the interval $[0, 1/2]$. Lemma 3.23 and Lemma 4.1 (applied with $N = 3$, $L = p - 1$) imply:

Theorem 13.7. *For any fixed $\alpha > \alpha_0$ there exists $\delta > 0$ such that for $n/2 \ge k \ge \alpha n$ the bound*

$$\sup_B \left| Q_k(B) - \frac{|B|}{p-1} \right| \ll |\mathcal{W}_k|^{2-\delta}$$

holds.

Let as before $r = \lfloor \log p \rfloor$. We define an *$s$-template* $\mathbf{T} = \{\mathcal{K}, \Sigma\}$ as a set $\mathcal{K}$ of s disjoint intervals $[m_i, m_i + l_i - 1] \subset [1, r]$, $i = 1, \ldots, s$ and a set Σ of s binary strings σ_i of length l_i for $i = 1, \ldots, s$.

The *length* of $\mathbf{T}$ is defined as the sum $k = l_1 + \ldots + l_s$.

We say that an integer u satisfies the s-template $\mathbf{T} = \{\mathcal{K}, \Sigma\}$ if its bit patterns on positions belonging to the intervals $[m_i, m_i + l_i - 1] \in \mathcal{K}$ coincide with the corresponding strings σ_i, $i = 1, \ldots, s$.

Given three s-templates $\mathbf{T}_1$, $\mathbf{T}_2$, $\mathbf{T}_3$ we denote by $M(\mathbf{T}_1, \mathbf{T}_2, \mathbf{T}_3)$ the number of pairs (x, y), $x, y = 1, \ldots, p-1$, such that $\lfloor g^x \rfloor_p$, $\lfloor g^y \rfloor_p$, $\lfloor g^{xy} \rfloor_p$ satisfy the templates $\mathbf{T}_1$, $\mathbf{T}_2$, $\mathbf{T}_3$, respectively.

Unfortunately it is not known how to derive the expected asymptotic formula for $M(\mathbf{T}_1, \mathbf{T}_2, \mathbf{T}_3)$ (even in the special case that the templates consist of a single interval each, not placed however at either end). Nevertheless, in [94] a lower bound is obtained which is non-trivial for a wide range of parameters. Moreover, if s is fixed then the bound is of the "correct order".

Theorem 13.8. *For any s-templates $\mathbf{T}_1$, $\mathbf{T}_2$, $\mathbf{T}_3$ of length k_1, k_2, k_3, the bound*

$$M(\mathbf{T}_1, \mathbf{T}_2, \mathbf{T}_3) \geq p^2 2^{-k_1-k_2-k_3-3s-6} + O\left(p^{23/12}\right)$$

holds.

Proof. We indicate only the main steps of the proof. For an s-template $\mathbf{T}$ of length k we denote by $\mathcal{U}$ the set of r-bit integers u which satisfy $\mathbf{T}$. Obviously $\mathcal{U}$ is of the form

$$\mathcal{U} = \left\{ A + \sum_{i=1}^{L} x_i 2^{t_i} \mid 0 \leq x_i \leq 2^{r_i} - 1,\ i = 1, \ldots, s \right\},$$

where $L \leq s+1$, A depends on the binary strings from Σ, $0 \leq r_i + t_i < t_{i+1}$, $i = 1, \ldots, L-1$, and $r_1 + \ldots + r_L = r - k$.

Let us put $b_i = h_i = 2^{r_i - 1}$, $i = 1, \ldots, L$ and define

$$B = A + \sum_{i=1}^{L} b_i 2^{t_i}, \qquad \mathcal{V} = \left\{ \sum_{i=1}^{s} x_i 2^{t_i} \mid 0 \leq x_i \leq h_i - 1,\ i = 1, \ldots, s \right\}.$$

We see that $B + v - w \in \mathcal{U}$ for all $v, w \in \mathcal{V}$ and each element u of $\mathcal{U}$ has at most $|\mathcal{V}|$ such representations. We also see that $|\mathcal{V}| = h_1 \ldots h_s = 2^{r-k-L} \geq 2^{r-k-s-1}$. We also remark that (3.2) implies

$$\sum_{a=0}^{p-1} \left| \sum_{v \in \mathcal{V}} \mathbf{e}_p(av) \right|^2 = \sum_{v,w \in \mathcal{V}} \sum_{a=0}^{p-1} \mathbf{e}_p\left(a(v-w)\right) = p|\mathcal{V}|. \tag{13.4}$$

Let $\mathcal{U}_j$, $\mathcal{V}_j$, B_j be defined in a similar way with respect to the template $\mathbf{T}_j$, $j = 1, 2, 3$.

From the above discussion we see that

$$M(\mathbf{T}_1, \mathbf{T}_2, \mathbf{T}_3) \geq \frac{W}{|\mathcal{V}_1||\mathcal{V}_2||\mathcal{V}_3|},$$

where W is the number of solutions $(x, y, v_1, v_2, v_3, w_1, w_2, w_3)$ of the system of congruences

$$\begin{aligned} g^x &\equiv B_1 + v_1 - w_1 \pmod p, \qquad v_1, w_1 \in \mathcal{V}_1, \\ g^y &\equiv B_2 + v_2 - w_2 \pmod p, \qquad v_2, w_2 \in \mathcal{V}_2, \\ g^{xy} &\equiv B_3 + v_3 - w_3 \pmod p, \qquad v_3, w_3 \in \mathcal{V}_3, \\ & \quad 1 \leq x, y \leq p-1. \end{aligned}$$

Using (3.2), as before, we write

$$\begin{aligned} W &= \frac{1}{p^3} \sum_{x,y=1}^{p-1} \sum_{v_1,w_1 \in \mathcal{V}_1} \sum_{v_2,w_2 \in \mathcal{V}_2} \sum_{v_3,w_3 \in \mathcal{V}_3} \\ &\quad \times \sum_{a_1,a_2,a_3=0}^{p-1} \mathbf{e}_p\big(a_1(g^x - B_1 - v_1 + w_1) + a_2(g^x - B_2 - v_2 + w_2) \\ &\qquad\qquad + a_3(g^{xy} - B_3 - v_3 + w_3)\big) \\ &= \frac{1}{p^3} \sum_{a_1,a_2,a_3=0}^{p-1} S_{a_1,a_2,a_3} \prod_{i=1}^{3} \sum_{v_i,w_i \in \mathcal{V}_i} \mathbf{e}_p\left(-a_i(B_i + v_i - w_i)\right), \end{aligned}$$

where the sums S_{a_1,a_2,a_3} are given by (3.9).

The term corresponding to $a_1 = a_2 = a_3 = 0$ in the last expression is equal to $(p-1)^2|\mathcal{V}_1|^2|\mathcal{V}_2|^2|\mathcal{V}_3|^2p^{-3}$.

To estimate the contribution

$$R = W - (p-1)^2|\mathcal{V}_1|^2|\mathcal{V}_2|^2|\mathcal{V}_3|^2p^{-3}$$

of the remaining terms of that sum, we apply Lemma 3.21 getting

$$R \ll p^{-13/12} \sum_{\substack{a_1,a_2,a_3=0 \\ \gcd(a_1,a_2,a_3,p)=1}}^{p-1} \prod_{i=1}^{3} \left| \sum_{v_i,w_i \in \mathcal{V}_i} \mathbf{e}_p\left(a_i(v_i - w_i)\right) \right|.$$

To simplify computation we add the term corresponding to $a_1 = a_2 = a_3 = 0$ back to the sum on the right-hand side. Using the formula (13.4) after simple evaluations we derive

$$R \ll p^{23/12}|\mathcal{V}_1||\mathcal{V}_2||\mathcal{V}_3|.$$

Therefore

$$M(\mathbf{T}_1, \mathbf{T}_2, \mathbf{T}_3) \geq |\mathcal{V}_1||\mathcal{V}_2||\mathcal{V}_3|p^{-1} + O\left(p^{23/12}\right).$$

Taking into account that

$$|\mathcal{V}_1||\mathcal{V}_2||\mathcal{V}_3| \geq 2^{3n-k_1-k_2-k_3-3s-3} \geq p^3 2^{-k_1-k_2-k_3-3s-6},$$

we obtain the desired estimate. □

Similar results, but without the assumption that g is a primitive root, are obtained in [94] as well. The bounds depend on the multiplicative order $t = \mathrm{ord}_p g$ of g and are non-trivial for $t \geq p^{3/4+\varepsilon}$.

It is well known that the Diffie–Hellman secret key is almost never used itself but rather some portion of its bits is used as a key K for some private key cryptosystem. The above results show that if this portion of bits is reasonably short compared to $\log p$, then the statistical distance between K and uniformly distributed binary strings of the same length is exponentially small.

Theorem 13.7 and other results of [195, 196] imply that this property also holds even if the exponents x and y are selected from arbitrary (but sufficiently large) sets $\mathcal{X}$ and $\mathcal{Y}$. For instance, they can selected in a way which helps to speed-up exponentiation.

More on the cryptographic interpretation of these results can be found in [94, 95], related in particular to some questions arising in [93, 159, 405].

For composite moduli, analogues of Lemmas 3.21 and 3.22, and thus of Theorems 13.4, 13.5 and 13.6, have been obtained in [189].

A related but somewhat different question about the distribution of the Diffie–Hellman secret key has been considered in [554]. Given a primitive root $g \in \mathbb{F}_p^*$, for $a \in \mathbb{F}_p^*$, we denote by $P(a)$ the probability of $g^{xy} = a$, where x, y are chosen uniformly and independently at random from the set $\{0, \dots, p-2\}$. This function been explicitly evaluated in [554]. More precisely

$$P(a) = \frac{1}{p-1} \prod_{i=1}^{\nu} \left((\gamma_i - \alpha_i + 1) \left(1 - \frac{1}{p_i}\right) + \frac{\delta_i}{p_i} \right), \tag{13.5}$$

where

$$p - 1 = p_1^{\gamma_1} \dots p_\nu^{\gamma_\nu} \qquad \text{and} \qquad t = p_1^{\alpha_1} \dots p_\nu^{\alpha_\nu}$$

are the prime number factorisation of $p-1$ and the multiplicative order t of a, respectively, and

$$\delta_i = \begin{cases} 1, & \text{if } \alpha_i = 0; \\ 0, & \text{if } \alpha_i \geq 1. \end{cases}$$

In particular

$$\min_{a \in \mathbb{F}_p^*} P(a) = \frac{1}{p-1} \prod_{i=1}^{\nu} \left(1 - \frac{1}{p_i}\right)$$

and

$$\max_{a \in \mathbb{F}_p^*} P(a) = \frac{1}{p-1} \prod_{i=1}^{\nu} \left(\gamma_i \left(1 - \frac{1}{p_i}\right) + 1\right).$$

These results imply a very nice bound

$$\log(p-1) \geq \mathcal{H}_p \geq \log(p-1) - \log_2 \tau(p-1)$$

on the entropy

$$\mathcal{H}_p = - \sum_{a \in \mathbb{F}_p^*} P(a) \log P(a)$$

of the distribution of the Diffie–Hellman secret key. Thus from (5.1) we derive

$$\mathcal{H}_p = \log p + O\left(\frac{\log p}{\log \log p}\right).$$

Chapter 14

Bit Security of the Diffie–Hellman Secret Key

We begin with presenting two results from [222] about the bit security of the Diffie–Hellman secret key which generalise Theorem 2 of [64]. We have already mentioned that the proof of Theorem 2 of [64] is not quite correct and it applies only to some special inputs. Using the bounds of exponential sums, namely Lemmas 3.15 and 3.16, allows us to complete the proof and also extend the result to more general settings. Accordingly, the bound of Lemma 3.24 has been used in [523] to obtain somewhat stronger results, see also Theorem 14.3 below.

For the Shamir message passing scheme, a similar correction of Theorem 3 of [64] has been obtained in [223], see also Chapter 18. Our technique also applies to other exponentiation based cryptosystems, such as the ElGamal cryptosystem, see Section 8.4 of [389] or Section 5.1 of [541], Matsumoto–Takachima–Imai key-agreement protocol, see Section 12.6 of [389] or Section 8.4.2 of [541] and other schemes mentioned in [64, 65].

For each integer $k \geq 1$ define the oracle $\mathcal{DH}_k$ as a "black box" which, given the values of $X = \lfloor g^x \rfloor_p$ and $Y = \lfloor g^y \rfloor_p$, outputs the value of $\mathrm{MSB}_{k,p}(g^{xy})$, where the function $\mathrm{MSB}_{k,p}$ is defined by (1.1).

Theorem 14.1. *Let* $k = \left\lceil \log^{1/2} p \right\rceil + \lceil \log\log p \rceil$. *For any* $\varepsilon > 0$, *sufficiently large* p *and any element* $g \in \mathbb{F}_p^*$ *of multiplicative order* $T \geq p^{1/3+\varepsilon}$, *there exists a probabilistic polynomial time algorithm which for any pair* $(a,b) \in [0, T-1]^2$, *given the values of* $A = \lfloor g^a \rfloor_p$ *and* $B = \lfloor g^b \rfloor_p$, *makes* $O\left(\log^{1/2} p\right)$ *calls of the oracle* $\mathcal{DH}_k$ *and computes* $\lfloor g^{ab} \rfloor_p$ *correctly with probability* $1 + O\left(p^{-\varepsilon/2}\right)$.

Proof. Given a pair $(a,b) \in [0, T-1]^2$ let us select an integer $u \in [0, T-1]$

uniformly at random. We compute

$$g_u = \lfloor Bg^u \rfloor_p,$$

thus $g_u \equiv g^{b+u} \pmod p$.

The probability that

$$\gcd(b+u, T) \ge Tp^{-1/3-\varepsilon/3} \tag{14.1}$$

is at most $\tau(T)T^{-1}p^{1/3+\varepsilon/3}$ where $\tau(T)$ is the number of positive integer divisors of T. Indeed, for any divisor $D|T$ with $D \ge Tp^{-1/3-\varepsilon/3}$ there are at most $T/D \le p^{1/3+\varepsilon/3}$ values of $s \in [0, T-1]$ with $\gcd(s,T) = D$.

Using the bound (5.1), we obtain that the probability of (14.1) is at most

$$O\left(T^{-1}p^{1/3+\varepsilon/2}\right) = O\left(p^{-\varepsilon/2}\right).$$

In the opposite case, when $\gcd(a+u, T) < Tp^{-1/3-\varepsilon/3}$, the multiplicative order of g_u is

$$T_u = \frac{T}{\gcd(b+u,T)} > p^{1/3+\varepsilon/3}.$$

Let $\alpha_u \equiv g^{a(b+u)} \pmod p$. Then

$$\mathrm{MSB}_{k,p}\left(\alpha_u g_u^x\right) = \mathrm{MSB}_{k,p}\left(g_u^{(a+x)}\right) = \mathrm{MSB}_{k,p}\left(g^{(a+x)(b+u)}\right).$$

Now we use the oracle $\mathcal{DH}_k$ with the input $\lfloor g^x A \rfloor_p$ and $\lfloor g^u B \rfloor_p$ to compute $\mathrm{MSB}_{k,p}(\alpha_u g_u^x)$ for an integer x chosen uniformly at random in the interval $[0, p-1]$. Because $T_u | p-1$, the values of residues of x modulo T_u are uniformly distributed in the interval $[0, T_u - 1]$ as well. From Lemmas 3.15 and 4.1 (applied with $N = 1$, $L = p-1$) we see that the sequence

$$g_u^x, \qquad x = 0, \ldots, T_u - 1,$$

is $p^{-\delta}$-homogeneously distributed modulo p for some $\delta > 0$ depending only on ε. Thus Lemma 6.3 can be applied. Therefore. one can construct a probabilistic polynomial time algorithm that:

- Selects a random $u \in [0, T-1]$.
- Applies algorithm $\mathcal{A}$ from Lemma 6.3 (now g_u plays the role of g in the conditions of Lemma 6.3). This algorithm makes $O\left(\log^{1/2} p\right)$ calls to the oracle $\mathcal{DH}_k$.
- If $T_u > p^{1/3+\varepsilon/3}$, the algorithm returns the correct value of α_u with probability $1 + O\left(p^{-1}\right)$.

Indeed, the only possible source of error is either the case $T_u \le p^{1/3+\varepsilon/3}$ or the probability error of the algorithm of Lemma 6.3. The probability of both events is $O\left(p^{-\varepsilon/2}\right)$.

Remarking that

$$g^{ab} \equiv \alpha_u A^{-u} \pmod p,$$

we obtain the desired result. □

It is easy to see that Theorem 14.1 is non-trivial for any $T \ge p^{1/3+\varepsilon}$. In a similar way, Lemma 3.16 produces a result which holds for almost all primes p and is non-trivial for $T \ge p^{\varepsilon}$.

Using the result of [304] one can show that the statement of Theorem 14.1 holds whenever $T \ge p^{1/4+\varepsilon}$.

Theorem 14.2. *Let r be a sufficiently large integer and let $k = \lceil r^{1/2} \rceil + \lceil \log r \rceil$. For any $\varepsilon > 0$ and for all primes $p \in [2^{r-1}, 2^r - 1]$, except at most $2^{(5/6+\varepsilon)r}$ of them, and any element $g \in \mathbb{F}_p^*$ of multiplicative order $T \ge p^{\varepsilon}$, the following statement holds: There exists a probabilistic polynomial time algorithm which for any pair $(a, b) \in [0, T-1]^2$, given the values of $A = \lfloor g^a \rfloor_p$ and $B = \lfloor g^b \rfloor_p$, makes $O\left(r^{1/2}\right)$ calls of the oracle $\mathcal{DH}_k$ and computes $\lfloor g^{ab} \rfloor_p$ correctly with probability $1 + O\left(p^{-\varepsilon/2}\right)$.*

It has also been shown in [64], that if the element $g \in \mathbb{F}_p^*$ is a small integer then an alternative method can be used to show the bit security of $O(1)$ bits of the corresponding Diffie–Hellman secret key. In particular, for $g = 2$, one has an analogue of Theorem 14.1 with the oracle $\mathcal{DH}_1$. However, for this result to be useful, one also needs to show that 2 is not "weak" as a generator of the Diffie–Hellman secret key, that is, that the multiplicative order of 2 is not smooth,, thus the Pohlig–Hellman algorithm cannot be used to solve efficiently the discrete logarithm problem in base 2, see Section 3.6.4 of [389] or Section 5.1 of [541]. Such results have recently been obtained in [466], showing that for the overwhelming majority of primes the Pohlig–Hellman algorithm is not efficient for finding the discrete logarithm to base 2 modulo p, and thus, based on our current state of art in solving the discrete logarithm problem, apparently it is as hard as for any other base. As in [466], we also remark, that besides stronger bit security results available in this case, using $g = 2$ as the base has an additional advantage of faster exponentiation.

Finally, we show that for large values of T, namely for $T \ge p^{1/2+\varepsilon}$, a weaker *diagonal* oracle $\mathcal{DDH}_k$ can be used, see [523].

For each integer k, we define the diagonal oracle $\mathcal{DDH}_k$ as a "black box" which, given the value of $X = \lfloor g^x \rfloor_p$, outputs the value of $\mathrm{MSB}_{k,p}\left(g^{x^2}\right)$.

As we have mentioned in Chapter 12, it has been shown in Section 5.3 of [375] that if there is an oracle computing $\left\lfloor g^{x^2} \right\rfloor_p$ from $X = \lfloor g^x \rfloor_p$, that is, the diagonal oracle $\mathcal{DDH}_k$ with $k = \lceil \log p \rceil$, then there is a deterministic polynomial time algorithm to compute $\lfloor g^{xy} \rfloor_p$ from $X = \lfloor g^x \rfloor_p$ and $Y = \lfloor g^y \rfloor_p$.

Several other security assumptions (in particular, similar to the Diffie–Hellman indistinguishability assumption) related to the pairs $\left(g^x, g^{x^2}\right)$ have recently been discussed in the literature, see [487].

Here we combine this result with our technique, in particular, we use Lemma 3.24, and show that the same statement holds for a diagonal oracle $\mathcal{DDH}_k$ with $k = O\left(\log^{1/2} p\right)$.

Theorem 14.3. *Let* $k = \left\lceil \log^{1/2} p \right\rceil + \lceil \log\log p \rceil$. *For any* $\varepsilon > 0$, *sufficiently large* p *and any element* $g \in \mathbb{F}_p^*$ *of multiplicative order* $T \geq p^{1/2+\varepsilon}$, *there exists a polynomial time algorithm which for any pair* $(a, b) \in [0, T-1]^2$, *given the values of* $A = \lfloor g^a \rfloor_p$ *and* $B = \lfloor g^b \rfloor_p$, *makes* $O\left(\log^{1/2} p\right)$ *calls of the diagonal oracle* $\mathcal{DDH}_k$ *and computes* $\lfloor g^{ab} \rfloor_p$ *correctly with probability* $1 + O\left(p^{-\varepsilon/2}\right)$

Proof. As we have just mentioned, it follows from Section 5.3 of [375] that it is enough to show that there exists an algorithm to compute $\left\lfloor g^{a^2} \right\rfloor_p$ from $A = \lfloor g^a \rfloor_p$.

Let $\alpha \equiv g^{a^2} \pmod p$. Then

$$\mathrm{MSB}_{k,p}\left(\alpha g^{x^2+2ax}\right) = \mathrm{MSB}_{k,p}\left(g^{(a+x)^2}\right),$$

thus we can use the diagonal oracle $\mathcal{DDH}_k$ with the input $\lfloor g^x A \rfloor_p$ to compute $\mathrm{MSB}_{k,p}\left(\alpha g^{x^2+2ax}\right)$ for an integer x chosen uniformly at random in the interval $[0, p-1]$. Because $T | p-1$, the values of residues of x modulo T are uniformly distributed in the interval $[0, T-1]$ as well.

From Lemmas 3.24 and 4.1 (applied with $N = 1$, $L = p-1$) we see that the sequence

$$g^{x^2+2ax}, \qquad x = 0, \ldots, T-1,$$

is $p^{-\delta}$-homogeneously distributed modulo p for some $\delta > 0$ depending only on ε. Thus Lemma 6.3 can be applied and the result follows. □

Finally, we show that a similar result holds for the *inversive* oracle $\mathcal{IDH}_k$ as well.

For each integer k define the inversive oracle $\mathcal{IDH}_k$ as a "black box" which, given the value of $X = \lfloor g^x \rfloor_p$ with $x \in \mathbb{Z}_T^*$, outputs the value of $\mathrm{MSB}_{k,p}\left(g^{x^{-1}}\right)$.

The mapping $g^x \to g^{x^{-1}}$ has recently been considered in the literature [356, 487]. In particular, it is shown in [487] that any algorithm which computes this mapping (for any g) can be used to compute the mapping $g^x \to g^{x^2}$ and thus, using the reduction of [375], to compute the Diffie–Hellman secret key.

However, we do not assume that the oracle $\mathcal{IDH}_k$ works for every g. Thus our result does not guarantee the recovering $\lfloor g^{ab} \rfloor_p$ from $A = \lfloor g^a \rfloor_p$ and $B = \lfloor g^b \rfloor_p$, but it just shows that it can be used to recover $\left\lfloor g^{a^{-1}} \right\rfloor_p$ from $A = \lfloor g^a \rfloor_p$. We also need the condition that T is prime. It is quite possible that an appropriate modification of our method can be used to eliminate this restriction (and thus extend our result to primitive roots g). On the other hand, in many cryptographic scenarios elements of prime multiplicative order play a key role.

Theorem 14.4. *Let $\varepsilon > 0$ and $1 \geq \rho > 1/2$ be fixed real numbers. Put*

$$k = \left\lceil \left(\frac{2\rho - 1}{5} - \varepsilon \right) \log p \right\rceil .$$

For any sufficiently large p and any element $g \in \mathbb{F}_p^*$ of multiplicative order $T \geq 0.6p^\rho$, where T is prime, there exists a polynomial time algorithm which for any $a \in \mathbb{Z}_T^*$, given the value of $A = \lfloor g^a \rfloor_p$, makes $O(\varepsilon^{-1})$ calls of the oracle $\mathcal{IDH}_k$ and computes $\left\lfloor g^{a^{-1}} \right\rfloor_p$ correctly with probability $1 + O(p^{-1})$.

Proof. Let $\mathcal{G} \subseteq \mathbb{F}_p^*$ be the multiplicative group generated by g. First of all we remark that when $v \in [1, T-2]$ is selected uniformly at random the value of $\lfloor g^v \rfloor_p$ is uniformly distributed in $\overline{\mathcal{G}} = \mathcal{G} \backslash \{1, g^{-1}\}$ and the cardinality $|\overline{\mathcal{G}}| = T - 2 \geq 0.5p^\rho$. Now making calls of the oracle $\mathcal{IDH}_k$ with

$$\lfloor A^{v_i} \rfloor_p = \lfloor g^{av_i} \rfloor_p \qquad \text{and} \qquad \lfloor A^{w_i} \rfloor_p = \lfloor g^{aw_i} \rfloor_p,$$

where $v_i = \lfloor (u_i + 1)^{-1} \rfloor_T$ and $w_i = \lfloor u_i^{-1} \rfloor_T$, we obtain approximations to $\left\lfloor g^{a^{-1}} g^{u_i a^{-1}} \right\rfloor_p$ and to $\left\lfloor g^{u_i a^{-1}} \right\rfloor_p$, $i = 1, \ldots, d$, for $d = \lceil 4\varepsilon^{-1}/5 \rceil$ values of $u_i \in [1, T-2]$ chosen uniformly and independently at random, we apply the algorithm of Lemma 6.10 to derive the result. □

As we have remarked in Chapter 6 using the full power of Lemma 6.1 or the results of [6] one can obtain slightly better estimates in Theorems 14.1, 14.2 and 14.3. It is also interesting to remark that if one assumes that the Diffie–Hellman key exchange protocol modulo p cannot be broken in less than about $\exp\left(C(\log p)^{1/3}\right)$ bit operations with some constant $C > 0$, that is, the presently known fastest (heuristic) algorithms [1, 4, 82, 352, 379, 389, 451, 462, 494, 508, 541] are essentially the best possible, then these results can be improved in a more substantial way. Indeed, allowing the algorithms of Theorems 14.1, 14.2 and 14.3 to run in the same time $\exp\left(C(\log p)^{1/3}\right)$ as needed for breaking the Diffie–Hellman protocol, one can use k of order about $\log^{1/3} p$.

Surprisingly enough, for the elliptic curve Diffie–Hellman key exchange protocol, a much stronger result has been obtained using more elementary considerations, see [63].

Finally, an analogue of Theorems 14.1 and 14.2 for an oracle giving k bits of the trace $\mathrm{Tr}_{\mathbb{F}_{p^m}/\mathbb{F}_p}(g^{xy})$, where $g \in \mathbb{F}_{p^m}^*$, has been obtained in [342, 521]. Some of these results are presented in Chapter 19.

Part IV

Other Cryptographic Constructions

Chapter 15

Security Against the Cycling Attack on the RSA and Timed-release Crypto

Here we give some applications of the results about the period of the power generator which are described in Chapter 23, see also [193].

The first application we discuss is the conclusion that the so-called *cycling attack* on the RSA cryptosystem has a negligible chance to be efficient. Despite the common belief that this should be the case, no rigorous proof of the statement has been given. The attack is based on the observation that the power generator (23.1) can be considered as a sequence of consecutive RSA encryptions starting with the "message" u_0. Thus, if the period is t, then after $t-1$ iterations of the encrypted message

$$u_1 \equiv u_0^e \pmod m$$

we obtain

$$u_t \equiv u_0 \pmod m$$

and if t is small then this is an efficient procedure. (Note that in RSA the encryption exponent e is co-prime to $\varphi(m)$, so the sequence (23.1) is purely periodic.) Even more, if $m = pl$ is a product of two primes p and l and t is small then, because it is very likely that the periods t_p and t_l of this sequence modulo p and l are distinct after at most $\min\{t_p, t_l\}$ iterations this attack may produce a complete factorisation of m. Indeed, it is very likely that

$$\gcd\left(u_{t_p} - u_0, m\right) = p \qquad \text{and} \qquad \gcd\left(u_{t_l} - u_0, m\right) = l.$$

This attack, as well as various ways of protecting against it, have been discussed in the literature, see [73, 372, 389, 476, 478]. In particular, so-called strong primes have been introduced (see the definition given in Chapter 5). Some heuristic arguments have recently been presented in [478] which show that randomly selected primes p and l are already likely to be strong against this attack so that it is not so necessary to make special choices. Theorem 23.4 implies a more precise and completely rigorous statement which basically means that for a random selection of parameters the expected complexity of this attack is about $m^{1/2}$, that is, of the same magnitude as the trial-division factorisation algorithm. Indeed, obviously $t_p(l-1) \geq t$ and $t_l(p-1) \geq t$; thus when t is of order m and $p \sim l \sim m^{1/2}$ we obtain that $\min\{t_p, t_l\}$ is of order $m^{1/2}$. In fact t_p and t_l can be studied independently using Lemma 5.1 directly, thus getting slightly more precise results.

Furthermore, in [249] some modifications of the cycling attack have been introduced. One of these is based on the properties of the sequence

$$u_n \equiv u_{n-1}^{e} u_{n-2}^{-1} \pmod m, \quad 0 \leq u_n \leq m-1, \qquad n = 2, 3, \dots , \tag{15.1}$$

with the *initial values* $u_0 = \vartheta^2$, $u_1 = \vartheta^e$, where ϑ is an integer co-prime to m, and the *exponent* $e \neq 0, \pm 2$. Thus, the sequence of exponents runs through the Lucas sequence $\alpha^j + \beta^j$ where α, β are the roots of the polynomial $x^2 - ex + 1$. The period of the sequence (15.1), and thus the success of the corresponding cycling attack, depends on the behaviour of the function $\Omega_D(n)$, defined in [249], where $D = e^2 - 4$. This function is defined in a somewhat similar way to the Carmichael function. Namely,

$$\Omega_D\left(2^k\right) = \begin{cases} 2^{k-1}, & \text{if } D \text{ is even}, \\ 3 \cdot 2^{k-1}, & \text{if } D \text{ is odd}. \end{cases}$$

For a prime $p \geq 3$,

$$\Omega_D\left(p^k\right) = \begin{cases} (p - (D/p))\, p^{k-1}, & \text{if } (D/p) \neq 0, \\ 2p^{k-1}, & \text{if } (D/p) = 0; \end{cases}$$

where (D/p) is the quadratic character of D modulo p. Finally, as in the case of the Carmichael function we put

$$\Omega_D(n) = \operatorname{lcm}\left(\Omega_D\left(p_1^{k_1}\right), \dots , \Omega_D\left(p_\nu^{k_\nu}\right)\right),$$

where

$$n = p_1^{k_1} \dots p_\nu^{k_\nu}$$

is the prime number factorisation of n.

The method of [192] can be used to obtain complete analogues of the results of Chapter 23 for the function $\Omega_D(n)$ and thus to also rule out the feasibility of this modified attack for randomly selected RSA moduli.

A variant of the cycling attack on elliptic curve analogues of the RSA cryptosystem has been considered in [288]. It would be interesting to obtain elliptic curve analogues of the results of [192, 193] and thus a rigorous proof that this attack is not feasible for almost all values of the parameters.

Our second application is related to the notion of *timed-release crypto* recently introduced in [477] which applies to the following scenario.

Assume one wants to send a message μ which is supposed to be read T seconds later. The simplest way to arrange this is to encrypt the message using any reliable (public or private key) cryptosystem using a secret key K and then, T seconds later, send the key K. However, this solution cannot be used if for some reason the sender is no longer on-line in T seconds, which may especially be true if the actual delay is several years. See the web site

`http://www.lcs.mit.edu/research/demos/cryptopuzzle0499`

for such an example.

In [477] the following elegant solution has been proposed. As before, use any (public/private key) cryptosystem and encrypt the message with a private key K, getting the encrypted message $E(\mu, K)$. Select an RSA modulus $m = pl$, an integer ϑ and an exponent e. We remark that in the original paper [477] only the value $e = 2$ has been considered but it may be more convenient to use a random value of e (to take advantage of the difference in the strength of Theorem 23.4 and Theorem 23.5).

Now the sender evaluates

$$U \equiv \vartheta^{e^s} \pmod m, \qquad 0 \le U < m,$$

by computing

$$f \equiv e^s \pmod{\varphi(m)}, \qquad 0 \le f \le \varphi(m),$$

in $O(\log s)$ modular multiplications and then by computing

$$\vartheta^{e^s} \equiv \vartheta^f \pmod m$$

in another $O(\log m)$ modular multiplications (using repeated squaring in both cases). Finally, the sender computes $L \equiv K + \vartheta^{e^s} \pmod m$, $0 \le L < m$, and sends the 6-tuple

$$[E(\mu, K), m, L, \vartheta, e, s].$$

To decrypt the message μ the receiver must recover K, that is, the receiver must compute $U \equiv \vartheta^{e^s} \pmod m$. It seems that the only way to compute U is to make s consecutive exponentiations

$$u_i \equiv u_{i-1}^e \pmod m$$

with $u_0 = \vartheta$, until it reaches $U = u_s$. Hopefully this is the only way without the knowledge of $\varphi(m)$. Moreover, there seems to be no advantage to the use of parallel computation, see the discussion of this issue in [477]. Thus, adjusting s and taking into account the expected computer performance one can predict the "breaking" time. Because there is no reasonable way to parallelize the procedure, this time does not depend on the total computational power, only on the performance of a single computer, which is more predictable.

However, if the period of the sequence $u_n \equiv \vartheta^{e^n} \pmod m$ is small, say $t < s$, then one can compute U in about t steps rather than in s steps, thus reading the message earlier than expected. This issue has been discussed at the end of Section 2.1 of [477]. Our results, namely Theorem 23.4 for a randomly selected e and, in weaker form, Theorem 23.5 for $e = 2$, provide rigorous support for the assumption that premature decryption will occur only extremely rarely.

Several modifications and applications of this scheme, which preserve the main idea, are given in [62, 367].

As in [275], now we show that if $\gcd(e, \varphi(m)) = 1$, then Lemma 6.10 implies a certain result about the bit security of this function.

Let $\mathcal{RSW}_\Delta$ denote an oracle which for any $v \in \mathbb{Z}_m^*$ outputs some V such that $\left|\lfloor v^{e^s} \rfloor_m - V\right| \leq \Delta$.

Theorem 15.1. *Let m be a sufficiently large integer. Let $\varepsilon > 0$ be a fixed real number. We define*

$$\Delta = m^{1/5-\varepsilon}.$$

Assume that the smallest prime divisor p of m satisfies the inequality

$$p \geq (m/\Delta)^{1/2}.$$

Given an oracle $\mathcal{RSW}_\Delta$, there exists a probabilistic polynomial time algorithm $\mathcal{A}$ such that for any $\vartheta \in \mathbb{Z}_m^*$ it makes the expected number $O(\varepsilon^{-1})$ of calls of the oracle $\mathcal{RSW}_\Delta$ and computes $\lfloor \vartheta^{e^s} \rfloor_m$ with probability at least $1 + O(m^{-1})$.

Proof. First of all we remark that because $\gcd(e, \varphi(m)) = 1$ when $v \in \mathbb{Z}_m^*$ is selected uniformly at random the values $\lfloor v^{e^s} \rfloor_m$ are uniformly distributed in $\mathbb{Z}_m^*$ as well. Now making calls of the oracle $\mathcal{RSW}_\Delta$ with $v = \vartheta$ and with $v = u_i$ and $v = \vartheta u_i$, $i = 1, \ldots, d$, for $d = \lceil 4\varepsilon^{-1}/5 \rceil$ values of $u_i \in \mathbb{Z}_m^*$, chosen uniformly and independently at random, we apply the algorithm of Lemma 6.10 to derive the result. □

Thus, Theorem 15.1 implies that $k = \lceil (4/5 - \varepsilon) \log m \rceil$ of the most significant bits of $\lfloor \vartheta^{e^s} \rfloor_m$ are as secure as the whole value.

We remark that in [477] this scheme is considered with $e = 2$. Here we consider a slightly different case of $\gcd(e, \varphi(m)) = 1$. If the 2-adic order of the Carmichael function $\lambda(m)$, is not too large (which is true for the overwhelming majority of the moduli), then the case $e = 2$ can be considered as well.

Chapter 16

The Insecurity of the Digital Signature Algorithm with Partially Known Nonces

Here we present the polynomial-time algorithm of [417] which recovers the *private key* of the signer if a small portion of bits of the so-called *nonce* in the Digital Signature Algorithm modulo a prime p is known for $O(\log^{1/2} p)$ signatures.

The main idea of this lattice reduction based attack has been outlined in the paper [414] which in turn sharpens and simplifies the approach of [277]. However, both papers [277] and [414] provide only heuristic results. Finally, it has been shown in [417] that using some new bounds of exponential sums this approach can be made rigorous.

Let p and $q \geq 3$ be prime numbers with $q|p-1$.

Let $g \in \mathbb{F}_p$ be a fixed element of multiplicative order q, that is $g^q = 1$ and $q \neq 1$ which is *publicly* known.

Let $\mathcal{M}$ be the set of messages to be signed and let $h : \mathcal{M} \to \mathbb{F}_q$ be an arbitrary function, usually called a *hash-function*.

The signer fixes an element $\alpha \in \mathbb{F}_q^*$ which is the *secret* key known only to the signer and publishes $A = \lfloor g^\alpha \rfloor_p$.

To sign a message $\mu \in \mathcal{M}$, the signer chooses a random integer $k \in \mathbb{F}_q^*$, usually called the *nonce*, which must be kept secret. The signer then defines the following two elements of $\mathbb{F}_q$:

$$r(k) = \left\lfloor \lfloor g^k \rfloor_p \right\rfloor_q \qquad \text{and} \qquad s(k, \mu) = \left\lfloor k^{-1} \left(h(\mu) + \alpha r(k)\right) \right\rfloor_q. \tag{16.1}$$

The pair $(r(k), s(k,\mu))$ is the signature on the message μ with a nonce k. This pair $(r(k), s(k,\mu))$ is sent to the verifier (together with the message μ, of course). Now the verifier can check the congruence

$$r(k)^{s(k,\mu)} \equiv g^{h(\mu)} A^{r(k)} \pmod p$$

together with the inequalities $0 < r(k) < q$ and $0 < s(k,\mu) < q$, see Section 11.5.1 of [389].

Clearly, knowledge of the exact value of only one value of k enables us to recover α.

Assume that we know the ℓ least significant bits of several nonces $k \in \mathbb{F}_q^*$. That is, for each nonce k, we are given an integer a such that $0 \le a \le 2^\ell - 1$ and $k - a = 2^\ell b$ for some integer $b \ge 0$. Given a message μ signed with the nonce k, the congruence

$$\alpha r(k) \equiv s(k,\mu)k - h(\mu) \pmod q,$$

can be rewritten for $s(k,\mu) \ne 0$ as

$$\alpha r(k) 2^{-\ell} s(k,\mu)^{-1} \equiv \left(a - s(k,\mu)^{-1} h(\mu)\right) 2^{-\ell} + b \pmod q. \tag{16.2}$$

Now define the following two elements

$$\begin{aligned} t(k,\mu) &= \lfloor 2^{-\ell} r(k) s(k,\mu)^{-1} \rfloor_q, \\ u(k,\mu) &= \lfloor 2^{-\ell} \left(a - s(k,\mu)^{-1} h(\mu)\right) \rfloor_q \end{aligned}$$

and remark that both $t(k,\mu)$ and $u(k,\mu)$ can easily be computed by the attacker from the publicly known information. Recalling that $0 \le b \le q/2^\ell$, we obtain

$$\lfloor \alpha t(k,\mu) - u(k,\mu) \rfloor_q < q/2^\ell. \tag{16.3}$$

Thus, the ℓ most significant bits of $\lfloor \alpha t(k,\mu) \rfloor_q$ are revealed. Collecting several relations of this kind for several pairs (k,μ), the problem of recovering the secret key α can be reduced to a variant of the hidden number problem. Therefore, if we can shown that the distribution of the multiplier $t(k,\mu)$ is close to uniform, then the algorithm of Lemma 6.3 can be applied to recover the secret key α.

For the hash function $h : \mathcal{M} \to \mathbb{F}_q$ we also denote by W the number of pairs $(\mu_1, \mu_2) \in \mathcal{M}^2$ with $h(\mu_1) = h(\mu_2)$. Thus, $W/|\mathcal{M}|^2$ is a probability of a *collision* and our results are non-trivial under a reasonable assumption that this probability is of order of magnitude close to $1/q$.

For an integer ℓ we define the oracle $\mathcal{DSA}_\ell$ which, for any given signature $(r(k), s(k,\mu))$, $k \in [0, q-1]$, $\mu \in \mathcal{M}$, generated by the Digital Signature Algorithm, returns the ℓ least significant bits of k.

Theorem 16.1. *Let Q be a sufficiently large integer. The following statement holds with $\vartheta = 1/3$ for all primes $p \in [Q, 2Q]$, and with $\vartheta = 0$ for all primes $p \in [Q, 2Q]$ except at most $Q^{5/6+\varepsilon}$ of them. For any $\varepsilon > 0$ there exists $\delta > 0$ such that for any element $g \in \mathbb{F}_p$ of multiplicative order $q \geq p^{\vartheta+\varepsilon}$ and any hash function h with*

$$W \leq \frac{|\mathcal{M}|^2}{q^{1-\delta}},$$

given an oracle $\mathcal{DSA}_\ell$ with $\ell = \left\lceil \log^{1/2} q \right\rceil + \lceil \log\log q \rceil$, there exists a probabilistic polynomial time algorithm which, with probability $1 + O(q^{-1})$, recovers the secret key α from $O\left(\log^{1/2} q\right)$ signatures $(r(k), s(k,\mu))$ with $k \in [0, q-1]$ and $\mu \in \mathcal{M}$ selected independently and uniformly at random.

Proof. Denote by $\mathcal{S}$ the set of pairs $(k,\mu) \in [1, q-1] \times \mathcal{M}$ with $s(k,\mu) \neq 0$ (that is, the set of pairs (k,μ) for which the congruence (16.2) holds and thus $t(k,\mu)$ is defined). Then we see from Lemma 4.5 that

$$|\mathcal{S}| = q|\mathcal{M}| \left(1 + O\left(q^{-\delta}\right)\right) \tag{16.4}$$

for all p and q satisfying the conditions of the theorem.

For each $\mu \in \mathcal{M}$ we denote by $\mathcal{K}_\mu$ the set of $k \in [1, q-1]$ for which $(k,\mu) \in \mathcal{S}$.

In order to use Lemma 4.1 to show that the sequence $t(k,\mu)$, $(k,\mu) \in \mathcal{S}$ is $2^{-\log^{1/2} q}$-homogeneously distributed modulo q we estimate the exponential sums

$$\sum_{(k,\mu)\in\mathcal{S}} \mathbf{e}_q\left(ct(k,\mu)\right), \qquad 1 \leq c \leq q-1.$$

We consider $c_0 \in \mathbb{F}_q^*$ corresponding to the largest (by absolute value) exponential sum

$$\sigma = \left| \sum_{(k,\mu)\in\mathcal{S}} \mathbf{e}_q\left(c_0 t(k,\mu)\right) \right| = \max_{\gcd(c,q)=1} \left| \sum_{(k,\mu)\in\mathcal{S}} \mathbf{e}_q\left(ct(k,\mu)\right) \right|.$$

Hence,

$$\sigma \leq \sum_{\mu\in\mathcal{M}} \left| \sum_{k\in\mathcal{K}_\mu} \mathbf{e}_q\left(c_0 t(k,\mu)\right) \right|.$$

For $\lambda \in \mathbb{F}_q$ we denote by $H(\lambda)$ the number of $\mu \in \mathcal{M}$ with $h(\mu) = \lambda$. We also define the integer $a \in [1, q-1]$ by the congruence $a \equiv 2^{-\ell} c_0 \pmod q$. Then

$$\sigma = \sum_{\lambda\in\mathbb{F}_q} H(\lambda) \left| \sum_{\substack{k=1 \\ \alpha r(k) \not\equiv -\lambda \pmod q}}^{q-1} \mathbf{e}_q\left(a \frac{kr(k)}{\lambda + \alpha r(k)}\right) \right|.$$

First of all we remark that

$$\sum_{\lambda \in \mathbb{F}_q} H(\lambda)^2 = W.$$

Then applying the Cauchy inequality, we obtain

$$\sigma^2 \le W \sum_{\lambda \in \mathbb{F}_q} \left| \sum_{\substack{k=1 \\ \alpha r(k) \not\equiv -\lambda \pmod q}}^{q-1} \mathbf{e}_q \left(a \frac{kr(k)}{\lambda + \alpha r(k)} \right) \right|^2 . \tag{16.5}$$

Furthermore,

$$\begin{aligned} &\sum_{\lambda \in \mathbb{F}_q} \left| \sum_{\substack{k=1 \\ \alpha r(k) \not\equiv -\lambda \pmod q}}^{q-1} \mathbf{e}_q \left(a \frac{kr(k)}{\lambda + \alpha r(k)} \right) \right|^2 \\ &= \sum_{\lambda \in \mathbb{F}_q} \sum_{\substack{k=1 \\ \alpha r(k) \not\equiv -\lambda \pmod q}}^{q-1} \\ &\qquad \times \sum_{\substack{m=1 \\ \alpha r(m) \not\equiv -\lambda \pmod q}}^{q-1} \mathbf{e}_q \left(a \left(\frac{kr(k)}{\lambda + \alpha r(k)} - \frac{mr(m)}{\lambda + \alpha r(m)} \right) \right) \\ &= \sum_{k,m=1}^{q-1} \sum_{\lambda \in \mathbb{F}_q}{}^{*} \mathbf{e}_q \left(a \left(\frac{kr(k)}{\lambda + \alpha r(k)} - \frac{mr(m)}{\lambda + \alpha r(m)} \right) \right), \end{aligned}$$

where, as in Lemma 3.2, the symbol $\sum^*$ means that the summation in the inner sum is taken over all $\lambda \in \mathbb{F}_q$ with

$$\lambda \not\equiv -\alpha r(k) \pmod q \qquad \text{and} \qquad \lambda \not\equiv -\alpha r(m) \pmod q.$$

It is easy to see that if $r(k) \ne r(m)$, then the rational function

$$F_{k,m}(X) = \frac{kr(k)}{X + \alpha r(k)} - \frac{mr(m)}{X + \alpha r(m)}$$

is not constant in $\mathbb{F}_q$. If $r(k) = r(m)$, then

$$F_{k,m}(X) = \frac{(k-m)r(k)}{X + \alpha r(k)}.$$

Thus it is constant only if $k = m$ or $r(k) = r(m) = 0$. From Lemma 4.5 we see that the number of such pairs is $O\left(q^{2-2\delta} + q\right)$ for some $\delta > 0$ for which we estimate the sum over λ trivially by q. For other pairs $(k,m) \in [1, q-1]^2$ we use

Lemma 3.2 getting

$$\sum_{\lambda \in \mathbb{F}_q} \left| \sum_{\substack{k=1 \\ \alpha r(k) \not\equiv -\lambda \pmod q}}^{q-1} \mathbf{e}_q\left(a \frac{kr(k)}{\lambda + \alpha r(k)}\right) \right|^2$$
$$= O\left(\left(q^{2-2\delta} + q\right) q + q^{5/2}\right) = O\left(q^{3-2\delta}\right)$$

(without loss of generality we may assume that $\delta < 1/4$). Substituting this estimate in (16.5) we obtain the bound

$$\max_{\gcd(c,q)=1} \left| \sum_{(k,\mu) \in \mathcal{S}} \mathbf{e}_q\left(ct(k,\mu)\right) \right| = O\left(W^{1/2} q^{3/2-\delta}\right). \tag{16.6}$$

From
$$W^{1/2} q^{1/2-\delta} |\mathcal{M}|^{-1} = O\left(q^{-\delta/2}\right) = o(2^{-\log^{1/2} q})$$

and Lemma 4.1 we conclude that for sufficiently large q the sequence $t(k,\mu)$, $(k,\mu) \in \mathcal{S}$ is $2^{-\log^{1/2} q}$-homogeneously distributed modulo q.

We choose $k \in [0, q-1]$ and $\mu \in \mathcal{M}$ independently and uniformly at random and ignore pairs $(k,\mu) \notin \mathcal{S}$. It follows from (16.4) that the expected number of choices in order to get d pairs $(k,\mu) \in \mathcal{S}$ is $d + O\left(dq^{-\delta}\right)$ for some $\delta > 0$ depending only on $\varepsilon > 0$.

Now, using the relation (16.3) and Lemma 6.3 we obtain the result. □

The papers [414, 417] also contain results of numerical experiments which imply that for practical values of the parameters even three least significant bits are enough to recover the private key. Certainly these results are better than the theoretical prediction which is not so surprising if one recalls that lattice reduction algorithms usually perform much better than their theoretical estimates guarantee.

Similar arguments work if, more generally, we are given consecutive bits at a known position. For instance, if we are given the ℓ most significant bits of k, that is, an element a such that $0 \le k - a \le q/2^\ell$, then we have

$$\lfloor \alpha T(k,\mu) - U(k,\mu) \rfloor_q < q/2^\ell$$

with

$$T(k,\mu) = \left\lfloor r(k) s(k,\mu)^{-1} \right\rfloor_q, \qquad U(k,\mu) = \left\lfloor \left(a - s(k,\mu)^{-1}\right) h(\mu) \right\rfloor_q.$$

So the result remains exactly the same for most significant bits. However, for oracles returning ℓ consecutive bits in the middle, one requires twice as many bits, see [417] for details.

In [166] the same method has been applied to a number of other similar signature schemes including the Nyberg–Rueppel scheme. However, although for many schemes the results are of the same strength as that of Theorem 16.1, for the Nyberg–Rueppel scheme the result is a little weaker. In particular, that result, although requiring the same number of bits of k and the same number of signatures, applies only to $q \geq p^{2/3+\varepsilon}$.

A different lattice reduction based attack on the Digital Signature Algorithm is given in [46]. It requires many more bits of the nonce k but may apply to fewer signatures; in some cases to just one signature. A certain elliptic curve cryptosystem, so-called MQV has been attacked in [325] by using similar ideas.

Generally, any leakage of information on k could prove dramatic. One can certainly ask how practical is our assumption that several bits of the nonce k become known. Here we outline several possible scenarios:

- One of the situations where this may happen is the case of using a biased pseudorandom number generator. Then, although bits of k are immediately leaked, one can try to "guess" them and then run the above attack.
- We also observe that if for efficiency reasons, one chooses a nonce k with fewer bits than q (to speed up the exponentiation), then our attack obviously applies. This improves a very different heuristic attack on very small values of $k = O(q^{1/2})$ that has recently been described in [318]. We remark that one can be tempted to use k of this size because standard "square-root" attacks, such as Shanks and Pollard attacks [389, 541], do not apply to inverting $r(k)$.
- Probably most of the danger comes from the *power* or *timing* analysis. In the signing algorithm, exponentiation is the only operation which is sensitive to the size of the nonce k. So, the attacker can just choose the signatures whose generation has taken less time than "on average". Those are likely to correspond to small values of k, thus most significant bits of k become known (they are simply zeros).

It is natural to assume that similar results can also be obtained for the ElGamal signature scheme. In fact, in Chapter 17 we show that the corresponding signature satisfies a much stronger uniform distribution property than what we have for the Digital Signature Algorithm. However, to apply our attack one needs to study the hidden number problem modulo a composite number, which has not yet been worked out.

Finally, the results and ideas of [417] which we have described in this chapter has recently been used in [78] to design an attack on another DSA-based cryptosystem. It is shown in [78] that in the above cryptosystem there is a way to extract all necessary information from the protocol itself, thus no additional "leakage" is assumed. In fact, the bound (16.6) allows us to make the attack of [78] rigorously proved and also to extend it to other small subgroups of $\mathbb{F}_p^*$ (not only those with a power of 2 elements as in [78]).

Chapter 17

Distribution of the ElGamal Signature

Here we present the result of [522] about the uniformity of distribution of pairs (u, v) generated by the ElGamal signature scheme which can be described in the following way. Let $\mathcal{M}$ be a finite set of messages to be signed and let $h : \mathcal{M} \to \mathbb{Z}_{p-1}$ be an arbitrary hash-function.

We assume that the primitive root g is *publicly* known.

Fix a certain element $\alpha \in \mathbb{Z}_{p-1}^*$ which is the *secret* key known only to the signer, and for an integer $k \in \mathbb{Z}_{p-1}^*$ called (as before) a *nonce* and a *message* $\mu \in \mathcal{M}$ we define the following two functions $v(k, \mu)$ by the conditions

$$u(k) = \lfloor g^x \rfloor_p \qquad \text{and} \qquad v(k, \mu) = \left\lfloor vk^{-1} \left(h(\mu) + \alpha u(k)\right)\right\rfloor_{p-1}$$

and call the pair $(u(k), v(k, \mu))$ the *ElGamal signature* of the message μ with nonce k, see [389, 541].

Clearly, the ElGamal signature scheme is very similar to the Digital Signature Algorithm. The most important distinction is that in the case of ElGamal signature there is no double reduction (modulo p and then modulo q) which occurs in the definition of $r(k)$ in the the Digital Signature Algorithm.

Let ρ be the probability of collision of the hash function h, that is,

$$\rho = \frac{W}{|\mathcal{M}|^2},$$

where W is the number of pairs $(\mu_1, \mu_2) \in \mathcal{M}^2$ with $h(\mu_1) = h(\mu_2)$.

Theorem 17.1. *For the discrepancy $D(\mathcal{S})$ of the set of $\varphi(p-1)|\mathcal{M}|$ points*

$$\mathcal{S} = \left\{ \left(\frac{u(k)}{p}, \frac{v(k, \mu)}{p-1} \right) \mid k \in \mathbb{Z}_{p-1}^*, \ \mu \in \mathcal{M} \right\}$$

the bound

$$D(\mathcal{S}) = O\left(\rho^{1/2}p^{\varepsilon}\right)$$

holds.

Proof. We define and estimate the sums

$$S(a,b) = \sum_{k\in \mathbb{Z}_{p-1}^*} \sum_{\mu\in\mathcal{M}} \mathbf{e}_p\left(au(k)\right) \mathbf{e}_{p-1}\left(bv(k,\mu)\right).$$

If $b=0$, then we have

$$S(a,0) = |\mathcal{M}| \sum_{k\in \mathbb{Z}_{p-1}^*} \mathbf{e}_p\left(au(k)\right) = |\mathcal{M}| \sum_{k\in \mathbb{Z}_{p-1}^*} \mathbf{e}_p\left(ag^k\right) \tag{17.1}$$

and thus Lemma 3.20 can be applied.
Now we assume that $a \in \mathbb{Z}_p$, $b \in \mathbb{Z}_{p-1}$ and $b \neq 0$.
We have

$$\begin{aligned} |S(a,b)| &= \left| \sum_{k\in \mathbb{Z}_{p-1}^*} \mathbf{e}_p\left(au(k)\right) \mathbf{e}_{p-1}\left(b\alpha k^{-1}u(k)\right) \sum_{\mu\in\mathcal{M}} \mathbf{e}_{p-1}\left(bk^{-1}h(\mu)\right) \right| \\ &\leq \sum_{k\in \mathbb{Z}_{p-1}^*} \left| \sum_{\mu\in\mathcal{M}} \mathbf{e}_{p-1}\left(bk^{-1}h(\mu)\right) \right| \\ &= \sum_{k\in \mathbb{Z}_{p-1}^*} \left| \sum_{\mu\in\mathcal{M}} \mathbf{e}_{p-1}\left(bkh(\mu)\right) \right|. \end{aligned}$$

Applying the Cauchy inequality, we obtain

$$|S(a,b)|^2 \leq \varphi(p-1) \sum_{k\in \mathbb{Z}_{p-1}^*} \left| \sum_{\mu\in\mathcal{M}} \mathbf{e}_{p-1}\left(bkh(\mu)\right) \right|^2.$$

Extending the summation over k to the whole set $\mathbb{Z}_{p-1}$ and using that

$$\left| \sum_{\mu\in\mathcal{M}} \mathbf{e}_{p-1}\left(bkh(\mu)\right) \right|^2 = \sum_{\mu_1,\mu_2\in\mathcal{M}} \mathbf{e}_{p-1}\left(bk\left(h(\mu_1)-h(\mu_2)\right)\right),$$

we derive

$$|S(a,b)|^2 \leq \varphi(p-1) \sum_{\mu_1,\mu_2\in\mathcal{M}} \sum_{k\in \mathbb{Z}_{p-1}} \mathbf{e}_{p-1}\left(bk\left(h(\mu_1)-h(\mu_2)\right)\right).$$

From (3.2) we conclude that

$$|S(a,b)|^2 \le \varphi(p-1)(p-1)V, \tag{17.2}$$

where V is the number of solutions of the congruence

$$b\,(h(\mu_1) - h(\mu_2)) \equiv 0 \pmod{p-1}, \qquad \mu_1, \mu_2 \in \mathcal{M},$$

or equivalently of the congruence

$$h(\mu_1) \equiv h(\mu_2) \pmod{(p-1)/d}, \qquad \mu_1, \mu_2 \in \mathcal{M},$$

where $d = \gcd(b, p-1)$. Therefore

$$V \le \sum_{j=0}^{d-1} U_j,$$

where U_j is the number of solutions of the congruence

$$h(\mu_1) \equiv h(\mu_2) + j(p-1)/d \pmod{p-1}, \qquad \mu_1, \mu_2 \in \mathcal{M}.$$

In particular $U_0 = W$. Using (3.2) we derive

$$\begin{aligned}
U_j &= \sum_{\mu_1,\mu_2\in\mathcal{M}} \frac{1}{p-1}\sum_{c=0}^{p-2} \mathbf{e}_{p-1}\left(c\,(h(\mu_1) - h(\mu_2) - j(p-1)/d)\right)\\
&= \frac{1}{p-1}\sum_{c=0}^{p-2} \mathbf{e}_{p-1}\left(-cj(p-1)/d\right) \sum_{\mu_1,\mu_2\in\mathcal{M}} \mathbf{e}_{p-1}\left(c\,(h(\mu_1) - h(\mu_2))\right)\\
&= \frac{1}{p-1}\sum_{c=0}^{p-2} \mathbf{e}_{p-1}\left(-cj(p-1)/d\right) \left|\sum_{\mu\in\mathcal{M}} \mathbf{e}_{p-1}\left(ch(\mu)\right)\right|^2\\
&\le \frac{1}{p-1}\sum_{c=0}^{p-2} \left|\sum_{\mu\in\mathcal{M}} \mathbf{e}_{p-1}\left(ch(\mu)\right)\right|^2 = U_0.
\end{aligned}$$

Therefore $V \le dW$ and from (17.2) we obtain the bound

$$|S(a,b)| \le \left(d\varphi(p-1)(p-1)W\right)^{1/2}. \tag{17.3}$$

Select $N = 2$ and $L = p-1$ in Lemma 4.1. If $b = 0$, then $a \ne 0$ and we apply Lemma 3.20 together with (17.1). For $b \ne 0$ we apply (17.3), getting

$$\begin{aligned}
D(\mathcal{S}) &\ll \frac{1}{p} + \frac{2^{\nu(p-1)} p^{1/2} \log p}{\varphi(p-1)} + \frac{((p-1)W)^{1/2}}{\varphi(p-1)^{1/2}|\mathcal{M}|} \sum_{d|p-1} d^{1/2}\\
&\qquad \times \sum_{0\le|a|<p-1} \sum_{\substack{0<|b|<p-1\\ d|b}} \frac{1}{\max\{|a|,1\}\max\{|b|,1\}}.
\end{aligned}$$

Obviously the first term never dominates, thus, substituting $b = cd$, we obtain

$$\begin{aligned} D(\mathcal{S}) &\ll \frac{2^{\nu(p-1)} p^{1/2} \log p}{\varphi(p-1)} + \frac{((p-1)W)^{1/2}}{\varphi(p-1)^{1/2}|\mathcal{M}|} \sum_{d|p-1} \frac{1}{d^{1/2}} \\ &\quad \times \sum_{0 \le |a| < p-1} \sum_{0 < |c| < (p-1)/d} \frac{1}{\max\{|a|, 1\} \max\{|c|, 1\}} \\ &\ll \frac{2^{\nu(p-1)} p^{1/2} \log p}{\varphi(p-1)} + \frac{(p-1)^{1/2} W^{1/2} \tau(p-1) \log^2 p}{\varphi(p-1)^{1/2}|\mathcal{M}|}. \end{aligned}$$

From (5.1) and (5.2) we obtain

$$D(\mathcal{S}) = O\left(p^{-1/2+\varepsilon} + \rho^{1/2} p^{\varepsilon}\right). \tag{17.4}$$

We remark that $\rho \ge 1/(p-1)$. Indeed, let $H(\lambda)$ be the number of messages $\mu \in \mathcal{M}$ with $h(\mu) = \lambda$. Then

$$\sum_{\lambda \in \mathbb{Z}_{p-1}} H(\lambda) = |\mathcal{M}| \qquad \text{and} \qquad \sum_{\lambda \in \mathbb{Z}_{p-1}} H(\lambda)^2 = W.$$

Thus from the Cauchy inequality we derive

$$|\mathcal{M}|^2 = \left(\sum_{\lambda \in \mathbb{Z}_{p-1}} H(\lambda) \right)^2 \le (p-1) \sum_{\lambda \in \mathbb{Z}_{p-1}} H(\lambda)^2 = (p-1)W.$$

Hence the first term in (17.4) never dominates and the result follows. □

First of all we note that under the reasonable assumption that the probability of collision is of "correct" order $\rho = O(1/p)$, then the bound of Theorem 17.1 is of the form $O\left(p^{-1/2+\varepsilon}\right)$.

On the other hand, Theorem 17.1 remains non-trivial even for very "weak" hash functions with probability of collision $\rho = O(p^{-\delta})$ for any $\delta > 0$.

Finally we remark that similar results can be obtained for several other generalisations of the ElGamal signature scheme, see Section 11.5.2 of [389].

Unfortunately it is not clear how to study the distribution of the pair $(r(k), s(k, \mu))$ arising in the Digital Signature Algorithm, see Chapter 16. Actually studying only the first component $r(k)$ already seems to be very hard.

Chapter 18

Bit Security of the RSA Encryption and the Shamir Message Passing Scheme

We have already mentioned that there exists a variety of very strong results (stronger than for almost all other cryptosystems) on bit security, see [182, 220] and references therein. Here we show that the methods of Chapter 6, more specifically Lemma 6.4, allow us to obtain several new results.

Informally, our result demonstrates that using RSA to send reasonably short strings, padded with random noise, is as secure against *Las Vegas* type attacks as the generic RSA encryption.

Let $m = pl$ be an RSA modulus, which is a product of two primes p and l. As usual, the public (encryption) exponent e and the private (decryption) exponent d, are chosen to satisfy

$$ed \equiv 1 \pmod{\varphi(m)}, \qquad 1 \le e, d < \varphi(m).$$

For each integer k and an arbitrary $A > 0$ we define the oracle $\mathcal{RSA}_{k,A}$ as a "black box" which, given the value of $y = \lfloor x^e \rfloor_m$, with $x \in \mathbb{Z}_m^*$, outputs the value of $\mathrm{MSB}_{k,m}(x)$ for at least $m/\log^A m$ values of $x \in \mathbb{Z}_m^*$, and outputs an error message for other $x \in \mathbb{Z}_m^*$.

That is, $\mathcal{RSA}_{k,A}$ is a Las Vegas type oracle which outputs some useful information polynomially often and never gives a wrong answer (but rather gives no answer at all).

Very roughly the main difference between $\mathcal{RSA}_{k,A}$ and other oracles considered in the literature can be outlined as follows. The oracle $\mathcal{RSA}_{k,A}$ provides some useful information for only very small number of inputs, while in other works the corresponding oracles give you some information (at least in a probabilistic sense) for all inputs. On the other hand, $\mathcal{RSA}_{k,A}$ returns many more bits that the aforementioned oracles, see [182, 220] for details.

Theorem 18.1. *Let* $k = \left\lceil \log^{1/2} m \right\rceil + \lceil \log\log m \rceil$ *and let* $ed \equiv 1 \pmod{\varphi(m)}$. *For any* $\varepsilon > 0$ *and sufficiently large* m, *there exists a probabilistic polynomial time algorithm which for any* $\alpha \in \mathbb{Z}_m^*$, *given the value of* $\beta = \lfloor \alpha^e \rfloor_m$ *makes the expected number of* $O\left(\log^{A+1/2} m\right)$ *calls of the oracle* $\mathcal{RSA}_{k,A}$ *and computes* α *correctly with probability* $1 + O\left(m^{-1+\varepsilon}\right)$.

Proof. Let $\mathcal{X} \subseteq \mathbb{Z}_m^*$ be the set of $x \in \mathbb{Z}_m^*$ for which, given $y = \lfloor x^e \rfloor_m$, the oracle $\mathcal{RSA}_{k,A}$ returns $\mathrm{MSB}_{k,m}(x)$. Given $\beta = \lfloor \alpha^e \rfloor_m$, let us select an integer $t \in \mathbb{Z}_m^*$ uniformly at random and feed $\lfloor \beta t^e \rfloor_m = \lfloor (\alpha t)^e \rfloor_m$ the oracle $\mathcal{RSA}_{k,A}$. After the expected number of $O\left(\log^{A+1/2} m\right)$ calls, we obtain at least $f = 4\left\lceil \log^{1/2} m \right\rceil$ integers $t_1, \ldots, t_f \in \mathbb{Z}_m^*$ for which $\lfloor \alpha t_1 \rfloor_m, \ldots, \lfloor \alpha t_f \rfloor_m \in \mathcal{X}$, and thus the oracle $\mathcal{RSA}_{k,A}$ returns $\mathrm{MSB}_{k,m}(\alpha t_1), \ldots, \mathrm{MSB}_{k,m}(\alpha t_f)$.

Lemma 6.4 asserts that the number of d-tuples $(t_1, \ldots, t_f) \in \left(\mathbb{Z}_m^*\right)^d$ for which its algorithm fails to recover $a \in \mathbb{Z}_m^*$ with $\gcd(a - \alpha, m) > 1$ from the values of $\mathrm{MSB}_{k,m}(\alpha t_1), \ldots, \mathrm{MSB}_{k,m}(\alpha t_f)$ is at most $O\left(m^{f-1}\right)$ (certainly $\mathbb{Z}_m^*$ has the required uniformity of distribution property, see Lemma 5.4). We have $|\mathcal{X}| \geq m/\log^A m$. Thus the probability that the algorithm succeeds is at least

$$\frac{|\mathcal{X}|^f + O\left(m^{f-1}\right)}{|\mathcal{X}|^f} = 1 + O\left(m^{-1} \log^{fA} m\right) = 1 + O\left(m^{-1/2}\right).$$

Now, we evaluate $b = \lfloor a^e \rfloor_m$ and $\gcd(b - \beta, m)$. Since $\gcd(a - \alpha, m) > 1$ we also have $\gcd(b - \beta, m) > 1$. If $\gcd(b - \beta, m) = m$ then we are done. Otherwise we obtain a nontrivial divisor of m, that is, either p or l, thus we can compute $\varphi(m)$ and find the decryption exponent d. □

Now assume we use RSA to pass a "short" message μ (for example to exchange a private key for a symmetric cryptosystem) of length $k \geq \left\lceil \log^{1/2} m \right\rceil + \lceil \log\log m \rceil$, which we pad with a random binary string π of length $h = \lfloor \log m \rfloor - k$. Assume that there is a Las Vegas algorithm which is able to recover the message μ for at least $m/\log^A m$ such pairs (μ, π). Then Theorem 18.1 claims that one can efficiently invert the RSA encryption.

Now we present the result of [223] on the bit security of the Shamir message passing scheme which is an analogue of Theorems 14.1 and 14.2. A similar result has been claimed in Theorem 3 of [64] but unfortunately the proof is not quite correct (Theorem 1 of [64] cannot be applied to multipliers arising in the proof of Theorem 3 because they are selected from a certain subset of $\mathbb{F}_p^*$ rather than from the whole set, as in Theorem 1).

The *Shamir message passing scheme* can be described in the following way, see [64], as well as Protocol 12.22 from [389].

To send a message $m \in [0, p-1]$ from *Alice* to *Bob*:

- *Alice* selects a random $a \in [0, p-2]$ with $\gcd(a, p-1) = 1$, computes $A = \lfloor m^a \rfloor_p$ and sends A to *Bob*.
- *Bob* selects a random $b \in [0, p-2]$ with $\gcd(b, p-1) = 1$, computes $B = \lfloor A^b \rfloor_p$ and sends B to *Alice*.
- *Alice* finds $u \in [0, p-2]$ satisfying the congruence $au \equiv 1 \pmod{p-1}$, computes $C = \lfloor B^u \rfloor_p$ and sends C to *Bob*.
- *Bob* finds $v \in [0, p-2]$ satisfying the congruence $bv \equiv 1 \pmod{p-1}$ and computes $m = \lfloor C^v \rfloor_p$.

For an integer $k \geq 1$ we suppose that we are given an oracle $\mathcal{SH}_k$ such that for any given values of X, Y and Z it outputs the k most significant bits of m if the triple (X, Y, Z) corresponds to a proper usage of the Shamir message passing scheme and an error message, otherwise.

More precisely, given X, Y and Z the oracle $\mathcal{SH}_k$ outputs:

- $\mathrm{MSB}_{k,p}(m)$, if

$$X = \lfloor m^x \rfloor_p, \qquad Y = \lfloor X^y \rfloor_p, \qquad Z = \lfloor Y^w \rfloor_p,$$

  where $xw \equiv 1 \pmod{p-1}$ for some $m \in [1, p-1]$ and $x, y \in [0, p-2]$ with $\gcd(xy, p-1) = 1$ and the function $\mathrm{MSB}_{k,p}$ is defined by (1.1);
- an error message, otherwise.

Theorem 18.2. *Assume that we are given an oracle $\mathcal{SH}_k$ as above, with*

$$k = \left\lceil \log^{1/2} p \right\rceil + \lceil \log \log p \rceil .$$

Then there exists a probabilistic polynomial time algorithm which computes the message m, for all except $O\left(p^{1/2+\varepsilon}\right)$ messages $m \in [1, p-1]$, from the values of $A = \lfloor m^a \rfloor_p$, $B = \lfloor A^b \rfloor_p$ and $C = \lfloor B^u \rfloor_p$, where $a, b \in [0, p-2]$ with $\gcd(ab, p-1) = 1$ and $au \equiv 1 \pmod{p-1}$, which uses the expected number of $O\left(\log^{1/2} p \log \log p\right)$ calls of the oracle $\mathcal{SH}_k$.

Proof. We exclude from consideration the messages $m \in [1, p-1]$ of multiplicative order less than $p^{1/2+\varepsilon/2}$. Obviously, the number E of such excluded messages does not exceed

$$E \le \tau(p-1)p^{1/2+\varepsilon/2}. \tag{18.1}$$

Indeed, for any divisor $D|p-1$ there are at most D values of $m \in [1, p-1]$ of multiplicative order D. Using the bound (5.1) we obtain from (18.1) that the exceptional set is of size $E = O\left(p^{1/2+\varepsilon}\right)$.

Let us consider a message m of multiplicative order $T \ge p^{1/2+\varepsilon/2}$.

We denote by $\mathcal{X}_b$ the set of integers $x \in [0, p-2]$ with $\gcd(bx+1, p-1) = 1$. In particular, because $\gcd(b, p-1) = 1$ we have $|\mathcal{X}_b| = \varphi(p-1)$.

For $x \in \mathcal{X}_b$ we put $a_x = a$ and define $b_x \in [1, p-2]$ from the congruence

$$b_x(bx+1) \equiv b \pmod{p-1}.$$

We also put

$$m_x = \lfloor m^{1+bx} \rfloor_p.$$

We remark that

$$A_x = \lfloor m_x^{a_x} \rfloor_p, \qquad B_x = \lfloor m_x^{a_x b_x} \rfloor_p, \qquad C_x = \lfloor m_x^{b_x} \rfloor_p$$

can be computed as

$$A_x = \lfloor AB^x \rfloor_p, \qquad B_x = B, \qquad C_x = C.$$

Although the value of b is not known, one can select elements $x \in \mathcal{X}_b$ uniformly and independently at random by querying the oracle $\mathcal{SH}_k$ with the triples (A_x, B_x, C_x) where the elements x are selected uniformly and independently at random in the interval $[0, p-2]$. If $\gcd(bx+1, p-1) > 1$ the oracle returns an error message, otherwise $x \in \mathcal{X}_b$. Now we choose $d = 2\left\lceil \log^{1/2} p \right\rceil$ elements $x_1, \ldots, x_d \in \mathcal{X}_b$ uniformly and independently at random. By the bound (5.1),

$$\frac{p-1}{|\mathcal{X}_b|} = \frac{p-1}{\varphi(p-1)} = O(\log\log p).$$

Hence the expected number of choices of $x \in [0, p-2]$ before we get d elements in $\mathcal{X}_b$ is $O(d \log\log p) = O\left(\log^{1/2} p \log\log p\right)$. We remark that these elements are independent and uniformly distributed in $\mathcal{X}_b$. Moreover, every output of the oracle provides k most significant bits of m_x.

Remarking that $m_x \equiv mB^x \pmod{p}$ and that $B \equiv m^{ab} \pmod{p}$ is of multiplicative order $T \ge p^{1/2+\varepsilon/2}$ (because $\gcd(ab, p-1) = 1$), we derive from the bound (5.2) and Lemmas 3.18 and 4.1 (applied with $N = 1$, $L = p-1$) that the sequence

$$B^x, \qquad x \in \mathcal{X}_b,$$

is $p^{-\delta}$-homogeneously distributed modulo p for some $\delta > 0$, depending only on ε. Therefore Lemma 6.3 applies and the result follows. □

We also remark that one can consider an oracle which instead of returning an error message for "inconsistent" inputs (A, B, C), returns just a random element from $\mathbb{F}_p$. In this case repeating each query twice one can easily distinguish between an $x \in \mathcal{X}_b$ and other values.

Similar results can be also be obtained for other exponentiation based cryptosystems, such as the ElGamal cryptosystem, see Section 8.4 of [389] or Section 5.1 of [541], Matsumoto–Takachima–Imai key-agreement protocol, see Section 12.6 of [389] or Section 8.4.2 of [541] and other schemes mentioned in [64, 65].

Chapter 19

Bit Security of the XTR and LUC Secret Keys

Here we obtain an analogue of Theorems 14.1, 14.2, 14.3 and 18.2 for the bit security of the key of the XTR key exchange protocol which has recently been introduced and studied in [332, 333, 334, 335, 551] and which is a computationally more efficient modification of some ideas of [76]. We also study the LUC key exchange protocol [47, 536]. This chapter is based on the results obtained in [342, 518], see also [521].

We do not give a description of the XTR key exchange protocol but rather present a brief outline of the XTR settings and give an explicit formula for the secret key, concentrating only on the details which are relevant to our results.

Let $g \in \mathbb{F}_{p^6}$ be of multiplicative order l dividing $p^2 - p + 1$.

We recall that the trace satisfies

$$\mathrm{Tr}_{\mathbb{F}_{p^2}/\mathbb{F}_p}\left(\mathrm{Tr}_{\mathbb{F}_{p^6}/\mathbb{F}_{p^2}}(u)\right) = \mathrm{Tr}_{\mathbb{F}_{p^6}/\mathbb{F}_p}(u)$$

for $u \in \mathbb{F}_{p^6}$.

The idea of the XTR key exchange protocol is based on the observation that for some specially selected element $g \in \mathbb{F}_{p^6}^*$, which we call *the XTR generator* of prime multiplicative order $l > 3$ such that

$$l | p^2 - p + 1, \tag{19.1}$$

one can efficiently compute $\mathrm{Tr}_{\mathbb{F}_{p^6}/\mathbb{F}_{p^2}}(g^{xy})$ from the values of x and $\mathrm{Tr}_{\mathbb{F}_{p^6}/\mathbb{F}_{p^2}}(g^y)$ or, alternatively, from the values of y and $\mathrm{Tr}_{\mathbb{F}_{p^6}/\mathbb{F}_{p^2}}(g^x)$. This allows us to reduce the size of messages to exchange. Namely, the communicating parties exchange only $\mathrm{Tr}_{\mathbb{F}_{p^6}/\mathbb{F}_{p^2}}(g^x)$ and $\mathrm{Tr}_{\mathbb{F}_{p^6}/\mathbb{F}_{p^2}}(g^y)$ (which can be encoded with about $2\log p$

bits each) rather than g^x and g^y (which require about $6 \log p$ bits each) in order to create a common XTR secret key $\mathrm{Tr}_{\mathbb{F}_{p^6}/\mathbb{F}_{p^2}}(g^{xy})$.

As it follows from Theorem 24 of [550] (see also [76, 332, 335]) any polynomial time algorithm to compute $\mathrm{Tr}_{\mathbb{F}_{p^6}/\mathbb{F}_{p^2}}(g^{xy})$ from g^x and g^y can be used to construct a polynomial time algorithm to compute g^{xy} from the same information.

Let ϑ be a fixed root of an irreducible quadratic polynomial over $\mathbb{F}_p$. We recall that any element $v \in \mathbb{F}_{p^2}$ can be represented by the pair

$$\left(\mathrm{Tr}_{\mathbb{F}_{p^2}/\mathbb{F}_p}(v), \mathrm{Tr}_{\mathbb{F}_{p^2}/\mathbb{F}_p}(\vartheta v)\right).$$

Thus $\mathrm{Tr}_{\mathbb{F}_{p^6}/\mathbb{F}_p}(g^{xy})$ is a part of the representation of $\mathrm{Tr}_{\mathbb{F}_{p^6}/\mathbb{F}_{p^2}}(g^{xy})$.

Thus the above results suggest that breaking XTR is not easier than breaking the classical Diffie–Hellman key exchange protocol.

Here we obtain one more result of this kind and show that even computing a small number of bits of the XTR secret key of $\mathrm{Tr}_{\mathbb{F}_{p^6}/\mathbb{F}_p}(g^{xy})$ from the values of $\mathrm{Tr}_{\mathbb{F}_{p^6}/\mathbb{F}_{p^2}}(g^x)$ and $\mathrm{Tr}_{\mathbb{F}_{p^6}/\mathbb{F}_{p^2}}(g^y)$ is as hard as breaking the classical Diffie–Hellman scheme. In fact we prove a stronger statement that computing a certain positive proportion of bits of $\mathrm{Tr}_{\mathbb{F}_{p^6}/\mathbb{F}_p}(g^{xy})$ from the values of g^x and g^y is as hard as computing g^{xy} from these values.

For a positive integer k, let $\mathcal{XTR}_k$ denote an oracle that, given the values of g^x and g^y, outputs $\mathrm{MSB}_{k,p}\left(\mathrm{Tr}_{\mathbb{F}_{p^6}/\mathbb{F}_p}(g^{xy})\right)$, where the function $\mathrm{MSB}_{k,p}$ is defined by (1.1).

We remark that the oracle $\mathcal{XTR}_k$ is assumed to have more information than actually available in the XTR protocol (where only the values $\mathrm{Tr}_{\mathbb{F}_{p^6}/\mathbb{F}_{p^2}}(g^x)$ and $\mathrm{Tr}_{\mathbb{F}_{p^6}/\mathbb{F}_{p^2}}(g^y)$ are given). This means that even if more information becomes available, breaking the XTR protocol is still as hard as finding the Diffie–Hellman secret key in the corresponding group.

Theorem 19.1. *Let p be a sufficiently large n-bit prime number and let the multiplicative order l of the XTR generator g satisfy the inequality $l \geq p^{3/2+\varepsilon}$ for some fixed $\varepsilon > 0$. Then there exists a polynomial time algorithm which, given the values of $A = g^a$ and $B = g^b$, for some $a, b \in [0, l-1]$, makes $O(\log^{1/2} p)$ calls of the oracle $\mathcal{XTR}_k$ with $k = \left\lceil 2\log^{1/2} p \right\rceil$ and computes g^{ab} correctly with probability at least $1 - p^{-1}$.*

Proof. The case $a = 0$ is trivial. Now assume that $1 \leq a \leq l - 1$. Then $g_0 = g^a$ is an element of multiplicative order l (because l is prime).

Select a random element $v \in [0, l-1]$. Applying the oracle $\mathcal{XTR}_k$ to A and $V = g^{b+v} = Bg^v$ we obtain

$$\mathrm{MSB}_{k,p}\left(\mathrm{Tr}_{\mathbb{F}_{p^6}/\mathbb{F}_p}\left(g^{a(b+v)}\right)\right) = \mathrm{MSB}_{k,p}\left(\mathrm{Tr}_{\mathbb{F}_{p^6}/\mathbb{F}_p}\left(g^{ab}t\right)\right)$$

where $t = g_0^v$.

Selecting $d = O(\log^{1/2} p)$ such elements $v_1, \ldots, v_d \in [0, l-1]$ uniformly and independently at random we can now apply Lemma 6.6 with $\alpha = g^{ab}$, $m = 6$, and the group $\mathcal{G}$ generated by g_0 (which coincides with the group generated by g). Indeed, we see that

$$\mathrm{Nm}_{\mathbb{F}_{p^6}/\mathbb{F}_{p^3}}(g) = g^{1+p^3} = g^{(p+1)(p^2-p+1)} = 1$$

thus $\mathcal{G} \in \mathcal{N}_{6,2}(p)$ and from Lemma 6.6 we obtain the desired result. □

Theorem 19.1, obtained in [342], improves a much weaker result of [521] whose method is similar to that used in [520] for the problem of finding an m-sparse polynomial over $\mathbb{F}_p$ from approximations to its values at polynomially many points $t \in \mathbb{F}_p$ selected uniformly at random, see Chapter 30.

Moreover, exactly the same statement holds for an oracle which outputs the value of $\mathrm{MSB}_{k,p}\left(\mathrm{Tr}_{\mathbb{F}_{p^6}/\mathbb{F}_p}(\omega g^{xy})\right)$ with any fixed $\omega \in \mathbb{F}_{p^6}^*$.

We have already mentioned that Theorem 12.8 has direct implication for the XTR and LUC key exchange protocols as well. For the sake of completeness, here we formulate the corresponding result in a more explicit form (only for XTR).

Let $0 < \varepsilon \le 1$ and let $\widetilde{\mathcal{XTR}}_\varepsilon$ denote an oracle that, given the values of g^x and g^y, for each $x \in [0, l-1]$ it returns $\mathrm{Tr}_{\mathbb{F}_{p^6}/\mathbb{F}_p}(g^{xy})$, for at least εt values of $y \in [0, l-1]$ and returns an error message for other values of $y \in [0, l-1]$.

Theorem 19.2. *Assume that*

$$1 \ge \varepsilon \ge 6l^{-1/5}.$$

Given an oracle $\widetilde{\mathcal{XTR}}_\varepsilon$, there exists a probabilistic algorithm which, given g^x and g^y, makes the expected number of at most $12\varepsilon^{-1}$ calls of the oracle $\widetilde{\mathcal{XTR}}_\varepsilon$, executes polynomial number $(\log p)^{O(1)}$ arithmetic operations in $\mathbb{F}_q$ per each call and returns g^{xy} for all pairs $(x, y) \in [0, l-1]^2$.

Obviously the same comment about more general oracles as after the proof of Theorem 12.8 applies to this result as well.

Let $g \in \mathbb{F}_{p^2}$ be a root of an irreducible quadratic polynomial of the form $f(X) = X^2 - AX + 1 \in \mathbb{F}_p[X]$. It is an easy matter to show that such elements exist. For example, for any root $\vartheta \in \mathbb{F}_{p^2}$ of an arbitrary irreducible quadratic polynomial over $\mathbb{F}_p$, $g = \vartheta^{p-1}$ is such an element.

In particular, $g \in \mathbb{F}_{p^2}$ is of multiplicative order dividing $p+1$.

In the LUC key exchange protocol the communicating parties exchange the values $\mathrm{Tr}_{\mathbb{F}_{p^2}/\mathbb{F}_p}(g^x)$ and $\mathrm{Tr}_{\mathbb{F}_{p^2}/\mathbb{F}_p}(g^y)$ and then, using some simple identities, compute the common secret key $\mathrm{Tr}_{\mathbb{F}_{p^2}/\mathbb{F}_p}(g^{xy})$. The name of this scheme comes from the fact that the exchanged values are the xth and yth elements of the corresponding Lucas sequence. Some simple properties of Lucas sequences help to design a kind of the repeated squaring algorithm, thus both parties can compute $\mathrm{Tr}_{\mathbb{F}_{p^2}/\mathbb{F}_p}(g^{xy})$ in a very efficient fashion. For more details we refer to [47, 536].

For a positive integer k we denote by $\mathcal{LUC}_k$ the oracle such that for any given values of g^x and g^y, it outputs $\mathrm{MSB}_{k,p}\left(\mathrm{Tr}_{\mathbb{F}_{p^2}/\mathbb{F}_p}(g^{xy})\right)$.

As before, we remark that the oracle $\mathcal{LUC}_k$ is assumed to have more information than actually available in the LUC protocol (where only the values $\mathrm{Tr}_{\mathbb{F}_{p^2}/\mathbb{F}_p}(g^x)$ and $\mathrm{Tr}_{\mathbb{F}_{p^2}/\mathbb{F}_p}(g^y)$ are given).

The following result has been obtained in [342].

Theorem 19.3. *Let p be a sufficiently large n-bit prime number and let the multiplicative order T of the LUC generator g satisfy the inequality $T \ge p^{1/2+\varepsilon}$ for some fixed $\varepsilon > 0$. Then there exists a polynomial time algorithm which, given the values of $A = g^a$ and $B = g^b$, for some $a, b \in [0, T-1]$, makes $O(\log^{1/2} p)$ calls of the oracle $\mathcal{LUC}_k$ with $k = \left\lceil 2\log^{1/2} p \right\rceil$ and computes g^{ab} correctly with probability at least $1 + O\left(p^{-\varepsilon/2}\right)$.*

Proof. The proof is similar to the proof of Theorem 14.1. Given a pair $(a, b) \in [0, T-1]^2$ let us select an integer $u \in [0, T-1]$ uniformly at random. We compute

$$g_u = Bg^u = g^{b+u}.$$

The probability that

$$\gcd(b+u, T) \ge Tp^{-1/2-\varepsilon/4} \qquad (19.2)$$

is at most $\tau(T)T^{-1}p^{1/2+\varepsilon/4}$ where $\tau(T)$ is the number of positive integer divisors of T. Indeed, for any divisor $D|T$ with $D \ge Tp^{-1/2-\varepsilon/4}$ there are at most $T/D \le p^{1/2+\varepsilon/4}$ values of $s \in [0, T-1]$ with $\gcd(s, T) = D$.

Using the bound (5.1), we obtain that the probability of (19.2) is at most

$$O\left(T^{-1}p^{1/2+2\varepsilon/4}\right) = O\left(p^{-\varepsilon/2}\right).$$

In the opposite case, when $\gcd(a+u, T) < Tp^{-1/2-\varepsilon/4}$, the multiplicative order of g_u is

$$T_u = \frac{T}{\gcd(b+u, T)} \ge p^{1/2+\varepsilon/4}.$$

Let $\alpha_u = g^{a(b+u)}$. For an integer x, applying the oracle $\mathcal{LUC}_k$ to $g^x A$ and $g^u B$ we obtain

$$\mathrm{MSB}_{k,p}\left(\mathrm{Tr}_{\mathbb{F}_{p^2}/\mathbb{F}_p}\left(g^{(a+x)(b+u)}\right)\right) = \mathrm{MSB}_{k,p}\left(\mathrm{Tr}_{\mathbb{F}_{p^2}/\mathbb{F}_p}\left(\alpha_u g_u^x\right)\right).$$

Choosing this integer x uniformly at random in the interval $[0, T-1]$ we see that because $T_u | p-1$, the values of residues of x modulo T_u are uniformly distributed in the interval $[0, T_u - 1]$ as well. Thus choosing $d = O(\log^{1/2} p)$ such elements $x_1, \ldots, x_d \in [0, T-1]$ uniformly and independently at random we can now recover α_u by applying the algorithm of Lemma 6.6 with the group $\mathcal{G}$ generated by g_u. Indeed, we see that

$$\mathrm{Nm}_{\mathbb{F}_{p^2}/\mathbb{F}_p}(g_u) = g_u^{1+p} = g^{(1+p)(b+u)} = 1$$

thus $\mathcal{G} \in \mathcal{N}_{2,1}(p)$. Finally, we can compute $\alpha = \alpha_u A^{-u}$. □

Some other results about algebraic properties of the XTR key exchange protocol are given in [324].

Chapter 20

Bit Security of NTRU

We recall that in the *NTRU cryptosystem* [267, 269], one selects integer parameters (N, q) and four sets $\mathcal{L}_f, \mathcal{L}_g, \mathcal{L}_\varphi, \mathcal{L}_m$ of polynomials f, g, φ, m in the ring $\mathcal{R} = \mathbb{Z}_q[X]/(X^N - 1)$. The coefficients of these polynomials are constrained by the choice of an additional parameter, p, a small integer or polynomial. In the original presentation of NTRU, see [267], the choice $p = 3$ is considered, and thus the polynomials f, g, φ, m are ternary. The authors of NTRU have since recommended [269] that to choose $p = X + 2$. This choice for p, along with other optimisations suggested in [269], leads to f, g, φ, m being constructed from binary polynomials. In particular this leads to the message representative polynomial m being binary.

We denote by $\odot$ the operation of multiplication in the ring $\mathcal{R}$.

We also recall the *key creation* procedure where the *receiver Alice* randomly selects polynomials $f \in \mathcal{L}_f$ and $g \in \mathcal{L}_g$ such that f has an inverse modulo q and modulo p (this issue has been discussed in [269, 535] together with heuristic estimates for a polynomial $f \in \mathcal{L}_f$ to satisfy these conditions).

Then *Alice* computes inverses f_q^* and f_p^* that satisfy

$$f \odot f_q^* \equiv 1 \pmod q, \qquad f \odot f_p^* \equiv 1 \pmod p,$$

and the product

$$h \equiv p \odot f_q^* \odot g \pmod q.$$

Alice publishes the polynomial h as her public key, retaining f as her private key. The polynomial f_p^* should also be stored for later use, and the polynomials f_q^*, g may be discarded. Note that the paper [269] recommends that f have the form $1 + p \odot F$ with $F \in \mathcal{R}$; in this case, f_p^* will always be 1.

Suppose the *sender Bob* wants to *encrypt* and send a secret message to *Alice*. *Bob* selects a message m from the set of plaintexts $\mathcal{L}_m$. Next, *Bob* selects a random "blinding" polynomial $\varphi \in \mathcal{L}_\varphi$ and uses the public key h to compute

$$e \equiv \varphi \odot h + m \pmod q. \tag{20.1}$$

Bob then transmits e to *Alice.*

We do not discuss the decryption procedure and thus the structure on the sets $\mathcal{L}_f, \mathcal{L}_g, \mathcal{L}_\varphi, \mathcal{L}_m$ which maximizes the chance of unambiguous decryption (rather we refer to [267, 269] for details). However the set $\mathcal{L}_m$ plays a crucial role and we recall that $\mathcal{L}_m$ consists of 2^N polynomials with $0, 1$-coefficients, that is

$$\mathcal{L}_m = \left\{a_0 + a_1 X + \ldots + a_{N-1} X^{N-1} \mid a_0, a_1 \ldots, a_{N-1} \in \{0,1\}\right\}.$$

Thus each coefficient of $m \in \mathcal{L}_m$ carries one bit of information. We also use the fact that the set $\mathcal{L}_\varphi$ is closed under multiplication by X^i for $i = 0, 1, \ldots, N-1$.

Here we present several results of [411] which imply that NTRU encryption has a very attractive bit security property. Namely, each coefficient of the encrypted message m is as secure as the whole message.

Our method is based on the following observation: Given an encryption e of a message m, we derive from (20.1) that

$$e_i = X^i e \equiv X^i \odot \varphi \odot h + X^i \odot m \pmod q \tag{20.2}$$

is a valid encryption of the message $m_i = X^i \odot m$ corresponding to the choice of the "blinding" polynomial $\varphi_i = X^i \odot \varphi \in \mathcal{L}_\varphi$. It is easy to see that m_i is just a cyclic shift of the message m and that since X^i is invertible modulo $X^N - 1$, φ_i is uniformly distributed in $\mathcal{L}_\varphi$. Thus given an oracle returning a prescribed coefficient of the message from a given encryption, one queries this oracle with the values of e_i, $i = 0, 1, \ldots, N-1$, to recover all bits of m one by one.

Here we show that the same idea can be used for much weaker oracles which return the value of some coefficient of the message only for a very small proportion of the messages.

We remark that our arguments rely on the fact that the sets $\mathcal{L}_f, \mathcal{L}_g, \mathcal{L}_\varphi, \mathcal{L}_m$ which guarantee unambiguous decryption are public. In particular, this means that any "guess" for a message m can be efficiently verified by an attacker by checking whether $(e - m) \odot h^{-1} = \varphi$ with some $\varphi \in \mathcal{L}_\varphi$. We note that for the parameter sets given in [267, 269], h (equivalently, g) is not invertible. However, because of the structure of φ, it is possible to use a "pseudo-inverse" of h to perform the check [415]. In fact, even in the settings of [267, 269], h maybe chosen to be invertible modulo $(X^N - 1)/(X - 1)$ which is quite enough for our approach. Nevertheless, in order to demonstrate the underlying ideas in the simplest possible situation, we always assume that h is invertible.

We say that an algorithm is polynomial time if its running time is bounded by $(N \log q)^{O(1)}$. In fact, typically q is of the same order as N, so this last expression can be replaced by simply $N^{O(1)}$. We do not evaluate the exact complexities (which depend on the accepted model of computation) but this is a very simple, although somewhat tedious task.

Assume that we are given an oracle $\mathcal{NTRU}_\gamma$ which for any sequence of k *distinct* and *valid* encryptions $e_1, \dots, e_k$, returns the leading coefficient of the corresponding messages $m_1, \dots, m_k$ for at least $\lfloor \gamma k \rfloor$ queries and returns the error message symbol $*$ for the rest of the queries.

It is important to remark that repeatedly querying this oracle with the same message cannot help to extract any new information because it maintains the required rate of correct outputs only for distinct messages. In the same way, we do not specify the behaviour of this oracle on invalid queries e which do not correspond to any valid message. For example, for such queries it can output 0 and 1 with probability $\gamma/2$ each, and output $*$ with probability $1-\gamma$. Thus this oracle cannot be used to distinguish between valid and invalid encryptions.

On the other hand, we assume that it is a *Las Vegas* type oracle which never gives a wrong answer, although sometimes may give no answer at all. One could argue that this model is qualitatively stronger than the "classical" model, though on the other hand, it is quantitatively quite comparable as it only returns correct answers for some (non-negligible) γ-fraction of all queries. Moreover, as will be seen, the results obtained are stronger, since the reconstruction algorithm for the message m can be made fully deterministic.

It is known [207] that when N is composite then the whole scheme is not secure, thus we consider only the case of prime N.

For this and other algorithms our general strategy is to create several valid encryptions, by "flipping" already known (or "guessed") bits of m. If all of them are distinct, then each time the oracle provides new information to us. On the other hand, if not all of them are distinct then, as the following statement shows, we can recover m without even using the oracle. The above can be summarised in the following statement which could be of independent interest.

Theorem 20.1. *Let e be any NTRU encryption of a message $m \in \mathcal{L}_m$ and let $\{r_1, \dots, r_k\} \subset \mathcal{R}$ be such that for all $i = 1, \dots, k$, the polynomial $e + r_i$ is a valid encryption of $m + r_i \in \mathcal{L}_m$. Then, either*

- *the polynomials $X^j \odot (m + r_i)$, $j = 0, \dots, N-1$, $i = 1, \dots, k\}$ are all distinct,*

or,

- *given the polynomials $r_1, \dots, r_k$ one can recover m from e in polynomial time.*

Proof. We remark that if

$$X^{j_1} \odot (m + r_1) = X^{j_2} \odot (m + r_2) \tag{20.3}$$

for two distinct pairs $(j_1, r_1) \neq (j_2, r_2)$, then

$$\left(X^{j_1} - X^{j_2}\right) \odot m = X^{j_2} \odot r_2 - X^{j_1} \odot r_1.$$

If $j_1 = j_2$ then

$$X^{j_1} \odot (r_2 - r_1) = 0,$$

and because X and $X^N - 1$ are relatively prime, we obtain $r_1 = r_2$ which contradicts our assumption. For $j_1 \neq j_2$ we remark that the greatest common divisor of $X^{j_1} - X^{j_2}$ and $X^N - 1$ is $X - 1$. Therefore, each two distinct pairs $(j_1, r_1) \neq (j_2, r_2)$ uniquely define m modulo $(X^N - 1)/(X - 1)$. Because m has $0, 1$ coefficients there could be at most 2 possible values for m which can be found and verified by checking whether $(e - m) \odot h^{-1} \odot p^{-1} \in \mathcal{L}_\varphi$. □

Theorem 20.2. *Let N be a prime. For any constant $A > 0$, given an oracle $\mathcal{NTRU}_\gamma$ with any*

$$\gamma \geq N^{-A},$$

there exists a deterministic polynomial time algorithm which, given a valid encryption e of a message $m \in \mathcal{L}_m$, makes $O(\gamma^{-2}N)$ calls to the oracle $\mathcal{NTRU}_\gamma$ and finds the message m.

Proof. The algorithm below calls the oracle $\mathcal{NTRU}_\gamma$ with valid encryptions of the form $X^j \odot (e + r)$ which correspond to some unknown messages of the form $X^j \odot (m + r)$ for some polynomial $r \in \mathcal{R}$ (chosen in such a way that $X^j \odot (m + r) \in \mathcal{L}_m$, and thus $X^j \odot (e + r)$ is its valid encryption).

By Theorem 20.1, in the algorithm below we always assume that each message (of the form $X^j \odot (m + r)$ where j and $r \in \mathcal{R}$ are known) it produces is distinct from all the previously generated messages, otherwise the above procedure finds m immediately. That is, each time a new message is produced we apply the above procedure (for this message and all previously used messages) and either find m or make sure this new message has never occurred before. Thus the corresponding encryptions are pairwise distinct as well.

We put $L = \lceil \log \gamma^{-1} \rceil$ and "guess" the L highest coefficients of m. For each guess we execute the below procedure. We remark that the total number of guesses is $2^L = O(N^A)$ and also it is enough to consider only the case when our guess is correct. Indeed, as we have already remarked, for each guess, the candidate for the message m produced by the below algorithm can easily be verified. Thus it is enough to find the right message m at least once (that is, for the right guess).

So we now assume that L highest coefficients $a_{N-1}, \ldots, a_{N-L+1}$ of m are known.

To find the coefficient a_{N-i}, $i = L+1, \ldots, N$, we consider the following 2^L valid messages (which are flips and cyclic shifts of m)

$$m_i^{(\nu)} = X^{i-1} \odot \left(m - \sum_{j=N-L}^{N-1} (a_j - \nu_j) X^j \right)$$

where $\nu = (\nu_{N-L}, \ldots, \nu_{N-1})$ runs over all L dimensional binary vectors, and compute the corresponding encryptions

$$e_i^{(\nu)} = X^{i-1} \odot \left(e - \sum_{j=N-L}^{N-1} (a_j - \nu_j) X^j \right).$$

For each $i = L+1, \ldots, N$, the algorithm makes 2^L calls of the oracle $\mathcal{NTRU}_\gamma$ with $e_i^{(\nu)}$. By the conditions on the oracle $\mathcal{NTRU}_\gamma$, because $2^L\gamma \geq 1$, it either returns correctly a_i at least once or we obtain an equation of the form (20.3).

To estimate the number of calls to the oracle $\mathcal{NTRU}_\gamma$, we remark that for each of the 2^L guesses the algorithm makes $O\left(2^L N\right)$ calls. Therefore the total number of calls is $O\left(2^{2L} N\right) = O\left(\gamma^{-2} N\right)$. □

We now consider the more classical *Monte Carlo* type oracles, which always return a result which however is correct only with with probability bounded away from $1/2$. However, we shall make one additional assumption.

Namely, assume that we are given an oracle $\widetilde{\mathcal{NTRU}}_\varepsilon$ which for any sequence of k *distinct* and *valid* encryptions $e_1, \ldots, e_k$, returns the leading coefficient of the corresponding messages $m_1, \ldots, m_k$ for at least $(0.5 + \varepsilon)k + O(1)$ queries and returns an arbitrary value for other messages. Thus, the added assumption is that the advantage of the oracle is in a sense uniformly distributed over the sample space.

As before we assume that the oracle $\widetilde{\mathcal{NTRU}}_\varepsilon$ is consistent in its replies, it maintains the required rate of correct outputs only for distinct messages. Thus, asked several times with the same value of e, it returns the same value (correct or not) for the leading coefficient of the corresponding message m. Again, as before, we do not specify the behaviour of this oracle on invalid queries e which do not correspond to any valid message. For example, for such queries it can output 1 with probability $0.5 + \varepsilon$ and 0 with probability $0.5 - \varepsilon$. As in the case of the previous oracle, this oracle cannot be used to distinguish valid from invalid encryptions either.

Theorem 20.3. *Let N be a prime. For any constant $A > 0$, given an oracle $\widetilde{\mathcal{NTRU}}_\varepsilon$ with any*

$$\varepsilon \geq N^{-A},$$

there exists a deterministic polynomial time algorithm which, given a valid encryption e of a message $m \in \mathcal{L}_m$, makes $O(\varepsilon^{-2}N)$ calls of the oracle $\widetilde{\mathcal{NTRU}}_\varepsilon$ and finds the message m.

Proof. Let C be a constant such that the oracle $\widetilde{\mathcal{NTRU}}_\varepsilon$ returns the correct answer for at least $(0.5+\varepsilon)k - C$ pairwise distinct and correct queries (that is, C is the implicit constant in $O(1)$ in the definition of $\widetilde{\mathcal{NTRU}}_\varepsilon$).

Put $L = \lceil \log(C\varepsilon^{-1} + 2) \rceil$. We start with "guessing" the L highest coefficients of m. In fact, we repeat the below procedure with all $2^L = O(N^A)$ possible "guesses" and remark that if our guess is correct, then the algorithm below finds the encrypted message correctly. Thus we just verify the obtained $2^L = O(N^A)$ results in order to find which one corresponds to the encrypted message.

As in the proof of Theorem 20.2 we remark that by Theorem 20.1, all messages which appear in our algorithm are of the form $X^j \odot (m+r)$ where j and $r \in \mathcal{R}$ are known (and pairwise distinct). Thus by checking polynomially many "suspects" we can assume that all valid encryptions used in our algorithm correspond to distinct messages.

Now, for each "guess" $(a_{N-L}, \ldots, a_{N-1})$ of the L highest coefficients of the message m we consider $2^L - 1$ "flips" of the L highest coefficients

$$m^{(\nu)} = m - \sum_{j=N-L}^{N-1} (a_j - \nu_j) X^j$$

and compute the corresponding encryptions

$$e^{(\nu)} = e - \sum_{j=N-L}^{N-1} (a_j - \nu_j) X^j$$

defined by an L-dimensional non-zero binary vector $\nu = (\nu_{N-L}, \ldots, \nu_{N-1})$.

Then for each $i = L+1, \ldots, N$ for which the corresponding coefficient a_{N-i} is still undefined, we query the oracle with $2^L - 1$ distinct encryptions $e_i^{(\nu)} = X^{i-1} \odot e^{(\nu)}$, getting (assuming our original guess of the L highest coefficients is correct) the value of a_{N-i} at least

$$\begin{aligned}(0.5+\varepsilon)(2^L-1) - C &= 0.5(2^L-1) + \varepsilon(2^L-1) - C \\ &\geq 0.5(2^L-1) + \varepsilon(C\varepsilon^{-1}+1) - C > 0.5(2^L-1)\end{aligned}$$

times. Thus we make the majority decision on the value of a_i. We remark that we make an odd number of queries thus the majority decision procedure is always correctly defined (even when our original guess of the coefficients is not correct). □

Finally, we can indeed consider the classical model where the oracle gives us the leading coefficient of m with probability $0.5 + \delta$, where the probability is over random choices of m, φ and the (possible) internal random choices of the oracle. This complicates things since there could for example, be certain messages for which the oracle has no advantage at all. Still, we can obtain a (weak) non-trivial result as follows. Denote such an oracle by $\widehat{\mathcal{NTRU}}_\delta$.

Theorem 20.4. *Let $0 < \alpha < 1$ be a constant. If there is an oracle $\widehat{\mathcal{NTRU}}_\delta$ with*

$$\delta \geq \frac{1}{2} - \frac{\alpha}{N},$$

there exists a deterministic polynomial time algorithm which, given a valid, random, encryption e of a random message $m \in \mathcal{L}_m$, makes N calls of the oracle $\widehat{\mathcal{NTRU}}_\delta$ and finds the message m with probability $1 - \alpha$.

Proof. Querying $\widehat{\mathcal{NTRU}}_\delta$ on $X^i \odot e$, gives us a_{N-i-1} except with probability at most α/N. Then, for randomly chosen m and φ, the probability that any of the N calls is wrong is then at most α. □

It is obvious that similar algorithms work for an oracle producing any other fixed coefficient of the message m rather than just the leading coefficient.

Finally, one can probably consider an oracle which is a combination of $\mathcal{NTRU}_\gamma$ and $\widetilde{\mathcal{NTRU}}_\varepsilon$, namely the oracle which for any sequence of k distinct and valid encryptions $e_1, \ldots, e_k$, returns $0, 1$ for at least $\gamma k + O(1)$ queries which is the correct value of the leading coefficient of the corresponding message for at least $(0.5 + \varepsilon)\gamma k + O(1)$ of them, while for the remaining queries it returns the error message symbol $*$.

It is easy to see that our approach applies to any choice of p, for example, to the original case $p = 3$, implying that finding any coefficient of the message m is as hard as finding the whole message.

Chapter 21

Distribution of the RSA and Exponential Pairs

Let $m = pl$ be a product of two distinct primes p and l and let $\mathcal{E}_m$ be the set of integers e, $1 \le e \le \varphi(m)$, with $\gcd(e, \varphi(m)) = 1$. Here we present some results of [514] about the distribution modulo m of the *RSA pairs* (x, x^e). For example such pairs arise in the *RSA signature* scheme, and thus studying their behaviour is a very natural question.

First of all we show that for almost all exponents $e \in \mathcal{E}_m$, this distribution is exponentially close to the uniform distribution, when x runs through the group of units $\mathbb{Z}_m^*$ modulo m. This result is an analogue of the results of Theorems 13.6 and 13.7. Then we also consider the case when x runs through all possible k-products of the form

$$x = \prod_{j=1}^{k} z_{i_j}, \qquad 1 \le i_1 < \ldots < i_k \le r,$$

for some fixed $z_1, \ldots, z_r \in \mathbb{Z}_m^*$. We show that for almost all r-element sequences $z_1, \ldots, z_r \in \mathbb{Z}_m^*$, the corresponding pairs (x, x^e) are uniformly distributed modulo m. Such pairs have been considered in [70] and provide a promising way to speed up the RSA encryption with precomputation.

Denote by D_e the discrepancy of the pairs of fractional parts

$$\left(\left\{ \frac{x}{m} \right\}, \left\{ \frac{x^e}{m} \right\} \right), \qquad x \in \mathbb{Z}_m^*.$$

Theorem 21.1. *Let $m = pl$ where p and l are two distinct primes. Then the bound*

$$\frac{1}{|\mathcal{E}_m|} \sum_{e \in \mathcal{E}_m} D_e = O(m^{-1/12} \log^2 m \log\log m)$$

holds.

Proof. Combining Lemma 4.1, in which we choose $N = 2$ and $L = \lfloor (m-1)/2 \rfloor$, with Lemma 3.26 and the bounds (3.16) and (3.17), we derive

$$\begin{aligned}\sum_{e \in \mathcal{E}_m} D_e = O\Bigg(1 + &\sum_{\substack{0<|a|+|b|<m \\ \gcd(a,b,m)=1}} \frac{m^{11/12}}{\max\{|a|,1\}\max\{|b|,1\}} \\ &+ \sum_{\substack{0<|a|+|b|<m \\ \gcd(a,b,m)=p}} \frac{ml^{-1/12}}{\max\{|a|,1\}\max\{|b|,1\}} \\ &+ \sum_{\substack{0<|a|+|b|<m \\ \gcd(a,b,m)=l}} \frac{mp^{-1/12}}{\max\{|a|,1\}\max\{|b|,1\}}\Bigg) \\ = O\left(1 + m^{11/12}\log^2 m + p^{-1}l^{11/12}\log^2 l + l^{-1}p^{11/12}\log^2 p\right)& \\ = O\left(m^{11/12}\log^2 m\right).&\end{aligned}$$

Recalling that $|\mathcal{E}_m| = \varphi(\varphi(m))$ and taking into account the bound (5.1) and the inequality $\varphi(m) \geq 0.25m$, we obtain the desired result. □

In particular, we see that for any $\delta > 0$ for a random exponent e chosen uniformly from $\mathcal{E}_m$ with probability at least $1 - \delta$ the bound

$$D_e = O\left(\delta^{-1} m^{-1/12} \log^2 m \log\log m\right)$$

holds.

Theorem 21.1 is somewhat related to the *message concealing* problem for the RSA cryptosystem which has been considered in [45] see also Section 8.2.3.viii of [389]. Namely, we say that a message $x \in \mathbb{Z}_m^*$ is not concealed by the RSA encryption modulo $m = pl$ with the encryption exponent e if $x^e \equiv x \pmod{m}$. Elementary number-theoretic arguments show, see [45], that there are exactly $\gcd(e-1, p-1)\gcd(e-1, l-1)$ unconcealed messages $x \in \mathbb{Z}_m^*$, which is very small for almost all choices of the encryption exponent $e \in \mathcal{E}_m$. Another interpretation of this is result is an exact formula for the number of fixed points of the RSA mapping.

For the discrete logarithm one can obtain bounds on the number of fixed points from the results of [92, 116, 565]. In fact for the discrete logarithm this problem is also known in the literature as the *Brizolis problem*, see Section F9 of [248]. In Chapter 27 we study fixed points of a certain subset sum mapping.

Theorem 21.1 shows that in fact for almost all encryption exponents $e \in \mathcal{E}_m$ the message $x \in \mathbb{Z}_m^*$ and its encryption $x^e \in \mathbb{Z}_m^*$ are not correlated and, statistically, behave as two uniformly and independently distributed random elements of $\mathbb{Z}_m^*$.

Here we show one more result related to the message concealing problem. Namely we consider the possibility of more general polynomial relations between x and x^e. To be more precise we consider polynomials which are not monomials modulo p and modulo l.

Theorem 21.2. *Let $m = pl$ be a product of two distinct primes p and l and let $e \in \mathcal{E}_m$ and let*

$$f(X) = \sum_{i=0}^{n} a_i X^i \in \mathbb{Z}[X]$$

be a polynomial of degree n such that for any $i, j = 0, \ldots, n$,

$$f(X) \not\equiv a_i X^i \pmod p \qquad \text{and} \qquad f(X) \not\equiv a_j X^j \pmod l.$$

Let T be the number of solutions of the congruence

$$f(x^e) \equiv x \pmod m, \qquad x \in \mathbb{Z}_m^*.$$

Then the bound

$$T \le 2n\varphi(m)^{1/2}$$

holds.

Proof. Obviously $T = T_p T_l$ where T_p and T_l are the numbers of solutions of the congruences

$$f(x^e) \equiv x \pmod p, \qquad x \in \mathbb{F}_p^* \qquad \text{and} \qquad f(x^e) \equiv x \pmod l, \qquad x \in \mathbb{F}_l^*,$$

respectively.

To estimate T_p we define $d \in [1, p-2]$ from the congruence $ed \equiv 1 \pmod{p-1}$. The congruence $f(x^e) \equiv x \pmod p$ has the same number of solutions $x \in \mathbb{F}_p^*$ as the congruence $f(x) \equiv x^d \pmod p$.

Obviously, T_p^2 is the number of solutions of the system of congruences

$$f(x) \equiv x^d \pmod p, \quad f(y) \equiv y^d \pmod p, \qquad x, y \in \mathbb{F}_p^*,$$

which after the substitution $y \equiv xz \pmod p$ reduces to the equivalent system

$$f(x) \equiv x^d \pmod p, \quad f(xz) \equiv (xz)^d \pmod p, \qquad x, z \in \mathbb{F}_p^*.$$

From here we derive $f(xz) \equiv z^d f(x) \pmod p$. For those values of z for which the polynomial $F_z(X) = f(Xz) - z^d f(X)$ is not identical to zero there are at most n corresponding values of x; for other values of z there are at most $p-1$ corresponding values of x. Thus,

$$T_p^2 \le (p-1)n + (p-1)Q,$$

where Q is the number of $z \in \mathbb{F}_p^*$ such that $F_z(X)$ is identical to zero modulo p. Taking into account that $f(X)$ contains at least two non-zero modulo p monomials, say $a_h X^h$ and $a_k X^k$, $a_h a_k \not\equiv 0 \pmod p$, $0 \le h < k \le n$, we find that this is possible only if

$$z^h \equiv z^d \pmod p \qquad \text{and} \qquad z^n \equiv z^d \pmod p.$$

In particular, $z^h \equiv z^k \pmod p$; therefore $Q \le k \le n$. Thus $T_p^2 \le 2(p-1)n$ and the result follows. □

We remark that if $f(X) \equiv a_n X^n \pmod m$, then the problem is essentially the original message concealing problem of [45]. Also, if $f(X) \equiv a_i X^i \pmod p$ or $f(X) \equiv a_j X^j \pmod l$ but $f(X) \not\equiv a_n X^n \pmod m$, then the problem is essentially equivalent to a combination of the original message concealing problem of [45] and our approach.

Similar arguments can be applied to congruences $h(x^e) \equiv x \pmod m$ with rational functions $h(X) = f(X)/g(X)$, $f(X), g(X) \in \mathbb{Z}[X]$.

Finally, using Lemma 2.6 one can obtain an analogue of Theorem 21.2 for sparse polynomials.

For integers $1 \le k \le r$ and an r-dimensional vector $\mathbf{z} = (z_1, \dots, z_r) \in (\mathbb{Z}_m^*)^r$ we denote by $D_{\mathbf{z},k,e}$ the discrepancy of the pairs of fractional parts

$$\left(\left\{\frac{x_{\mathbf{z}}(\mathbf{u})}{m}\right\}, \left\{\frac{x_{\mathbf{z}}^e(\mathbf{u})}{m}\right\}\right), \qquad \mathbf{u} = (u_1, \dots, u_r) \in \mathfrak{B}_{r,k},$$

where as before

$$x_{\mathbf{z}}(\mathbf{u}) = \prod_{j=1}^{r} z_j^{u_j}$$

and $\mathfrak{B}_{r,k}$ is defined by (1.5) and (1.6).

Using Lemma 3.27 and the bounds (3.18) and (3.19), in the same way as we have used Lemma 3.26 and the bounds (3.16) and (3.17) in the proof of Theorem 21.1, we obtain the following statement:

Theorem 21.3. *Let $m = pl$ where p and l are two distinct primes. Then the bound*

$$\frac{1}{|\mathbb{Z}_m^*|^r |\mathcal{E}_m|} \sum_{\mathbf{z} \in (\mathbb{Z}_m^*)^r} \sum_{e \in \mathcal{E}_m} D_{\mathbf{z},k,e} = O\left(\left(|\mathfrak{B}_{r,k}|^{-1/2} + m^{-1/12}\right) \log^2 m \log\log m\right)$$

holds.

In particular, we see that for any $\delta > 0$ for a random vector $\mathbf{z}$ and a random exponent e chosen uniformly and independently from $(\mathbb{Z}_m^*)^r$ and $\mathcal{E}_m$ with probability at least $1-\delta$ the bound

$$D_{\mathbf{z},k,e} = O\left(\delta^{-1}\left(|\mathfrak{B}_{r,k}|^{-1/2} + m^{-1/12}\right)\log^2 m \log\log m\right)$$

holds.

Another important cryptographic construction is exponentiation in a finite field. In particular, it is a part of the Digital Signature Algorithm, of the Diffie–Hellman key exchange protocol and of many other cryptographic constructions.

Let us fix an element $g \in \mathbb{F}_p^*$ of multiplicative order $\mathrm{ord}_p g = t$. We consider the distribution of the *exponential pairs* (s, g^s) in the Cartesian product $\mathbb{Z}_t \times \mathbb{F}_p$, when s runs through

- the whole set $\mathbb{Z}_t$ as in many standard schemes;
- all k-sums $s = z_{i_1} + \ldots + z_{i_k}$, $1 \le i_1 < \ldots < i_k \le r$, where $z_1, \ldots, z_r \in \mathbb{Z}_t$ are selected at random as in the exponentiation with precomputation scheme which has been recently introduced in [70], see also [420].

These results are similar to the above results for the RSA pairs and have been obtained in [516].

Denote by D the discrepancy of the pairs of fractional parts

$$\left(\left\{\frac{s}{t}\right\}, \left\{\frac{g^s}{p}\right\}\right), \qquad s \in \mathbb{Z}_t.$$

Theorem 21.4. *The bound*

$$D = O(B(t,p)\log^2 t)$$

holds, where $B(t,p)$ is given by (3.8).

Proof. Select $L = t-1$ in Lemma 4.1. Applying Lemma 3.15 to the sums $S(a,c)$ given by (3.7) and the identity

$$S(a,0) = 0 \tag{21.1}$$

which holds with any integer a such that $\gcd(a,t) = 1$, we obtain the desired result. □

For integers $1 \le k \le r$ and an r-dimensional vector $\mathbf{z} = (z_1, \ldots, z_r) \in \mathbb{Z}_t^r$ we denote by $D_{\mathbf{z},k}$ the discrepancy of the pairs of fractional parts

$$\left(\left\{\frac{s_{\mathbf{z}}(\mathbf{u})}{t}\right\}, \left\{\frac{g^{s_{\mathbf{z}}(\mathbf{u})}}{p}\right\}\right), \qquad \mathbf{u} = (u_1, \ldots, u_r) \in \mathfrak{B}_{r,k},$$

where

$$s_{\mathbf{z}}(\mathbf{u}) = \sum_{j=1}^{r} z_j u_j \tag{21.2}$$

and $\mathfrak{B}_{r,k}$ is defined by (1.5) and (1.6).

Theorem 21.5. *The bound*

$$\frac{1}{t^r} \sum_{\mathbf{z}\in \mathbb{Z}_t^r} D_{\mathbf{z},k} = O\left(\left(|\mathfrak{B}_{r,k}|^{-1/2} + B(t,p)t^{-1}\right) \log^2 t\right)$$

holds, where $B(t,p)$ is given by (3.8).

Proof. We define

$$W(a,c) = \sum_{\mathbf{z}\in \mathbb{Z}_t^r} \left| \sum_{\mathbf{u}\in\mathfrak{B}_{r,k}} \mathbf{e}_t\left(as_{\mathbf{z}}(\mathbf{u})\right) \mathbf{e}_p\left(cg^{s_{\mathbf{z}}(\mathbf{u})}\right)\right|.$$

Using the Cauchy inequality and changing the order of summation, we derive

$$\begin{aligned} &W(a,c)^2 \\ &\le t^r \sum_{\mathbf{z}\in \mathbb{Z}_t^r} \left| \sum_{\mathbf{u}\in\mathfrak{B}_{r,k}} \mathbf{e}_t\left(as_{\mathbf{z}}(\mathbf{u})\right) \mathbf{e}_p\left(cg^{s_{\mathbf{z}}(\mathbf{u})}\right)\right|^2 \\ &= t^r \sum_{\mathbf{u},\mathbf{v}\in\mathfrak{B}_{r,k}} \sum_{\mathbf{z}\in \mathbb{Z}_t^r} \mathbf{e}_t\left(as_{\mathbf{z}}(\mathbf{u}) - as_{\mathbf{z}}(\mathbf{v})\right) \mathbf{e}_p\left(cg^{s_{\mathbf{z}}(\mathbf{u})} - cg^{s_{\mathbf{z}}(\mathbf{v})}\right). \end{aligned}$$

The contribution to this sum of each pair with $\mathbf{u} = \mathbf{v}$ is t^r. For each pair $\mathbf{u},\mathbf{v} \in \mathfrak{B}_{r,k}$ with $\mathbf{u} \neq \mathbf{v}$ we can find i and j, $1 \le i < j \le r$, with $u_i = v_j = 1$ and $u_j = v_i = 0$. Without loss of generality we may assume that $i = 1$, $j = 2$. In this case $s_{\mathbf{z}}(\mathbf{u}) = A + z_1$ and $s_{\mathbf{z}}(\mathbf{v}) = B + z_2$, where A and B depend only on $\mathbf{u}$, $\mathbf{v}$ and $z_3, \ldots, z_r$. Therefore

$$\begin{aligned} &\sum_{\mathbf{z}\in \mathbb{Z}_t^r} \mathbf{e}_t\left(as_{\mathbf{z}}(\mathbf{u}) - as_{\mathbf{z}}(\mathbf{v})\right) \mathbf{e}_p\left(cg^{s_{\mathbf{z}}(\mathbf{u})} - cg^{s_{\mathbf{z}}(\mathbf{v})}\right) \\ &\qquad = \sum_{a_3,\ldots,a_r\in\mathbb{Z}_t} \mathbf{e}_t\left(a(A-B)\right) \sum_{z_1\in\mathbb{Z}_t} \mathbf{e}_t\left(az_1\right) \mathbf{e}_p\left(cg^A g^{z_1}\right) \\ &\qquad\qquad \times \sum_{z_2\in\mathbb{Z}_t} \mathbf{e}_t\left(-az_2\right) \mathbf{e}_p\left(-cg^B g^{z_2}\right). \end{aligned}$$

If $\gcd(c,p) = 1$, then from Lemma 3.15 we see that each inner sum is $O\left(B(t,p)\right)$ thus

$$\max_{a,c\in\mathbb{Z}^2,\ \gcd(c,p)=1} W(a,c) = O\left(t^r|\mathfrak{B}_{r,k}|^{1/2} + t^{r-1}|\mathfrak{B}_{r,k}|B(t,p)\right).$$

Also, for an integer a with $\gcd(a,t) = 1$ we have from (21.1) that

$$W(a,0) \le t^r |\mathfrak{B}_{r,k}|^{1/2}.$$

Selecting $L = t - 1$ in Lemma 4.1 and applying the above estimates, we obtain the desired result. □

In particular, we see that for any $\delta > 0$ for a random vector $\mathbf{z}$ chosen uniformly at random from $\mathbb{Z}_t^r$ with probability at least $1 - \delta$ the bound

$$D_{\mathbf{z},k} = O\left(\delta^{-1}\left(|\mathfrak{B}_{r,k}|^{-1/2} + B(t,p)t^{-1}\right)\log^2 t\right)$$

holds.

One easily verifies that the above bounds are non-trivial for $t \ge p^{1/3+\varepsilon}$ with any fixed $\varepsilon > 0$ and sufficiently large p.

Accordingly, using Lemma 3.16 instead of Lemma 3.15, one obtains bounds which are non-trivial for $t \ge p^{\varepsilon}$ with any fixed $\varepsilon > 0$ and almost all sufficiently large p. Using the improvement of Lemma 3.15 from [304], one can obtain a nontrivial bound for $t \ge p^{1/4+\varepsilon}$ and any prime p.

Chapter 22

Exponentiation and Inversion with Precomputation

As in Chapter 21 we define $s_{\mathbf{z}}(\mathbf{u})$ by (21.2).

For an arbitrary set $\mathcal{B} \in \mathfrak{B}_r$, the uniformity of distribution (in a very strong sense) of $g^{s_{\mathbf{z}}(\mathbf{u})}$, $\mathbf{u} \in \mathcal{B}$, for almost all $\mathbf{z} \in \mathbb{Z}_t^r$ has been obtained in [419].

With $\mathcal{B} = \mathfrak{B}_{r,k}(h)$ this result has been applied to evaluating the security of a certain algorithm for exponentiation with precomputation. This scheme generalises that of [70] which corresponds to the case $h = 2$. However it has turned out that selecting h as a certain function of k and r one can achieve an asymptotically growing speed-up. It has also been applied to designing a new provably secure "server assisted" exponentiation scheme.

Given a vector $\mathbf{z} = (z_1, \dots, z_r) \in \mathbb{Z}_t^r$, a residue $c \in \mathbb{Z}_t$ and a set $\mathcal{B} \subseteq \mathbb{Z}_t^r$, we denote by $N_{\mathbf{z}}(\mathcal{B}, c)$ the number of solutions of the congruence

$$s_{\mathbf{z}}(\mathbf{u}) \equiv c \pmod t, \qquad \mathbf{u} \in \mathcal{B}.$$

Theorem 22.1. *The identity*

$$\frac{1}{t^r} \sum_{\mathbf{z} \in \mathbb{Z}_t^r} \sum_{c \in \mathbb{Z}_t} \left(N_{\mathbf{z}}(\mathcal{B}, c) - \frac{|\mathcal{B}|}{t} \right)^2 = \frac{t-1}{t} |\mathcal{B}|$$

holds.

Proof. From (3.2) we have

$$N_{\mathbf{z}}(\mathcal{B}, c) = \frac{1}{t} \sum_{\mathbf{u} \in \mathcal{B}} \sum_{\lambda=0}^{t-1} \mathbf{e}_t \left(\lambda \left(s_{\mathbf{z}}(\mathbf{u}) - c \right) \right).$$

The contribution of the term corresponding to $\lambda = 0$ is $|\mathcal{B}|/t$. Therefore

$$\sum_{c\in\mathbb{Z}_t}\left(N_{\mathbf{z}}(\mathcal{B},c)-\frac{|\mathcal{B}|}{t}\right)^2$$

$$=\sum_{c\in\mathbb{Z}_t}\left(\frac{1}{t}\sum_{\lambda=1}^{t-1}\mathbf{e}_t\left(-\lambda c\right)\sum_{\mathbf{u}\in\mathcal{B}}\mathbf{e}_t\left(\lambda s_{\mathbf{z}}(\mathbf{u})\right)\right)^2$$

$$=\frac{1}{t^2}\sum_{c\in\mathbb{Z}_t}\sum_{\lambda,\eta=1}^{t-1}\mathbf{e}_t\left(-c(\lambda+\eta)\right)\sum_{\mathbf{u},\mathbf{v}\in\mathcal{B}}\mathbf{e}_t\left(\lambda s_{\mathbf{z}}(\mathbf{u})+\eta s_{\mathbf{z}}(\mathbf{v})\right)$$

$$=\frac{1}{t^2}\sum_{\lambda,\eta=1}^{t-1}\sum_{\mathbf{u},\mathbf{v}\in\mathcal{B}}\mathbf{e}_t\left(\lambda s_{\mathbf{z}}(\mathbf{u})+\eta s_{\mathbf{z}}(\mathbf{v})\right)\sum_{c\in\mathbb{Z}_t}\mathbf{e}_t\left(-\lambda c-\eta c\right).$$

The sum over c vanishes if $\lambda+\eta\not\equiv 0 \pmod{t}$ and is equal to t otherwise. Hence,

$$\sum_{c\in\mathbb{Z}_t}\left(N_{\mathbf{z}}(\mathcal{B},c)-\frac{|\mathcal{B}|}{t}\right)^2 = \frac{1}{t}\sum_{\lambda=1}^{t-1}\sum_{\mathbf{u},\mathbf{v}\in\mathcal{B}}\mathbf{e}_t\left(\lambda\left(s_{\mathbf{z}}(\mathbf{u})-s_{\mathbf{z}}(\mathbf{v})\right)\right) = \frac{1}{t}\sum_{\lambda=1}^{t-1}\left|\sum_{\mathbf{u}\in\mathcal{B}}\mathbf{e}_t\left(\lambda s_{\mathbf{z}}(\mathbf{u})\right)\right|^2.$$

We have

$$\sum_{\mathbf{z}\in\mathbb{Z}_t^r}\left|\sum_{\mathbf{u}\in\mathcal{B}}\mathbf{e}_t\left(\lambda s_{\mathbf{z}}(\mathbf{u})\right)\right|^2=\sum_{\mathbf{u},\mathbf{v}\in\mathcal{B}}\sum_{\mathbf{z}\in\mathbb{Z}_t^r}\mathbf{e}_t\left(\lambda\left(s_{\mathbf{z}}(\mathbf{u})-s_{\mathbf{z}}(\mathbf{v})\right)\right).$$

If $\mathbf{u}=\mathbf{v}$ the contribution of the inner sum is t^r. If $\mathbf{u}\neq\mathbf{v}$, then, assuming without loss of generality that $v_r\neq u_r$, we obtain

$$\sum_{\mathbf{z}\in\mathbb{Z}_t^r}\mathbf{e}_t\left(\lambda\left(s_{\mathbf{z}}(\mathbf{u})-s_{\mathbf{z}}(\mathbf{v})\right)\right)$$

$$=\sum_{(z_1,\dots,z_{r-1})\in\mathbb{Z}_t^{r-1}}\mathbf{e}_t\left(\lambda\sum_{j=1}^{r-1}z_j(u_j-v_j)\right)\sum_{z_r=0}^{t-1}\mathbf{e}_t(\lambda z_r(u_r-v_r))=0.$$

Thus

$$\sum_{\mathbf{z}\in\mathbb{Z}_t^r}\left|\sum_{\mathbf{u}\in\mathcal{B}}\mathbf{e}_t\left(\lambda s_{\mathbf{z}}(\mathbf{u})\right)\right|^2=t^r|\mathcal{B}|,$$

and the desired result follows. □

In particular, using the Cauchy inequality we obtain

$$\frac{1}{t^r} \sum_{\mathbf{z}\in \mathbb{Z}_t^r} \sum_{c\in \mathbb{Z}_t} \left| N_{\mathbf{z}}(\mathfrak{B}_{r,k}(h), c) - \frac{|\mathfrak{B}_{r,k}(h)|}{t} \right| \le t^{1/2} |\mathfrak{B}_{r,k}(h)|^{1/2}$$

which has been used as the main tool in [419]. It has been interpreted as the statistical distance between the output of exponentiation with precomputation in a subgroup of $\mathbb{F}_p^*$ and the uniform distribution.

Several other results of this kind have recently been obtained in [134] where some applications to fast generation of points on elliptic curves have been given as well.

Similar results about the distribution modulo q of the products of the form

$$\kappa = t_1 \dots t_k t_{k+1}^{-1} \dots t_{2k}^{-1} \pmod q, \qquad t_i \in [1, T],\ i = 1, \dots, 2k,$$

where T is sufficiently small relative to q, are given in [330]. They are motivated by the algorithm of [327] of fast generation of pairs (κ, λ) with $\lambda\kappa \equiv 1 \pmod q$. This algorithm can be used (and actually has been designed) for speeding up the Digital Signature Algorithm, see Chapter 16 for a description of this algorithm.

The algorithm from [327] works as follows, in a special partial case. Given a prime q and two more integer parameters $T \ge 2$ and $m \ge 1$:

- Select independently and uniformly at random $2m$ integer numbers $t_1, \dots, t_{2m} \in [1, T]$;
- For $i = 1, \dots, 2m$, compute $u_i = \lfloor q^{-1} \rfloor_{t_i}$ and $w_i = (qu_i - 1)/t_i$;
- For $i = 1, \dots, 2m$, using the identity $t_i^{-1} \equiv w_i \pmod q$, compute $v_i = \lfloor t_i^{-1} \rfloor_q$;
- Compute and output

$$\kappa = \lfloor t_1 \dots t_m v_{m+1} \dots v_{2m} \rfloor_q \quad \text{and} \quad \lambda = \lfloor v_1 \dots v_m t_{m+1} \dots t_{2m} \rfloor_q.$$

It is easy to see that $\lambda = \lfloor \kappa^{-1} \rfloor_q$. The efficiency of the algorithm is based on the observation that for each arithmetic operation it performs one of the operands is of size T. Furthermore, once $\lfloor q \rfloor_{t_i}$ has been computed, the inversion required for the computation of u_i involves only numbers of size $\le T$. Thus if the bit length of T is essentially smaller than the bit length of q and m is reasonably small this algorithm is faster than the standard inversion modulo q using the *Extended Euclidean Algorithm*. The efficiency of this algorithm (and its slightly more general form described in [327]) has been numerically verified, see [327, 330].

However, it has remained an open question whether this new way of generating k and $\lfloor k^{-1} \rfloor_q$ undermines the security of the Digital Signature Algorithm. In [327] some heuristic arguments in support of the security of the new scheme are given.

Here we present the results of [330] which show that using bounds of character sums one can establish rigorous security results for the above scheme (for some values of the parameters T and m). In fact we show that the distribution of the value of κ is exponentially close to the uniform distribution. Therefore any algorithm attacking this modification immediately implies an attack on the original scheme with exponentially close probabilities of success.

More precisely, for $k \in \mathbb{F}_q^*$, let $P_{m.T}(k)$ be the probability that the output κ of the above algorithm equals k. We use some known bounds of exponential sums to prove that for a wide range of parameters T and m the *statistical distance*

$$\Delta(m,T) = \sum_{k\in\mathbb{F}_q^*} \left| P_{m.T}(k) - \frac{1}{q-1} \right| \tag{22.1}$$

is exponentially small, namely

$$\Delta(m,T) < q^{-\delta}$$

for some constant $\delta > 0$. The range of parameters allowed by this general result do, however, not seem to be of much practical value. We show that under the assumption of the Extended Riemann Hypothesis an essentially stronger result can be obtained that allows parameter choices in a more realistic and practical range.

We stress that the uniformity of distribution of the nonce k is absolutely essential. Indeed, it has been shown in the series of papers [277, 414, 417], see also Chapter 16, that the knowledge of some bits of k can be used to break the Digital Signature Algorithm (that is, recovering the private key α) in polynomial time.

Let $\mathcal{X}$ be the set of multiplicative characters of the multiplicative group $\mathbb{F}_q^*$, see Section 1 of Chapter 5 of [343]. We denote by $\mathcal{X}^*$ the subset of non-trivial characters.

We define

$$\sigma(T) = \max_{\chi\in\mathcal{X}^*} \left| \sum_{t=1}^{T} \chi(t) \right| .$$

Theorem 22.2. *For any integers $T \geq 2$ and $m \geq 1$ the bound*

$$\Delta(m,T) < q^{1/2}\sigma(T)^{2m-1}T^{-2m+1/2}$$

holds for the statistical distance $\Delta(m,T)$ given by (22.1).

Proof. Let $N_{m,T}(k)$ be the number of sequences $t_1, \dots, t_{2m} \in [1,T]$ for which $t_1 \dots t_m t_{m+1}^{-1} \dots t_{2m}^{-1} \equiv k \pmod q$. Then $P_{m,T}(k) = N_{m.T}(k)T^{-2m}$.

From the following well-known identity

$$\sum_{\chi\in\mathcal{X}} \chi(z) = \begin{cases} q-1, & \text{if } z = 1, \\ 0, & \text{otherwise,} \end{cases}$$

which holds for any $z \in \mathbb{F}_q^*$ (see Theorem 5.4 of [343]), we derive

$$N_{m,T}(k) = \frac{1}{q-1} \sum_{t_1,\dots,t_{2m}=1}^{T} \sum_{\chi \in \mathcal{X}} \chi(t_1 \dots t_m t_{m+1}^{-1} \dots t_{2m}^{-1} k^{-1}).$$

We remark that $\chi(\lambda^{-1}) = \overline{\chi(\lambda)}$ for $\lambda \in \mathbb{F}_q^*$ and that $z\bar{z} = |z|^2$, where $\bar{z}$ denotes the conjugate of a complex number z. Therefore, changing the order of summation, separating the term $T^{2m}/(q-1)$ which corresponds to the trivial character, and noting that k^{-1} runs through $\mathbb{F}_q^*$ together with k we obtain

$$\left| N_{m,T}(k) - \frac{T^{2m}}{q-1} \right| = \frac{1}{q-1} \sum_{\chi \in \mathcal{X}^*} \chi(k) \left| \sum_{t=1}^{T} \chi(t) \right|^{2m}.$$

Therefore

$$\sum_{k \in \mathbb{F}_q^*} \left| N_{m,T}(k) - \frac{T^{2m}}{q-1} \right|^2$$
$$= \frac{1}{(q-1)^2} \sum_{k \in \mathbb{F}_q^*} \left(\sum_{\chi \in \mathcal{X}^*} \chi(k) \left| \sum_{t=1}^{T} \chi(t) \right|^{2m} \right)^2$$
$$= \frac{1}{(q-1)^2} \sum_{k \in \mathbb{F}_q^*} \sum_{\chi_1,\chi_2 \in \mathcal{X}^*} \chi_1(k)\chi_2(k) \left| \sum_{t=1}^{T} \chi_1(t) \right|^{2m} \left| \sum_{t=1}^{T} \chi_2(t) \right|^{2m}$$
$$= \frac{1}{(q-1)^2} \sum_{\chi_1,\chi_2 \in \mathcal{X}^*} \left| \sum_{t=1}^{T} \chi_1(t) \right|^{2m} \left| \sum_{t=1}^{T} \chi_2(t) \right|^{2m} \sum_{k \in \mathbb{F}_q^*} \chi_1(k)\chi_2(k).$$

Using that the product of two characters is a character as well and the identity

$$\sum_{k \in \mathbb{F}_q^*} \chi(k) = \begin{cases} q-1, & \text{if } \chi = \chi_0, \\ 0, & \text{otherwise,} \end{cases}$$

where χ_0 is the trivial character (see Theorem 5.4 of [343]), we see that the inner sum vanishes unless

$$\chi_2(k) = \chi_1(k)^{-1} = \chi_1(k^{-1}) = \overline{\chi_1(k)}, \qquad k \in \mathbb{F}_q^*$$

in which case it is equal to $q-1$. Therefore

$$\sum_{k \in \mathbb{F}_q^*} \left| N_{m,T}(k) - \frac{T^{2m}}{q-1} \right|^2 = \frac{1}{q-1} \sum_{\chi \in \mathcal{X}^*} \left| \sum_{t=1}^{T} \chi(t) \right|^{2m} \left| \sum_{t=1}^{T} \overline{\chi(k)} \right|^{2m}$$
$$= \frac{1}{q-1} \sum_{\chi \in \mathcal{X}^*} \left| \sum_{t=1}^{T} \chi(t) \right|^{4m} \le \frac{\sigma(T)^{4m-2}}{q-1} \sum_{\chi \in \mathcal{X}^*} \left| \sum_{t=1}^{T} \chi(t) \right|^2.$$

We have

$$\frac{1}{q-1}\sum_{\chi\in\mathcal{X}^*}\left|\sum_{t=1}^{T}\chi(t)\right|^2 < \frac{1}{q-1}\sum_{\chi\in\mathcal{X}}\left|\sum_{t=1}^{T}\chi(t)\right|^2 = T.$$

Hence

$$\begin{aligned}\sum_{k\in\mathbb{F}_q^*}\left|P_{m.T}(k)-\frac{1}{q-1}\right|^2 &\le T^{-4m}\sum_{k\in\mathbb{F}_q^*}\left|N_{m.T}(k)-\frac{T^{2m}}{q-1}\right|^2\\ &\le \sigma(T)^{4m-2}T^{-4m+1}.\end{aligned}$$

From the Cauchy inequality we obtain the desired result. □

One can now use the known bound of character sums to get concrete estimates of $\Delta(m,T)$.

In particular, as in [330] here we use the *Burgess bound* in the form of Lemma 3.10 and the conditional, under the Extended Riemann Hypothesis, bound of [230].

Theorem 22.3. *For any $\varepsilon > 0$ and $A \ge 0$ there exists some $m_0(\varepsilon, A) > 0$ such that for any integers $T \ge 2$ and $m \ge 1$ satisfying the inequalities*

$$T \ge q^{1/4+\varepsilon} \qquad \textit{and} \qquad m \ge m_0(\varepsilon, A)$$

the statistical distance $\Delta(m,T)$ given by (22.1) satisfies the inequality $\Delta(m,T) \le q^{-A}$.

Proof. From Theorem 22.2 and Lemma 3.10 we obtain that for some $\gamma > 0$ depending only on ε the bound

$$\Delta(m,T) < q^{1/2}T^{-1/2}q^{-\gamma(2m-1)} \le q^{3/8-\gamma(2m-1)}$$

holds. Choosing

$$m_0(\varepsilon, A) = \frac{8A+3+8\gamma}{16\gamma}$$

we obtain the desired result. □

Unfortunately the range of parameters allowed by Theorem 22.3 does not seem to be of any practical value. However under the Extended Riemann Hypothesis an essentially stronger result can be obtained.

Theorem 22.4. *Assume the Extended Riemann Hypothesis. Then for any $A > 0$ and any integers $T \ge 2$ and $m \ge 1$ such that*

$$v = \frac{\ln T}{\ln\ln q} \to \infty \qquad \textit{and} \qquad m \ge (2A+1)\frac{\ln q}{v\ln v} + 1$$

for sufficiently large q, the statistical distance $\Delta(m,T)$ given by (22.1) satisfies the bound $\Delta(m,T) \le q^{-A}$.

Proof. From Theorem 22.2 and Lemma 3.11 we obtain the bound

$$\begin{aligned}\Delta(m,T) &< q^{1/2}T^{-1/2}v^{-(v/2+o(v))(2m-1)} \le q^{1/2}v^{-v(2m-1)/3} \\ &\le q^{1/2-(4A+2)/3} \le q^{-A},\end{aligned}$$

provided that q is large enough. □

In particular, if q is about n bits long and T is selected about ℓ bits long with $\ell \ge \ln n^{1+\varepsilon}$, then for m of order $n/\ln\ell$ the algorithm of [327] generates a secure sequence of pairs $\kappa, \lambda = \lfloor \kappa^{-1} \rfloor_q$. Thus the values of T used in this algorithm can be rather small.

Part V

Pseudorandom Number Generators

Chapter 23

RSA and Blum–Blum–Shub Generators

Let ϑ, m and e be integers such that $\gcd(\vartheta, m) = 1$ and $e \geq 2$. Then one can define the sequence (u_n) by the recurrence relation

$$u_n \equiv u_{n-1}^e \pmod m, \quad 0 \leq u_n \leq m-1, \qquad n = 1, 2, \ldots, \tag{23.1}$$

with the *initial value* $u_0 = \vartheta$ (an integer co-prime to m) and *exponent* e.

This sequence is known as the *power generator* of pseudorandom numbers and has many applications to cryptography, see [50, 138, 319, 541].

In two special cases $\gcd(e, \varphi(m)) = 1$ and $e = 2$ this sequence is known as the *RSA generator* and as the *Blum–Blum–Shub generator*, respectively.

We remark that for cryptographic application, usually only some bits of each element u_n are revealed and used. Sometimes, the names of RSA and Blum–Blum–Shub generators, are associated only with these, truncated, sequences.

Although various properties of the *power generator* have been studied in a number of papers [50, 73, 137, 138, 182, 254, 319, 389, 396, 410, 541] very few unconditional results are known. One such result is due to [137]. It is shown in that paper that the rightmost bit of the Blum–Blum–Shub generator takes values 0 and 1 almost equally often, provided that the period is large enough. In this chapter we present several new results about this generator that have recently been obtained in [191, 192, 194, 236, 511].

In particular, we prove that if the period of the RSA generator or the Blum–Blum–Shub generator is large enough, then the elements of these sequences are uniformly distributed modulo m and a positive proportion of the rightmost bits is uniformly distributed as well. The result is supported by estimates from [192], we present here as Theorems 23.1 and 23.4. showing that typically the period is close to m.

We obtain the aforementioned results in the most important case for applications when $m = pl$ where p and l are distinct primes. Such numbers are called *Blum integers* (sometimes given with certain additional conditions such as that $p \equiv l \equiv 3 \pmod 4$) and that p and l are of the same magnitude). However, the same results hold for m prime, as is easy to see from our proof, and similar, but somewhat weaker, results can be obtained for arbitrary composite moduli as well. This, although it requires some additional work and adjusting the technique, can be done without involving any essentially new ideas, see [189].

It is obvious that the sequence (23.1) eventually becomes periodic with some period t. We always assume that this sequence is *purely periodic*, that is, that $u_n = u_{n+t}$ beginning with $n = 0$. It is easy to see that if $\gcd(e, \varphi(m)) = 1$, then this is always the case, otherwise we consider a shift of the original sequence.

Moreover if $\gcd(e, \varphi(m)) = 1$, this period is given by $t = \mathrm{ord}_T e$, where $T = \mathrm{ord}_m \vartheta$. Therefore, the largest possible value of t, over all possible choices of ϑ and e with $\gcd(e, \varphi(m)) = 1$, is $\lambda(\lambda(m))$, where as before $\lambda(m)$ is the Carmichael function.

The period of the power generator has recently been studied in the paper [192] whose results we present here.

First of all we show that for almost all choices of the initial value ϑ and the exponent e with $\gcd(e, \varphi(m)) = 1$ the corresponding period is close to its largest possible value $\lambda(\lambda(m))$. Then we obtain a lower bound on $\lambda(\lambda(m))$ which shows that for almost all products $m = pl$ this value is close to m. Then we combine both these statements. We also obtain similar but weaker results for the Blum–Blum–Shub generator.

Theorem 23.1. *For any positive integer m and any numbers $K_1, K_2 \geq 1$, let W denote the number of pairs of integers ϑ, e with $1 \leq \vartheta \leq m$, $1 \leq e \leq \lambda(m)$ and $\gcd(\vartheta, m) = \gcd(e, \lambda(m)) = 1$, such that the period of the power generator given by (23.1) is at most $\lambda(\lambda(m))/K_1K_2$. Then*

$$W \leq \varphi(m)\varphi(\lambda(m)) \left(\frac{\tau(\lambda(m))}{K_1} + \frac{\tau(\lambda(\lambda(m)))}{K_2} \right).$$

Proof. We apply Lemma 5.2 first with $M = m$ and $K = K_1$. So the number of values of ϑ in $[1, m]$ with $T = \mathrm{ord}_m \vartheta > \lambda(m)/K_1$ is at least $\varphi(m)(1 - \tau(\lambda(m))/K_1)$. For each such ϑ we again apply Lemma 5.2, now with $M = T$ and $K = K_2$. We deduce that the number of choices for e in $[1, T]$ with $\mathrm{ord}_T e > \lambda(T)/K_2$ is at least $\varphi(T)(1 - \tau(\lambda(T))/K_2)$. Thus, there are at least $\varphi(\lambda(m))(1 - \tau(\lambda(T))/K_2)$ choices of $e \in [1, \lambda(m)]$ that are co-prime to $\lambda(m)$ and such that $\mathrm{ord}_T e > \lambda(T)/K_2$.

Note that Lemma 5.3 implies that if ϑ, e are chosen as above, then the period $\mathrm{ord}_T e$ of the power generator satisfies

$$\mathrm{ord}_T e > \frac{\lambda(T)}{K_2} \geq \frac{\lambda(\lambda(m))T}{\lambda(m)K_2} \geq \frac{\lambda(\lambda(m))\lambda(m)}{\lambda(m)K_1K_2} = \frac{\lambda(\lambda(m))}{K_1K_2}.$$

Since $\tau(\lambda(T)) \leq \tau(\lambda(\lambda(m)))$, the result follows. □

Using the bound (5.1), we can obtain that for any $\varepsilon > 0$ and sufficiently large m, depending on the choice of ε, the number of pairs of integers ϑ, e in the range $1 \leq \vartheta \leq m$, $1 \leq e \leq \lambda(m)$ with $\gcd(\vartheta, m) = \gcd(e, \varphi(m)) = 1$, and such that the period t of the power generator given by (23.1) satisfies $t \leq \lambda(\lambda(m))/\lambda(m)^{\varepsilon}$ is at most $\varphi(m)\varphi(\lambda(m))/\lambda(\lambda(m))^{\varepsilon/3}$.

Here we apply Lemma 5.1 to obtain a lower bound for the largest possible period of the power generator.

Theorem 23.2. *For Q sufficiently large and for any $\Delta \geq 2(\log\log Q)^3$ the number of pairs $(p, l) \in \mathcal{P}^2$, $1 < p < l \leq Q$, with*

$$\lambda(\lambda(pl)) < Q^2 \exp(-\Delta)$$

is at most $Q^2 \exp\left(-0.16(\Delta \log \Delta)^{1/3}\right)$.

Proof. Fix $\Delta \geq 2(\log\log Q)^3$ and put

$$D = \exp\left(0.16(\Delta \log \Delta)^{1/3}\right).$$

The number W of pairs $(p, l) \in \mathcal{P}^2$, $1 < p < l \leq Q$, with $\gcd(p-1, l-1) \geq D$ satisfies

$$W \leq \sum_{d \geq D} \pi(Q; d, 1)^2.$$

We use the estimate (5.4) for $d < Q^{1/2}$ together with (5.1) and just $\pi(Q; d, 1) \leq Q/d$ for $d \geq Q^{1/2}$, getting

$$W \ll \sum_{Q^{1/2} \geq d \geq D} \frac{Q^2}{\varphi(d)^2 (\log Q)^2} + \sum_{d \geq Q^{1/2}} \frac{Q^2}{d^2} \ll \frac{Q^2(\log\log D)^2}{D(\log Q)^2} + Q^{3/2}.$$

Let $R(n)$ be the number of solutions of the equation $n = \lambda(pl)$ in pairs $(p, l) \in \mathcal{P}^2$, $1 < p < l \leq Q$, with $\gcd(p-1, l-1) < D$. Obviously $R(n) \leq D\tau(n)$. Indeed, we have at most $\tau(n)$ possibilities for $p-1$ and at most D possibilities for $\gcd(p-1, l-1)$. These two choices determine l.

Put $N = Q^2$. We say that n is *exceptional* if $\lambda(n) \le N \exp(-\Delta)$. Because we have $\Delta \ge (\log\log N)^3$, Lemma 5.1 may be applied. Let $\mathcal{E}$ denote the set of exceptional $n \le N$. Applying Lemma 5.1 and the Cauchy inequality and (5.3) we obtain

$$\begin{aligned}\sum_{n\in\mathcal{E}} R(n) &\le |\mathcal{E}|^{1/2}\left(\sum_{n\in\mathcal{E}} R(n)^2\right)^{1/2} \ll |\mathcal{E}|^{1/2} D\left(\sum_{n\le N} \tau(n)^2\right)^{1/2}\\ &\ll ND\exp\left(-0.34\,(\Delta\log\Delta)^{1/3}\right).\end{aligned}$$

Noting that the cardinality of the set $\mathcal{Q}$ of these pairs $(p,l) \in \mathcal{P}^2$, $1 < p < l \le Q$, for which $\lambda(pl)$ is exceptional is at most

$$|\mathcal{Q}| \le W + \sum_{n\in\mathcal{E}} R(n),$$

after simple calculations we derive the result. □

Now we show, using sieve methods, that for a large number of pairs $(p,l) \in \mathcal{P}^2$ we have $\lambda(\lambda(pl)) \gg pl$. Specifically we prove the following result.

Theorem 23.3. *There exist positive constants c_1 and c_2 such that, for more than $c_1Q^2/(\log Q)^4$ pairs $(p,l) \in \mathcal{P}^2$, $1 < p < l \le Q$, we have*

$$\lambda(\lambda(pl)) > c_2Q^2.$$

Proof. This proof contains a real parameter $\alpha > 0$ which will eventually be fixed. Throughout we assume that Q is chosen sufficiently large in terms of α. We require both upper and lower bound sieve results.

The upper bound we require is as follows. Let $\beta > 0$ and define

$$P = \prod_{p<Q^\beta} p.$$

Let $1 \le k \le Q^{1-\beta}$ be an integer satisfying $\gcd(k,P) = 1$. The number of primes $Q^\beta \le p < Q$ in the arithmetic progression $p \equiv 1 \pmod k$ and also satisfying $\gcd((p-1)/2, P) = 1$ is majorised by the number of positive integers $m \le Q$, $m \equiv 1 \pmod k$ for which $\gcd(m(m-1)/2, P) = 1$. Bounding this latter set above by means of the ("two-dimensional" upper bound) sieve, for example Theorem 5.1 of the standard reference [250], we see that this number is $O\left(Q/k\beta^2(\log Q)^2\right)$.

Using the lower bound sieve, see for example Theorem 8.3 of [250], and estimating the error term with the aid of the Bombieri–Vinogradov theorem, we may fix $\alpha > 0$ sufficiently small that, in every interval $(Q/2, Q)$ there are at least $cQ/\alpha(\log Q)^2$

primes p such that $(p-1)/2$ is free of primes less than Q^α where the constant $c > 0$ is absolute. We may assume, by discarding no more than $Q^{1-\alpha}$ of our primes, that $p-1$ is also square-free. By possibly making α somewhat smaller (and using the above upper bound sieve to remove the primes and products of two primes) we can also demand that $(p-1)/2$ be the product of three or more primes, the product of any two of which is thus no more than $Q^{1-\alpha}$. To remove the primes we merely take $k = 1$ and $\beta = 1/2$. To remove the products of two primes we group in accordance with the smaller of the two, calling that one k, apply the upper bound, again with $\beta = 1/2$, and then sum over k. The result follows from (5.6) in view of the fact that as α decreases $\log(1/\alpha)$ grows more slowly than $1/\alpha$.

Consider now the integers pl in the interval $\left(Q^2/4, Q^2\right)$ formed by pairing distinct primes p, l as above. There are at least $c^2Q^2/3\alpha^2(\log Q)^4$ of them.

By removing $O\left(Q^{2-\alpha}\right)$ of these pairs, we may assume that

$$\gcd\left(\frac{p-1}{2}, \frac{l-1}{2}\right) = 1$$

for all pairs under consideration. Thus, for such a pair (p, l) we may write

$$\frac{p-1}{2}\frac{l-1}{2} = q_1 \dots q_r, \tag{23.2}$$

where $q_1, \dots, q_r \in \mathcal{P}$ are distinct and each satisfies $Q^\alpha \le q_j < Q^{1-2\alpha}$, $j = 1, \dots, r$. In particular, $r < 2/\alpha$. Moreover we have

$$\lambda\left(\lambda(pl)\right) = \lambda\left(\operatorname{lcm}(p-1, l-1)\right) = \operatorname{lcm}(q_1 - 1, \dots, q_r - 1)$$

and for all large Q this satisfies

$$\lambda\left(\lambda(pl)\right) \ge \frac{(q_1-1)\dots(q_r-1)}{D(p;l)^{r^2/2}} \ge \frac{Q^2}{17D(p;l)^{r^2/2}},$$

where

$$D(p;l) = \max_{1\le i<j\le r} \gcd(q_i - 1, q_j - 1).$$

For given d, we next apply our upper bound sieve result to bound the number of pairs $(p, l) \in \mathcal{P}^2$, $1 < p < l \le Q$, for which

$$q_i \equiv q_j \equiv 1 \pmod d$$

for some $1 \le i < j \le r$, where $q_1, \dots, q_r$ are defined by (23.2). We choose $\beta = \alpha$. To bound the number of occurrences with q_i and q_j both coming from the same member of the pair, we apply the bound once with $k = q_iq_j < Q^{1-\alpha}$. For the case where each of the two comes from a different member of the pair, we apply the bound twice, once with $k = q_i$ and once with $k = q_j$. The number of pairs in question for both cases is bounded by

$$N_d \ll \frac{Q^2}{\alpha^4 (\log Q)^4} \left(\sum_{\substack{q\in\mathcal{P},\ Q^\alpha \le q < Q^{1-\alpha} \\ q\equiv 1 \pmod d}} \frac{1}{q} \right)^2.$$

For $d > \log Q$ we use (5.5) and for smaller d we use (5.6). These give

$$\sum_{\substack{q\in\mathcal{P},\ Q^\alpha \le q < Q^{1-\alpha} \\ q\equiv 1 \pmod d}} \frac{1}{q} \ll \frac{\log(1/\alpha)\log d}{\varphi(d)}.$$

Thus the number of pairs $(p, l) \in \mathcal{P}^2$, $1 < p < l \le Q$, in our set which satisfy $D(p; l) > D$ is bounded above by

$$\begin{aligned} \sum_{d>D} N_d &\ll \frac{Q^2}{\alpha^4 (\log Q)^4} \sum_{d>D} \left(\frac{\log(1/\alpha)\log d}{\varphi(d)} \right)^2 \\ &\ll \frac{Q^2}{(\log Q)^4} \frac{(\log(1/\alpha))^2}{\alpha^4} \frac{(\log D)^2}{D}. \end{aligned}$$

Provided D is chosen sufficiently large in terms of α, say $D = \alpha^{-3}$, and α is sufficiently small, this latter set makes a small contribution compared to those for which $D(p; l) \le D$ and the result follows. □

We remark that the constants c_1 and c_2 in Theorem 23.3 can be effectively evaluated.

Now we study the period of the sequence (u_n) given by (23.1) with $m = pl$ when the primes p and l, the initial value ϑ and the exponent e are all selected at random.

Theorem 23.4. *For Q sufficiently large and for any $\Delta \ge 6 (\log\log Q)^3$, and $\delta > 0$. the following statement holds for all pairs $(p, l) \in \mathcal{P}^2$, $1 < p < l \le Q$, except at most $Q^2 \exp\left(-0.16 (\Delta \log \Delta)^{1/3}\right)$ of them. For all pairs (ϑ, e) with*

$$1 \le \vartheta \le m-1, \qquad 1 \le e \le \lambda(m), \qquad \gcd(\vartheta, m) = \gcd(e, \lambda(m)) = 1,$$

where $m = pl$, except at most $m\lambda(m)^{1-\delta/2}$ of them, the period t of the sequence (u_n) given by (23.1) satisfies

$$t \ge m^{1-\delta}.$$

Proof. First of all we recall that the above period $t = \mathrm{ord}_T e$, where $T = \mathrm{ord}_m \vartheta$.

Let $\mathcal{R}$ be the set of the pairs $(p,l) \in \mathcal{P}^2$, $1 < p < l \le Q$, for which

$$\lambda(\lambda(pl)) \ge Q^2 \exp(-\Delta).$$

It follows from the above estimates and from Theorem 23.2 that

$$|\mathcal{R}| \ge \frac{1}{2}\pi(Q)^2 - Q^2 \exp\left(-0.16\,(\Delta \log \Delta)^{1/3}\right).$$

Let us fix some pair $(p,l) \in \mathcal{R}$ and put $m = pl$. We apply Theorem 23.1 with $K_1 = K_2 = \lambda(m)^{1-\delta/3}$. Together with (5.1) it follows that, except for at most

$$\varphi(m)\varphi(\lambda(m))\lambda(m)^{-\delta/2} \le m\lambda(m)^{1-\delta/2}$$

of them, the period t of the corresponding power generator is at least $\lambda(\lambda(m))\lambda(m)^{1-2\delta/3} \ge m^{1-\delta}$. □

One can certainly obtain a more precise statement than that of Theorem 23.4. In fact, Theorem 23.1 can be slightly improved as well.

A similar result also holds for p and l which are described in Theorem 23.3.

It would be of great interest to be able to give a result of the strength of Theorem 23.4 but where now e is kept fixed while the other parameters p, l, ϑ vary. This seems a much harder question and is quite reminiscent of the Gauss–Artin problem on primitive roots. Using the same ideas as in Theorem 23.3 we are able to show that quite often the period is reasonably large. Because of its special interest we consider the case when $e = 2$, that is the Blum–Blum–Shub generator.

Theorem 23.5. *Given $\varepsilon > 0$, there exist positive constants c, γ such that for Q sufficiently large, there are more than $cQ^2/(\log Q)^4$ pairs $(p,l) \in \mathcal{P}^2$, $p < l \le Q$, such that for all integers ϑ with*

$$1 \le \vartheta \le m-1 \qquad \text{and} \qquad \gcd(\vartheta, m) = 1,$$

where $m = pl$, except at most $m^{1-\gamma}$ of them, the period t of the sequence (u_n) given by (23.1) with $e = 2$ satisfies

$$t \ge cQ^{1-\varepsilon}.$$

Proof. We select the pairs $(p,l) \in \mathcal{P}^2$, $1 < p < l \le Q$, which gave us large values of $\lambda(\lambda(pl))$ as described in the proof of Theorem 23.3 and then eliminate those pairs for which

$$\mathrm{ord}_{q_1 \dots q_r} 2$$

is small. Here $q_1, \dots, q_r$ are given by (23.2) and also satisfy the condition that $\gcd(q_i - 1, q_j - 1)$ is bounded for all $i \ne j$. The number of primes q up to a bound B for which $\mathrm{ord}_q 2 < q^{1/2-\varepsilon/2}$ is at most $B^{1-\varepsilon}$, since these primes divide the product of $2^j - 1$ for j up to $B^{1/2-\varepsilon/2}$ and each factor $2^j - 1$ clearly has fewer than j prime factors. Thus, the number of primes $p \le Q$ with $p-1$ divisible by a prime $q > Q^\alpha$ with $\mathrm{ord}_q 2 < q^{1/2-\varepsilon/2}$ is $O\left(Q^{1-\alpha\varepsilon}\right)$. Hence, we may assume that the pairs p, l produced in Theorem 23.3 all have, for each i, $\mathrm{ord}_{q_i} 2 \ge q_i^{1/2-\varepsilon/2}$. Then

$$\begin{aligned} \mathrm{ord}_{q_1\dots q_r} 2 &= \mathrm{lcm}\,(\mathrm{ord}_{q_1} 2, \dots, \mathrm{ord}_{q_r} 2) \gg \mathrm{ord}_{q_1} 2 \cdot \ldots \cdot \mathrm{ord}_{q_r} 2 \\ &\ge (q_1 \dots q_r)^{1/2-\varepsilon/2} \gg (pl)^{1/2-\varepsilon/2} \gg Q^{1-\varepsilon}. \end{aligned}$$

But, by Lemma 5.2 the number of choices of ϑ up to m that are co-prime to m and with $\mathrm{ord}_m \vartheta < q_1 \dots q_r$ is $O\left(m/Q^\alpha\right)$. Hence, $\mathrm{ord}_m \vartheta$ is either $q_1 \dots q_r$ or $2q_1 \dots q_r$ (since $\lambda(m) = 2q_1 \dots q_r$) for all ϑ up to m and co-prime to m, except at most $O\left(m^{1-\alpha}\right)$ of them. For any choice of ϑ, the period t of the power generator with $e = 2$ is the order of 2 modulo the largest odd divisor of $\mathrm{ord}_m \vartheta$. So, but for few exceptional choices of ϑ, this period is $\mathrm{ord}_{q_1\dots q_r} 2$ and, as we have already seen, $\mathrm{ord}_{q_1\dots q_r} 2 \gg Q^{1-\varepsilon}$. This completes the proof. □

Several important cryptographic applications of these results have been given in [193], they are also discussed in Chapter 15.

Now we study the distribution of the power generator. In particular we show, under some natural conditions, that if the period $t \ge m^{3/4}$, then this generator produces a uniformly distributed sequence. As we have seen, typically the period is of order m; in this case the discrepancy of the power generator is $O(m^{-1/8+\varepsilon})$.

According to a standard principle, we can bound the discrepancy by bounding the corresponding exponential sums. Thus we define exponential sums

$$S_a = \sum_{n=1}^{t} \mathbf{e}_m(au_n).$$

The following estimate is due to [194]. It improves and generalises the first non-trivial estimate of such sums from [191]. The proof is based on a combination of some ideas from [236, 511] with a general approach to exponential sums with non-linear pseudorandom numbers given by (26.7) which has recently been developed in [128, 234, 245, 246, 431, 432, 433, 434, 436, 437]. This approach has also been used in [429, 440, 441, 442] to study several other pseudorandom numbers generated by similar methods. Outlines of several recent achievements can be found in [435, 438].

For $\nu \ge 1$, an integer parameter, we define

$$\alpha(\nu) = \frac{2\nu+1}{2\nu(\nu+1)} \qquad \text{and} \qquad \beta(\nu) = \frac{3\nu+2}{4\nu(\nu+1)}. \tag{23.3}$$

Theorem 23.6. *Let $m = pl$ where p and l are two distinct primes with*

$$\gcd(p-1, l-1) = \Delta.$$

If the sequence (u_n), given by (23.1), is purely periodic with period t, then for any integer $\nu \ge 1$ the bound

$$\max_{\gcd(a,m)=1} |S_a| \ll \Delta^{\alpha(\nu)} t^{1-\alpha(\nu)} m^{\beta(\nu)}$$

holds with $\alpha(\nu)$, $\beta(\nu)$ given by (23.3).

Proof. Let $T = \operatorname{ord}_m \vartheta$ be the multiplicative order of ϑ modulo m. Since (u_n) is purely periodic with period t,

$$\vartheta \equiv u_0 \equiv u_t \equiv \vartheta^{e^t} \pmod m.$$

Hence $e^t \equiv 1 \pmod T$. Therefore we conclude that $\gcd(e, T) = 1$ and that $t = \operatorname{ord}_T e$ is the multiplicative order of e modulo T.

We also remark that for any integers $\lambda \ge 0$ and $\mu \ge 1$ the sequence $(u_{\lambda+\mu n})$, is purely periodic with period $t/\gcd(\mu, t)$. Indeed, put

$$\rho = \vartheta^{e^\lambda} \qquad \text{and} \qquad f = e^\mu,$$

thus

$$u_{\lambda+\mu n} \equiv \rho^{f^n} \pmod m, \qquad n = 1, 2, \ldots .$$

We also see that

$$\operatorname{ord}_m \rho = T \qquad \text{and} \qquad \operatorname{ord}_T f = \frac{t}{\gcd(\mu, t)}.$$

Denote by t_p and t_l the periods of (u_n) modulo p and l, respectively.

First of all we consider the case $\gcd(t_p, t_l) = 1$; thus $t = t_p t_l$. Moreover in this case we can find integers q_p and q_l such that

$$q_p \equiv 1 \pmod{t_p}, \qquad q_p \equiv 0 \pmod{t_l},$$

and

$$q_l \equiv 0 \pmod{t_p}, \qquad q_l \equiv 1 \pmod{t_l}.$$

We can also find integers Q and R such that $Ql + Rp = 1$. Then we have

$$\begin{aligned} S_a &= \sum_{n=0}^{t_p-1} \sum_{k=0}^{t_l-1} \mathbf{e}_m\left(a\vartheta^{e^{nq_p+kq_l}}\right) = \sum_{n=0}^{t_p-1} \sum_{k=0}^{t_l-1} \mathbf{e}_m\left(a(Ql+Rp)\vartheta^{e^{nq_p+kq_l}}\right) \\ &= \sum_{n=0}^{t_p-1} \sum_{k=0}^{t_l-1} \mathbf{e}_p\left(aQ\vartheta^{e^{nq_p+kq_l}}\right) \mathbf{e}_l\left(aR\vartheta^{e^{nq_p+kq_l}}\right). \end{aligned}$$

Remarking that

$$\vartheta^{e^{nq_p+kq_l}} \equiv \vartheta^{e^{nq_p}} \pmod p \qquad \text{and} \qquad \vartheta^{e^{nq_p+kq_l}} \equiv \vartheta^{e^{kq_l}} \pmod l$$

we derive

$$\begin{aligned} S_a &= \sum_{n=0}^{t_p-1} \mathbf{e}_p\left(aQ\vartheta^{e^{nq_p}}\right) \sum_{k=0}^{t_l-1} \mathbf{e}_l\left(aR\vartheta^{e^{kq_l}}\right) \\ &= \sum_{n=0}^{t_p-1} \mathbf{e}_p\left(A\vartheta^{e^n}\right) \sum_{k=0}^{t_l-1} \mathbf{e}_l\left(B\vartheta^{e^k}\right), \end{aligned}$$

where $A = aQ$ and $B = aR$. We are going to apply Lemma 4.4 with modulus $M = T_p = \text{ord}_p\vartheta$. It is easy to see that the multiplicative order of e modulo T_p is just $\tau = t_p$. For some $\delta > 0$ and $h \geq T_p^\delta$ to be chosen later we select r as in Lemma 4.4. Let $\mathcal{L}$ denote the set of k which satisfy the corresponding congruence. Put $L = |\mathcal{L}|$. Then,

$$\sum_{n=0}^{t_p-1} \mathbf{e}_p\left(A\vartheta^{e^n}\right) = \frac{1}{L}\sum_{k\in\mathcal{L}}\sum_{n=0}^{t_p-1} \mathbf{e}_p\left(A\vartheta^{e^{n+k}}\right).$$

By the Hölder inequality we have

$$\begin{aligned} \left|\sum_{n=0}^{t_p-1} \mathbf{e}_p\left(A\vartheta^{e^n}\right)\right|^{2\nu} &\leq L^{-2\nu}t_p^{2\nu-1}\sum_{n=0}^{t_p-1}\left|\sum_{k\in\mathcal{L}} \mathbf{e}_p\left(A\vartheta^{e^{n+k}}\right)\right|^{2\nu} \\ &= L^{-2\nu}t_p^{2\nu-1}\sum_{n=0}^{t_p-1}\left|\sum_{k\in\mathcal{L}} \mathbf{e}_p\left(A\vartheta^{e^n e^k}\right)\right|^{2\nu}. \end{aligned}$$

Let $d = (p-1)/T_p$. Obviously, the powers ϑ^{e^n}, $n = 0, \dots, t_p - 1$, are pairwise distinct and are each of the form x^d modulo p. Moreover, for each n there are precisely d values of $x = 1, \dots, p-1$ which give rise to this value of ϑ^{e^n}. Thus by replacing ϑ^{e^n} by x^d we may write this last sum over n as d^{-1} times the corresponding sum over x where x runs over a certain subset of the non-zero residue classes modulo p. Using positivity we may majorize this last sum by the same sum over all classes modulo p. In this way we obtain

$$\begin{aligned}
&\left|\sum_{n=0}^{t_p-1} \mathbf{e}_p\left(A\vartheta^{e^n}\right)\right|^{2\nu} \\
&\quad \le L^{-2\nu} d^{-1} t_p^{2\nu-1} \sum_{x=0}^{p-1} \left|\sum_{k\in\mathcal{L}} \mathbf{e}_p\left(Ax^{de^k}\right)\right|^{2\nu} \\
&\quad = L^{-2\nu} d^{-1} t_p^{2\nu-1} \sum_{j_1,\dots,j_\nu\in\mathcal{L}} \ \sum_{k_1,\dots,k_\nu\in\mathcal{L}} \\
&\qquad \times \sum_{x=0}^{p-1} \mathbf{e}_p\left(A\left(x^{de^{j_1}} + \ldots + x^{de^{j_\nu}} - x^{de^{k_1}} - \ldots - x^{de^{k_\nu}}\right)\right) \\
&\quad = L^{-2\nu} d^{-1} t_p^{2\nu-1} \sum_{j_1,\dots,j_\nu\in\mathcal{L}} \ \sum_{k_1,\dots,k_\nu\in\mathcal{L}} \\
&\qquad \times \sum_{x=0}^{p-1} \mathbf{e}_p\left(A\left(x^{dre^{j_1}} + \ldots + x^{dre^{j_\nu}} - x^{dre^{k_1}} - \ldots - x^{dre^{k_\nu}}\right)\right)
\end{aligned}$$

because $\gcd(r, T_p) = 1$. For the case that $(k_1, \dots, k_\nu)$ is a permutation of $(j_1, \dots, j_\nu)$ we must use the trivial bound and this gives a contribution $O(L^\nu p)$. In case this does not happen (there are at most $L^{2\nu}$ ways) the inner sum above is a character sum with a polynomial of degree at most dh. By Lemma 3.1 each of these terms contributes at most $dhp^{\frac{1}{2}}$. Thus

$$\left|\sum_{n=0}^{t_p-1} \mathbf{e}_p\left(A\vartheta^{e^n}\right)\right|^{2\nu} \ll L^{-2\nu} d^{-1} t_p^{2\nu-1} \left(L^\nu p + L^{2\nu} dhp^{\frac{1}{2}}\right)$$

and so

$$\begin{aligned}
\left|\sum_{n=0}^{t_p-1} \mathbf{e}_p\left(A\vartheta^{e^n}\right)\right| &\ll t_p^{1-1/2\nu}\left(L^{-1/2}p^{1/2\nu}d^{-1/2\nu} + h^{1/2\nu}p^{1/4\nu}\right) \\
&\ll t_p^{1-1/2\nu}\left(L^{-1/2}T_p^{1/2\nu} + h^{1/2\nu}p^{1/4\nu}\right).
\end{aligned}$$

By Lemma 4.4 we have $L \gg t_p h/T_p$. We substitute this in, use the trivial bound $T_p \le p$ and equalize by choosing

$$h = \left\lceil p^{(2\nu+1)/2(\nu+1)} t_p^{-\nu/(\nu+1)} \right\rceil \ge p^{1/2(\nu+1)},$$

which thus satisfies $h \ge T_p^\delta$ with $\delta = 1/2(\nu+1)$. After a simple computation we obtain

$$\left|\sum_{n=0}^{t_p-1} \mathbf{e}_p\left(A\vartheta^{e^n}\right)\right| \ll t_p^{1-\alpha(\nu)} p^{\beta(\nu)}$$

and similarly

$$\left|\sum_{k=0}^{t_l-1} \mathbf{e}_l\left(B\vartheta^{e^k}\right)\right| \ll t_l^{1-\alpha(\nu)} l^{\beta(\nu)}$$

from which the result follows, provided that $\gcd(t_p, t_l) = 1$.

In the general case we put $\mu = \gcd(t_p, t_l)$ and remark that

$$S_a = \sum_{n=0}^{t-1} \mathbf{e}_m\left(au_n\right) = \sum_{\lambda=1}^{\mu} \sum_{n=0}^{t/\mu-1} \mathbf{e}_m\left(au_{\lambda+n\mu}\right).$$

From the above observation about the period of $(u_{\lambda+n\mu})$ one easily verifies that this sequence has relatively prime periods modulo p and l (because they are divisors of t_p/μ and t_l/μ, respectively). By the above bound we obtain

$$S_a \ll \mu(t/\mu)^{1-\alpha(\nu)} m^{\beta(\nu)}. \tag{23.4}$$

Now from $\mu \le \gcd(p-1, l-1) = \Delta$ the result follows. □

It is easy to check that for a given $\delta > 0$ one may by selecting sufficiently large ν, obtain a bound which is non-trivial for all $t \ge \Delta m^{3/4+\delta/2}$. If also $\gcd(p-1, l-1) \ll m^{\delta/2}$, then it suffices to take $t \ge m^{3/4+\delta}$. Under these conditions an asymptotically optimal choice for small δ is $\nu = \lceil 1/2\delta \rceil$ which gives the bound $O(t^{1-c\delta^2})$ with an absolute constant $c > 0$. For sufficiently small δ, this holds with any value $c < 2/3$. On the other hand, for large values of t the choice $\nu = 1$ becomes optimal, yielding the bound

$$\max_{\gcd(a,m)=1} |S_a| \ll \Delta^{3/4} t^{1/4} m^{5/8}.$$

In the most interesting case that $t \gg m^{1-\delta/2}$ and $\Delta \ll m^{\delta/2}$ with a small δ, selecting $\nu = 1$ we obtain

$$\max_{\gcd(a,m)=1} |S_a| \ll t^{7/8+\delta}. \tag{23.5}$$

Moreover, as the results of [192], which we have presented in Theorems 23.1 and 23.4, show, for a random choice of the parameters of the generator this case occurs almost always.

We note that although the condition $\gcd(p-1, l-1) \ll m^{\varepsilon}$ is satisfied for almost all pairs of primes p and l, in fact our method works without any restriction on $\gcd(p-1, l-1)$. Obviously, the parameter μ in the proof of Theorem 23.6 can be estimated from the inequality $t \le m/\mu$ which, being substituted in (23.4) leads to the estimate

$$\max_{\gcd(a,m)=1} |S_a| \ll t^{1-(2\nu+1)/\nu(\nu+1)} m^{(7\nu+4)/4\nu(\nu+1)}$$

which holds for any p and l. One can also modify the scheme a little to prove results under the natural assumption that the primes p and l are about the same order. For example, for the Blum integers $m = pl$ with $p < l \le p^{1+\varepsilon}$ and $p \equiv l \equiv 3 \pmod 4$ we have

$$\max_{\gcd(a,m)=1} |S_a| \ll t^{1-(3\nu+1)/2\nu(2\nu+1)} m^{(5\nu+2)/4\nu(2\nu+1)+\varepsilon}.$$

Several more statements of this kind can be proved as well, however in the case $\gcd(p-1, l-1) \ll m^{\varepsilon}$ the bound of Theorem 23.6 supersedes them all.

To study the distribution of s-tuples $(u_n, \dots, u_{n+s-1})$ we need to estimate more general sums. For an integer vector $\mathbf{a} = (a_0, \dots, a_{s-1}) \in \mathbb{Z}^s$ we define the exponential sum

$$S_{\mathbf{a}} = \sum_{n=1}^{t} \mathbf{e}_m \left(\sum_{i=0}^{s-1} a_i u_{n+i} \right).$$

Although in the next result we estimate these sums only for the special case of the Blum–Blum–Shub generator, it can be extended to any power generator with a small value of e.

Theorem 23.7. *Let $m = pl$ where p and l are two distinct primes with*

$$\gcd(p-1, l-1) = \Delta.$$

If the sequence (u_n), given by (23.1) with $e = 2$, is purely periodic with period t, then for any integer $\nu \ge 1$ and any dimension $s \ge 1$, the bound

$$\max_{\gcd(a_0,\dots,a_{s-1},m)=1} |S_{\mathbf{a}}| \ll 2^{s\nu/(\nu+1)} \Delta^{\alpha(\nu)} t^{1-\alpha(\nu)} m^{\beta(\nu)}$$

holds with $\alpha(\nu)$, $\beta(\nu)$ given by (23.3).

Proof. We put

$$h = \left\lceil 2^{-2s\nu/(\nu+1)} p^{(2\nu+1)/2(\nu+1)} t_p^{-\nu/(\nu+1)} \right\rceil \gg p^{1/2(\nu+1)}$$

and proceed as in Theorem 23.6 getting the sum

$$\sum_{x=0}^{p-1} \mathbf{e}_p \left(A \sum_{i=0}^{s-1} \sum_{\eta=1}^{\nu} a_i \left(x^{dr2^{j_\eta+i}} - x^{dr2^{k_\eta+i}} \right) \right),$$

which is an exponential sum with a polynomial of degree at most $2^{s-1}h$. Continuing as in the proof of Theorem 23.6, after simple calculations we obtain the desired result. □

We remark that both Theorem 23.6 and Theorem 23.7 apply to prime moduli m as well. In fact in this case even stronger results can be obtained, see [187]. For arbitrary composite moduli, analogues of these results are obtained in [189]. In particular the results of [187, 189] imply some uniformity of distribution results of the residues of the doubly exponential function g^{e^x} which occur in some cryptographic constructions, for example, see [91].

Now we show that a positive proportion of the leftmost and rightmost bits of (u_n) are uniformly distributed. Let σ be a binary string of length k. Denote by $L(\sigma)$ the number of $n = 1, \dots, t$, such that σ is the string of the k least significant bits of u_n.

Theorem 23.8. *Let $m = pl$ where p and l are two distinct odd primes with*

$$\gcd(p-1, l-1) = \Delta.$$

If the sequence (u_n), given by (23.1), is purely periodic with period t, then for any integer $\nu \geq 1$ the bound

$$\sup_{\sigma} \left| L(\sigma) - t2^{-k} \right| \ll \Delta^{\alpha(\nu)} t^{1-\alpha(\nu)} m^{\beta(\nu)} \log m$$

holds with $\alpha(\nu)$, $\beta(\nu)$ given by (23.3) and where the supremum is taken over all binary strings σ of length k.

Proof. We denote by $\overline{\sigma}$ the integer whose bit representation coincides with σ and put $K = 2^k$, $H = \lfloor (m - 1 - \overline{\sigma})/K \rfloor$. We remark that $L(\sigma)$ is equal to the number $W(\sigma)$ of solutions of the congruence

$$u_n \equiv Kx + \overline{\sigma} \pmod m, \qquad 1 \leq n \leq t,\ 0 \leq x \leq H.$$

Thus, using (3.2) we write

$$\begin{aligned} W(\sigma) &= \frac{1}{m} \sum_{n=1}^{t} \sum_{x=0}^{H} \sum_{a=0}^{m-1} \mathbf{e}_m \left(a \left(u_n - Kx - \overline{\sigma} \right) \right) \\ &= \frac{1}{m} \sum_{a=0}^{m-1} \mathbf{e}_m \left(-a\overline{\sigma} \right) S_a \sum_{x=0}^{H} \mathbf{e}_m(-aKx). \end{aligned}$$

The term corresponding to $a = 0$ equals

$$t(H+1)m^{-1} = t2^{-k} + O(1),$$

which gives the main term of the desired formula, apart from an admissible error. To estimate the contribution R of the remaining terms we apply Theorem 23.6

getting

$$R \ll \Delta^{\alpha(\nu)} t^{1-\alpha(\nu)} m^{-1+\beta(\nu)} \sum_{\substack{a=1 \\ \gcd(a,m)=1}}^{m-1} \left| \sum_{x=0}^{H} \mathbf{e}_m(-aKx) \right|$$
$$+ m^{-1} \sum_{a=1}^{l-1} |S_{ap}| \left| \sum_{x=0}^{H} \mathbf{e}_l(-aKx) \right| + m^{-1} \sum_{a=1}^{p-1} |S_{al}| \left| \sum_{x=0}^{H} \mathbf{e}_p(-aKx) \right| .$$

We still need to estimate the sums $|S_{ap}|$ and $|S_{al}|$. Let t_p be the period of the sequence (u_n) modulo p so that $t \le t_p(l-1)$. It is easy to check that the bound of Theorem 23.6 applies to the case when m is prime as well, and moreover the factor related to the $\gcd(p-1, l-1)$ does not appear in this case. Therefore,

$$S_{al} = \frac{t}{t_p} \sum_{n=1}^{t_p} \mathbf{e}_p(au_n) \ll l t_p^{1-\alpha(\nu)} p^{\beta(\nu)}$$

if $\gcd(a,p) = 1$. A similar estimate also holds for S_{ap}.

Remarking that $\gcd(K,m) = 1$, we see that $-aK$ can be replaced by just a. Therefore, applying (3.4), after some simple calculations we obtain the desired estimate. □

Virtually the same proof yields the same result for the most significant bits. One simply replaces $Kx + \overline{\sigma}$ by $x + M\overline{\sigma}$ for suitable M in the above congruence for u_n. Since the most significant bits of a number are the ones most responsible for locating it as a point on the line, this case may also be formulated somewhat differently. Namely, from Lemma 4.1 (applied with $N = 1$, $L = m - 1$) one easily derives the following result.

Theorem 23.9. *Let $m = pl$ where p and l are two distinct odd primes with*

$$\gcd(p-1, l-1) = \Delta.$$

If the sequence (u_n), given by (23.1), is purely periodic with period t, then for any integer $\nu \ge 1$ for the discrepancy D of the sequence

$$\left\{ \frac{u_n}{m} \right\}, \qquad n = 1, \ldots, t,$$

the bound

$$D \ll \Delta^{\alpha(\nu)} t^{-\alpha(\nu)} m^{\beta(\nu)} \log m$$

holds with $\alpha(\nu)$, $\beta(\nu)$ given by (23.3).

Under the condition $\gcd(p-1.l-1) \ll m^{\varepsilon}$ the error term in Theorem 23.8 is of the form $t^{1-(2\nu+1)/2\nu(\nu+1)}m^{(3\nu+2)/4\nu(\nu+1)+\varepsilon}$ and the error term in Theorem 23.9 is of the form $t^{-(2\nu+1)/2\nu(\nu+1)}m^{(3\nu+2)/4\nu(\nu+1)+\varepsilon}$.

With very little change we can use Theorem 23.7 to also study the multidimensional distribution of the Blum–Blum–Shub generator. Let $\Sigma = (\sigma_0, \dots, \sigma_{s-1})$ be a collection of s binary strings of length k. Denote by $L(\Sigma)$ the number of $n = 1, \dots, t$, such that σ_i is the string of the k least significant bits of u_{n+i}, $i = 0, \dots, s-1$. Combining Theorem 23.7 with the same arguments used in the proof of Theorem 23.8 we obtain:

Theorem 23.10. *Let $m = pl$ where p and l are two distinct odd primes with*

$$\gcd(p-1, l-1) = \Delta.$$

There exists an absolute constant $C > 0$ such that if the sequence (u_n), given by (23.1) with $e = 2$, is purely periodic with period t, then for any integer $\nu \ge 1$ and every integer s the bound

$$\sup_{\Sigma} \left| L(\Sigma) - t2^{-ks} \right| \ll \Delta^{\alpha(\nu)} t^{1-\alpha(\nu)} m^{\beta(\nu)} (C \log m)^s$$

holds with $\alpha(\nu)$, $\beta(\nu)$ given by (23.3), where the supremum is taken over all collections Σ of s binary strings of length k.

Accordingly, Theorem 23.7 and Lemma 4.1 yield:

Theorem 23.11. *Let $m = pl$ where p and l are two distinct primes with*

$$\gcd(p-1, l-1) = \Delta.$$

There exists an absolute constant $C > 0$ such that if the sequence (u_n), given by (23.1) with $e = 2$, is purely periodic with period t, then for any integer $\nu \ge 1$ and every integer s for the discrepancy D of the sequence

$$\left(\left\{ \frac{u_n}{m} \right\}, \dots, \left\{ \frac{u_{n+s-1}}{m} \right\}, \right), \qquad n = 1, \dots, t,$$

the bound

$$D \ll \Delta^{\alpha(\nu)} t^{-\alpha(\nu)} m^{\beta(\nu)} (C \log m)^s$$

holds with $\alpha(\nu)$, $\beta(\nu)$ given by (23.3).

We see that if $\gcd(p-1, l-1) \ll m^{\delta/2}$ and $t \ge m^{3/4+\delta}$, then, for sufficiently large ν depending on δ, the bounds of Theorems 23.10 and 23.11 are non-trivial, that is of the form $o(t)$ and $o(1)$, respectively, for all $s \le c\delta^2 \log m / \log\log m$ with an absolute constant $c > 0$. In particular, under this condition Theorem 23.11 gives the statement of uniform distribution of s-tuples produced by the Blum–Blum–Shub generator.

Let us consider the binary sequence (ξ_n) where ξ_n is the rightmost bit of u_n. We note that all cryptographic applications make use of this sequence instead of the original sequence (u_n), see [50, 137, 138, 182, 254, 319, 389, 541]. On the other hand, although as we have seen the period t of the sequence (u_n) admits a reasonably simple number-theoretic characterization via the Carmichael function, the period τ of the sequence (ξ_n) does not seem to be easy to evaluate. It is obvious that τ satisfies the property $\tau|t$. Also, a certain lower bound for the Blum–Blum–Shub generator is valid for some very special moduli m, see Section 14.8 of [138]. The paper [194] provides the first general non-trivial lower bound on τ which we present here.

First of all we note that using Theorem 23.10 one can easily derive that if, for some fixed $\delta > 0$, the primes p and l and the period t satisfy the conditions $\gcd(p-1, l-1) \ll m^{\delta/2}$ and $t \ge m^{3/4+\delta}$, then τ is exponentially large in the bit size of m. Indeed, taking $k = 1$ in Theorem 23.10, we note that $L(\Sigma) > 0$ provided that $s \le c(\delta) \log m / \log\log m$ for some constant $c(\delta) > 0$. For such s there are at least 2^s distinct s-tuples of the form $(\xi_n, \dots, \xi_{n+s-1})$, so that, choosing s as large as permitted, we find

$$\tau \ge 2^s \gg 2^{c(\delta) \log m / \log\log m}.$$

Now we refine this bound by using a modified approach which allows us to show that all $L(\Sigma) > 0$ for a slightly larger value of s by counting, instead of $L(\Sigma)$ itself, a weighted sum over the integers n which contribute to it.

Theorem 23.12. *Let $m = pl$ where p and l are distinct odd primes with*

$$\gcd(p-1, l-1) = \Delta.$$

If the sequence (u_n), given by (23.1) with $e = 2$, is purely periodic with period $t \ge \Delta m^{3/4+\delta}$ for some fixed $\delta > 0$, then there exists a constant $\gamma(\delta) > 0$ such that the bound

$$\tau \gg m^{\gamma(\delta)}$$

holds.

Proof. Using Theorem 23.7 with sufficiently large ν, we derive that there exists $\eta > 0$ such that

$$S_{\mathbf{a}} \ll 2^s t^{1-\eta} \tag{23.6}$$

for any vector $\mathbf{a} = (a_0, \dots, a_{s-1})$ with $\gcd(a_0, \dots, a_{s-1}, m) = 1$. Let t_p and t_l be the periods of the sequence (u_n) modulo p and l, respectively. It is also easy to see that the same considerations also give the bounds

$$S_{l\mathbf{a}} \ll 2^s l t_p^{1-\eta} \qquad \text{and} \qquad S_{p\mathbf{a}} \ll 2^s p t_l^{1-\eta} \tag{23.7}$$

for vectors with $\gcd(a_0, \dots, a_{s-1}, p) = 1$ and $\gcd(a_0, \dots, a_{s-1}, l) = 1$, respectively.

Let $\sigma = (\sigma_1, \dots, \sigma_s)$ be a binary string of length $s \geq 1$. Denote by $Q(\sigma)$ the number of $n = 1, \dots, t$, such that $(\xi_n, \dots, \xi_{n+s-1}) = \sigma$. Put $H = \lfloor m/4 \rfloor$. Denote by $W(\sigma)$ the number of solutions of the system of congruences

$$u_{n+i} \equiv 2(H + x_i - y_i) + \sigma_i \pmod{m}, \qquad 0 \leq i \leq s-1,$$

where

$$1 \leq n \leq t, \qquad 0 \leq x_0, \dots, x_{s-1}, y_0, \dots, y_{s-1} \leq H-1.$$

It is obvious that if $W(\sigma) > 0$, then $Q(\sigma) > 0$.

Using (3.2), we obtain

$$\begin{aligned} W(\sigma) &= \frac{1}{m^s} \sum_{n=1}^{t} \sum_{x_0, \dots, x_{s-1}=0}^{H-1} \sum_{y_0, \dots, y_{s-1}=0}^{H-1} \\ &\qquad \times \sum_{a_0, \dots, a_{s-1}=0}^{m-1} \mathbf{e}_m \left(\sum_{i=0}^{s-1} a_i \left(u_{n+i} - 2x_i + 2y_i - 2H - \sigma_i \right) \right) \\ &= \frac{1}{m^s} \sum_{a_0, \dots, a_{s-1}=0}^{m-1} \mathbf{e}_m \left(- \sum_{i=0}^{s-1} a_i \left(2H + \sigma_i \right) \right) S_{\mathbf{a}} \\ &\qquad \times \sum_{x_0, \dots, x_{s-1}=0}^{H-1} \sum_{y_0, \dots, y_{s-1}=0}^{H-1} \mathbf{e}_m \left(-2 \sum_{i=0}^{s-1} a_i \left(x_i - y_i \right) \right) \\ &= \frac{1}{m^s} \sum_{a_0, \dots, a_{s-1}=0}^{m-1} \mathbf{e}_m \left(- \sum_{i=0}^{s-1} a_i \left(2H + \sigma_i \right) \right) S_{\mathbf{a}} \\ &\qquad \times \prod_{i=0}^{s-1} \left| \sum_{x_i=0}^{H-1} \mathbf{e}_m \left(2a_i x_i \right) \right|^2 . \end{aligned}$$

The term corresponding to $a_0 = \dots = a_{s-1} = 0$ is equal to $tH^{2s}m^{-s}$. To estimate the contribution of the terms with $\gcd(a_0, \dots, a_{s-1}, m) = 1$ we apply (23.6) together with the identity (m is odd)

$$\sum_{a_0, \dots, a_{s-1}=0}^{m-1} \prod_{i=0}^{s-1} \left| \sum_{x_i=0}^{H-1} \mathbf{e}_m \left(2a_i x_i \right) \right|^2 = \left(\sum_{a=0}^{m-1} \left| \sum_{x=0}^{H-1} \mathbf{e}_m \left(ax \right) \right|^2 \right)^s = m^s H^s.$$

To estimate the contribution of the terms with $\gcd(a_0, \dots, a_{s-1}, m) = p$ and $\gcd(a_0, \dots, a_{s-1}, m) = l$ we use (23.7) together with the analogue of the above identity modulo l and p, respectively. This gives

$$W(\sigma) = tH^{2s}m^{-s} + O(2^s t^{1-\eta} H^s) = tH^s \left(H^s m^{-s} + O(2^s t^{-\eta}) \right).$$

We can assume that $m \geq 15$, thus

$$H^s m^{-s} \geq \left(\frac{m-3}{4m}\right)^s \geq 5^{-s}.$$

Therefore there is a constant $\gamma(\delta) > 0$ such that $W(\sigma) > 0$ provided that we have $s \leq \gamma(\delta) \log m$, and from this the result follows. □

It is easy to see that if $t \ll m \ll t^{1+\varepsilon}$, then the bound (23.5) implies that $\tau \geq m^{1/24-\varepsilon}$.

This also implies a lower bound for the linear complexity L of the sequence (ξ_n) over $\mathbb{F}_2$. We recall that the linear complexity of a periodic sequence is defined as the smallest possible order of a linear recurrence relation which this sequence satisfies, see [138, 389, 483, 484]. It is clear that all L-tuples $(\xi_n, \dots, \xi_{n+L-1})$, $n = 1, \dots, \tau$, are pairwise distinct. Hence $\tau \leq 2^L$ and from Theorem 23.12 we derive the bound

$$L \geq \gamma(\delta) \log m. \tag{23.8}$$

There is another way to guarantee that the period of (ξ_n) is large enough. Indeed, this period is obviously a divisor of the period of (u_n). Thus, if $\lambda(\lambda(pl))$ has no small prime divisors (except 2) then obviously the period of (ξ_n) is not too small (or is just 2 which can easily be detected and avoided). If one is willing to accept heuristic results then, as a simple consequence of the well-known conjecture about prime k-tuples, see [19], there are in every interval $(x, 2x)$ with large x, at least $c_1 x/(\log x)^3$ primes p such that $q = (p-1)/2$ and $r = (q-1)/2$ are both also prime. That is, we request that r, $q = 2r+1$, and $p = 4r+3$ are prime. It follows on pairing such primes p that there are at least $c_2 Q^2/(\log Q)^6$ pairs of primes (p, l) with $Q/2 < p < l \leq Q$ for which $\lambda(\lambda(pl)) = 2rs$ where $r = p/4 + O(1)$ and $s = l/4 + O(1)$ are primes. Thus the period of (ξ_n) is either 2 or at least $2^{-5/2} m^{1/2}$. However, obtaining unconditional results about the arithmetic structure of the values of the Carmichael function (as is the case for the Euler function) is a notoriously hard problem.

Now we present the lower bounds of [511] on the linear complexity of the power generator considered in the ring $\mathbb{Z}_m$ which are much stronger than (23.8).

First of all we consider the case of prime moduli.

Theorem 23.13. *Let $m = p$ be prime. Assume that the sequence (u_n), given by (23.1) with $m = p$, is purely periodic with period t. Then, for the linear complexity L_p of this sequence the bound*

$$L_p \geq \frac{t^2}{p-1}$$

holds.

Proof. Let τ be the largest positive integer for which the powers e^x, $x = 1, \dots, \tau$, are pairwise distinct modulo $p-1$. Obviously $\tau \ge t$.

It is easy to see that there exists a, $0 \le a \le p-1$ such that the number T of solutions of the congruence

$$e^x \equiv a + y \pmod{p-1}, \qquad 1 \le x \le \tau,\ 0 \le y \le t-1,$$

satisfies

$$T \ge \frac{\tau t}{p-1} \ge \frac{t^2}{p-1}.$$

Let $j_1, \dots, j_T$ be the corresponding values of x and let $k_1, \dots, k_T$ be the corresponding values of y.

Assume that $L_p \le T-1$. Remarking that

$$u_{n+j_i} \equiv \vartheta^{e^{n+j_i}} \equiv u_n^{e^{j_i}} \equiv u_n^{a+k_i} \pmod p, \qquad n = 1, 2, \dots,\ i = 1, \dots, T,$$

from Lemma 2.1 we see that there exist integers $c_1, \dots, c_T$, not all equal to zero modulo p, such that

$$\sum_{i=1}^{T} c_i u_n^{a+k_i} \equiv 0 \pmod p, \qquad n = 1, 2, \dots.$$

Because $u_n \not\equiv 0 \pmod p$, $n = 1, 2, \dots$, we conclude that the non-zero modulo p polynomial

$$f(X) = \sum_{i=1}^{T} c_i X^{k_i}$$

of degree

$$\deg f \le \max_{1 \le i \le T} k_i \le t-1$$

has t distinct zeros u_n, $n = 1, \dots, t$, modulo p, which is impossible. Hence $L_p \ge T$ and the desired result follows. □

Thus, when t is of order p, that is, in the case that is most interesting for applications, the linear complexity is $L_p = t + O(t^{1/2})$, that is, it is very close to its largest possible value t.

We use Theorem 23.13 to estimate the linear complexity modulo a Blum integer.

Theorem 23.14. *Let $m = pl$, where p and l are two distinct primes. Assume that the sequence (u_n), given by (23.1), is purely periodic with period t. Then for the linear complexity L_m of this sequence the bound*

$$L_m \ge t\varphi(m)^{-1/2}$$

holds.

Proof. Let t_p be the period of the sequence (u_n) modulo p and let t_l be the period of the sequence (u_n) modulo l. Obviously $t \le t_p t_l$. Therefore

$$\frac{t_p^2 t_l^2}{(p-1)(l-1)} \ge \frac{t^2}{\varphi(m)}.$$

Without loss of generality we may assume that

$$\frac{t_p^2}{p-1} \ge t\varphi(m)^{-1/2}.$$

Using that L_m is not smaller than the linear complexity modulo p, from Theorem 23.13 we derive the desired statement. □

It is easy to see from the Chinese Remainder Theorem that the linear complexity modulo a composite number m is equal to the largest linear complexity modulo all prime power divisors of m. So for a Blum integer $m = pl$ with $p \sim l \sim m^{1/2}$ the linear complexity L_m does not exceed

$$L_m = \max\{L_p, L_l\} \le \max\{t_p, t_l\} \le \max\{p, l\} = (1 + o(1))m^{1/2},$$

where L_p and t_p are the linear complexity and the period modulo p and L_l and t_l are the linear complexity and the period modulo l. Thus, in the case that is most interesting for applications, when t is of order m, Theorem 23.14 and the above inequality imply that L_m is of the largest possible order $m^{1/2}$.

The results of Theorems 23.13 and 23.14 demonstrate that the power generator does not contain any hidden linear structure. In particular, this rules out a possibility of the lattice reduction attack on this generator and even its truncated versions, see [197, 287, 313, 319] where such attacks have been described on the linear congruential generator

$$x_n \equiv ax_{n-1} + b \pmod m, \quad 0 \le x_n \le m-1, \qquad n = 1, 2, \ldots,$$

where a and b are integers with $\gcd(a, m) = 1$.

Lower bounds on the linear complexity profile $L_m(N)$ are also known, see [236]. It is shown in [236] that

$$L_m(N) \ge \frac{\min\{N^2, 4t\}}{4\max\{p-1, l-1\}}, \qquad 1 \le N \le t.$$

Provided that $p \sim l$ and $t \ge \varphi(m)^{1/2+\varepsilon}$, this bound is non-trivial for values of N beginning with $N \ge \varphi(m)^{1/4+\varepsilon}$. An alternative method for estimation of the linear complexity profile of the Blum–Blum–Shub generator has been proposed in [236] as well. This method leads to the estimate

$$L_m(N) \ge \min\left\{\frac{(8N+1)^{1/2} - 3}{2}, \frac{t}{\varphi(m)^{1/2}}\right\}$$

on the linear complexity profile of the Blum–Blum–Shub generator. For $t \geq \varphi(m)^{1/2+\varepsilon}$, the bound is non-trivial for all $N \geq 5$.

The distribution of the power generator modulo a power of a fixed prime number has been studied in [188].

Finally we remark that several new results about the uniformity distribution of elements of very general non-linear pseudorandom number generators have recently been obtained in [128, 234, 245, 246, 431, 432, 433, 434, 436, 437], see also the surveys [435, 438]. We present some of these results in Chapter 26.

Chapter 24

Naor–Reingold Function

Let p and l be prime numbers with $l|p-1$.

Select an element $g \in \mathbb{F}_p^*$ of multiplicative order l. Then for each r-dimensional vector $\mathbf{a} = (a_1, \dots, a_r) \in \mathbb{F}_l^r$ one can define the function

$$f_{\mathbf{a}}(x) = g^{a_1^{x_1} \dots a_r^{x_r}} \in \mathbb{F}_p, \tag{24.1}$$

where $x = x_1 \dots x_r$ is the bit representation of an integer x, $0 \le x \le 2^r - 1$, with some extra leading zeros if necessary.

In [405] the function $f_{\mathbf{a}}(x)$ has been proposed as an efficient pseudo-random function (for a randomly chosen vector $\mathbf{a} \in \mathbb{F}_l^r$). It is shown in [405] that this function can be computed in parallel by threshold circuits of bounded depth and polynomial size (see also [311]) and has some very desirable security property, provided certain standard cryptographic assumptions hold.

Here we show that the Naor–Reingold pseudorandom function has several more useful features, which comes as an additional bonus to the aforementioned cryptographic properties of $f_{\mathbf{a}}(x)$. Namely, we prove that for almost all vectors $\mathbf{a} \in \mathbb{F}_l^r$, the sequence $f_{\mathbf{a}}(x)$, $x = 0, 1, \dots, 2^r - 1$, is asymptotically uniformly distributed, provided that $l \ge p^{1/3+\varepsilon}$ with any fixed $\varepsilon > 0$. We also obtain a lower bound on its linear complexity. These results have been obtained in [513, 517]. Several other results, including the generalisation of the above construction to elliptic curves and bounds of the linear and non-linear complexity can be found in [22, 235, 512, 528].

We denote by $D_{l,p,g}(\mathbf{a})$ the discrepancy of the set

$$\{f_{\mathbf{a}}(x) \mid x = 0, 1, \dots, 2^r - 1\}.$$

Theorem 24.1. *Let $r = \lfloor \log p \rfloor$. Then the bound*

$$\frac{1}{l^r} \sum_{\mathbf{a} \in \mathbb{F}_l{}^r} D_{l,p,g}(\mathbf{a})^2 \ll \Delta(l, p)$$

holds, where

$$\Delta(l,p) = \begin{cases} p^{1-\gamma}l^{-1}\log^2 p, & \text{if } l \ge p^{\gamma}, \\ pl^{-2}\log^2 p, & \text{if } p^{\gamma} > l \ge p^{2/3}, \\ p^{1/2}l^{-5/4}\log^2 p, & \text{if } p^{2/3} > l \ge p^{1/2}, \\ p^{1/4}l^{-3/4}\log^2 p, & \text{if } p^{1/2} > l \ge p^{1/3}, \end{cases}$$

and $\gamma = 2.5 - \log 3 = 0.9150\ldots$.

Proof. We may assume that p is large enough, in particular that $r \ge 3$. From Lemma 4.1 with $L = p-1$ and the Cauchy inequality we conclude that

$$\begin{aligned} \sum_{\mathbf{a}\in\mathbb{F}_l{}^r} D_{l,p,g}(\mathbf{a})^2 &\ll \sum_{\mathbf{a}\in\mathbb{F}_l{}^r} \left(\frac{1}{p} + \frac{1}{2^r}\sum_{h=1}^{p-1}\frac{1}{h}\left|\sum_{x=0}^{2^r-1} \mathbf{e}_p\left(hf_{\mathbf{a}}(x)\right)\right|\right)^2 \\ &\ll \sum_{\mathbf{a}\in\mathbb{F}_l{}^r} \left(\frac{1}{p^2} + \frac{1}{2^{2r}}\left(\sum_{h=1}^{p-1}\frac{1}{h}\left|\sum_{x=0}^{2^r-1} \mathbf{e}_p\left(hf_{\mathbf{a}}(x)\right)\right|\right)^2\right) \\ &\ll \frac{l^r}{p^2} + \frac{1}{2^{2r}}\sum_{\mathbf{a}\in\mathbb{F}_l{}^r}\sum_{j=1}^{p-1}\frac{1}{j}\sum_{h=1}^{p-1}\frac{1}{h}\left|\sum_{x=0}^{2^r-1} \mathbf{e}_p\left(hf_{\mathbf{a}}(x)\right)\right|^2. \end{aligned}$$

Therefore

$$\sum_{\mathbf{a}\in\mathbb{F}_l{}^r} D_{l,p,g}(\mathbf{a})^2 \ll \frac{l^r}{p^2} + \frac{\log p}{2^{2r}}\sum_{h=1}^{p-1}\frac{1}{h}W_h, \tag{24.2}$$

where

$$W_h = \sum_{\mathbf{a}\in\mathbb{F}_l{}^r}\left|\sum_{x=0}^{2^r-1} \mathbf{e}_p\left(hf_{\mathbf{a}}(x)\right)\right|^2.$$

Replacing the square of the inner sum by a double sum and changing the order of summation we obtain

$$W_h = \sum_{x,y=0}^{2^r-1}\sum_{\mathbf{a}\in\mathbb{F}_l{}^r} \mathbf{e}_p\left(h\left(f_{\mathbf{a}}(x) - f_{\mathbf{a}}(y)\right)\right).$$

If $x = y$ the inner sum is equal to l^r.

Now we consider the case $x \ne y$. We say that $x \succ y$ if $x_i \ge y_i$, $i = 1,\ldots,r$, where $x = x_1\ldots x_r$ and $y = y_1\ldots y_r$ are the bit representations of x and y.

We also say that integers x and y are *comparable* if either $x \succ y$ or $y \succ x$.

If $x \ne y$ and $x \succ y$ we fix i, $1 \le i \le r$, with $x_i = 1$, $y_i = 0$.

We see that the term $f_{\mathbf{a}}(y)$ does not depend on a_i.

Let the vector $(z_1, \dots, z_{r-1})$ be formed by all the bits of x except x_i, that is, $z_k = x_k$ if $1 \le k < i$ and $z_k = x_{k+1}$ if $i \le k \le r-1$. Therefore,

$$\left| \sum_{\mathbf{a} \in \mathbb{F}_l{}^r} \mathbf{e}_p \left(h \left(f_{\mathbf{a}}(x) - f_{\mathbf{a}}(y) \right) \right) \right| \le \sum_{\mathbf{b} \in \mathbb{F}_l{}^{r-1}} \left| \sum_{r=0}^{l-1} \mathbf{e}_p \left(h \vartheta_{\mathbf{b},x}^r \right) \right|,$$

where $\mathbf{b} = (b_1, \dots, b_{r-1})$ and

$$\vartheta_{\mathbf{b},x} = g^{b_1^{z_1} \dots b_{r-1}^{z_{r-1}}}.$$

We see that if

$$b_1 \dots b_{r-1} \not\equiv 0 \pmod{l}$$

then, because l is prime, $\vartheta_{\mathbf{b},x}$ is of multiplicative order l. Hence the bound of Lemma 3.15 applies to the inner sum. For other $O\left(rl^{r-2}\right)$ vectors $\mathbf{b}$ we estimate the inner sum trivially by l.

It is easy to see that there are

$$\sum_{k=0}^{r} \binom{r}{k} 2^k = 3^r$$

pairs of (x, y), $0 \le x, y \le 2^r - 1$, with $x \succ y$. Thus this part of the sum can be estimated as

$$\left| \sum_{\substack{x,y=0 \\ x \ne y,\ x \succ y}}^{2^r-1} \sum_{\mathbf{a} \in \mathbb{F}_l{}^r} \mathbf{e}_p \left(h \left(f_{\mathbf{a}}(x) - f_{\mathbf{a}}(y) \right) \right) \right| \ll 3^r \left(rl^{r-1} + l^{r-1} B(l,p) \right).$$

The case $x \ne y$ and $y \succ x$ can be considered quite analogously.

Finally, let us consider pairs of x and y which are not comparable. In this case there are i and j, $1 \le i, j \le r$, with $x_i = y_j = 1$ and $x_j = y_i = 0$. We see that the term $f_{\mathbf{a}}(y)$ does not depend on a_i and the term $f_{\mathbf{a}}(x)$ does not depend on a_j.

Let the vector $(z_1, \dots, z_{r-2})$ be formed by all the bits of x except x_i and x_j, that is,

- $z_k = x_k$ if $1 \le k < I$;
- $z_k = x_{k+1}$ if $I \le k < J - 1$;
- $z_k = x_{k+2}$ if $J - 1 \le k \le r - 2$;

where $I = \min\{i, j\}$ and $J = \max\{i, j\}$.

We also form the vector $(w_1, \dots, w_{r-2})$ in a similar way from all the bits of y except y_i and y_j.

Therefore,

$$\left|\sum_{\mathbf{a}\in\mathbb{F}_l{}^r} \mathbf{e}_p\left(h\left(f_{\mathbf{a}}(x)-f_{\mathbf{a}}(y)\right)\right)\right| \le \sum_{\mathbf{b}\in\mathbb{F}_l{}^{r-2}} \left|\sum_{r=0}^{l-1} \mathbf{e}_p\left(h\lambda_{\mathbf{b},x}^r\right)\right| \left|\sum_{s=0}^{l-1} \mathbf{e}_p\left(h\mu_{\mathbf{b},y}^s\right)\right|,$$

where $\mathbf{b} = (b_1, \ldots, b_{r-2})$,

$$\lambda_{\mathbf{b},x} = g^{b_1^{x_1}\ldots b_{r-2}^{x_{r-2}}} \qquad \text{and} \qquad \mu_{\mathbf{b},y} = g^{b_1^{w_1}\ldots b_{r-2}^{w_{r-2}}}.$$

We see that if

$$b_1 \ldots b_{r-2} \not\equiv 0 \pmod l$$

then, because l is prime, $\lambda_{\mathbf{b},x}$ and $\mu_{\mathbf{b},y}$ are both of multiplicative order l. Hence the bound of Lemma 3.15 applies to both inner sums. For other $O(nl^{r-3})$ vectors $\mathbf{b}$ we estimate the inner sums trivially by l each.

Therefore, for each pair of x and y which are not comparable the bound

$$\left|\sum_{\mathbf{a}\in\mathbb{F}_l{}^r} \mathbf{e}_p\left(h\left(f_{\mathbf{a}}(x)-f_{\mathbf{a}}(y)\right)\right)\right| \ll rl^{r-1} + l^{r-2}B(l,p)^2$$

holds.

Putting everything together and taking into account that $2^r = O(p)$ and $3^r = O(p^\alpha)$, where $\alpha = \log 3$, we derive

$$\begin{aligned} W_h &\ll 2^r l^r + 3^r\left(rl^{r-1} + l^{r-1}B(l,p)\right) + 2^{2r}\left(rl^{r-1} + l^{r-2}B(l,p)^2\right) \\ &\ll pl^r + rp^\alpha l^{r-1} + p^\alpha l^{r-1}B(l,p) + rp^2 l^{r-1} + p^2 l^{r-2} B(l,p)^2. \end{aligned}$$

It is easy to see that the terms including $B(l,p)$ dominate all other terms. Thus

$$W_h \ll p^\alpha l^{r-1} B(l,p) + p^2 l^{r-2} B(l,p)^2. \tag{24.3}$$

Combining (24.2) and (24.3), we derive

$$\begin{aligned} \frac{1}{l^r}\sum_{\mathbf{a}\in\mathbb{F}_l{}^r} D_{l,p,g}(\mathbf{a})^2 &\ll \frac{1}{p^2} + \frac{\log p}{2^{2r}} \sum_{h=1}^{p-1} \frac{p^\alpha l^{-1}B(l,p) + p^2 l^{-2}B(l,p)^2}{h} \\ &\ll \left(p^{\alpha-2}l^{-1}B(l,p) + l^{-2}B(l,p)^2\right)\log^2 p. \end{aligned}$$

Remarking that the first term in the numerator dominates if and only if $l \ge p^\gamma$, we obtain the desired result. □

In particular, if the vector $\mathbf{a} \in \mathbb{F}_l{}^r$ is chosen uniformly at random, then for any $\delta > 0$ with probability at least $1-\delta$

$$D_{l,p,g}(\mathbf{a}) \ll \delta^{-1}\Delta(l,p)^{1/2}.$$

The obtained in [304] improvement of Lemma 3.15 allows us to derive a nontrivial bound on $\Delta(l,p)$ starting with $l \ge p^{1/4+\varepsilon}$.

Lemma 3.16 can be used in a similar way to produce the following result which essentially tells us that even elements of very small period are likely to produce uniformly distributed sequences.

Theorem 24.2. *Let Q be a sufficiently large integer. For any $\varepsilon > 0$ there exists $\delta > 0$ such that for all primes $p \in [Q, 2Q]$, except at most $Q^{5/6+\varepsilon}$ of them, and any element $g_{p,T} \in \mathbb{F}_p$ of multiplicative order $T \ge p^\varepsilon$ the bound*

$$\frac{1}{l^r} \sum_{\mathbf{a} \in \mathbb{F}_l{}^r} D_{l,p,g_p}(\mathbf{a})^2 \le p^{-\delta}$$

holds, where $r = \lfloor \log p \rfloor$.

Now we obtain a lower bound on the linear complexity of the Naor–Reingold pseudorandom function. This result has recently been obtained in [517] and it improves the previously known bound from [235].

Theorem 24.3. *Assume that for some $\gamma > 0$,*

$$r \ge (1+\gamma)\log l.$$

Then for any δ with $\gamma/4 > \delta > 0$ and sufficiently large l, the linear complexity, $L_\mathbf{a}$, of the sequence $f_\mathbf{a}(x)$, $x = 0, \dots, 2^r - 1$, satisfies

$$L_\mathbf{a} \ge \begin{cases} l^{1-2\delta}, & \text{if } \gamma \ge 1, \\ l^{\gamma-2\delta}, & \text{if } \gamma < 1, \end{cases}$$

for all except possibly at most $O\left((l-1)^{r-\delta}\right)$ vectors $\mathbf{a} \in (\mathbb{F}_l^*)^r$.

Proof. Let

$$s = \lfloor \log l \rfloor \qquad \text{and} \qquad t = r - s.$$

We also put

$$h = \left\lceil \frac{l^{-\delta}(l-1)2^s}{l+2^s} \right\rceil.$$

Therefore

$$h \ge l^\delta \left(l2^{-t/2} + 1 \right) \log l \tag{24.4}$$

for $\gamma > 4\delta$ provided that l is large enough.

For $\mathbf{b} \in (\mathbb{F}_l^*)^m$, as in Lemma 4.6, let $T_\mathbf{b}(h)$ denote the number of distinct values taken by the products of the form

$$\mathbf{b}^\mathbf{z} = b_1^{z_1} \dots b_m^{z_m}, \qquad \mathbf{z} = (z_1, \dots, z_m) \in \mathfrak{B}_m,\ \mathbf{z} \ne 0,$$

among the elements of the set $\{1, \dots, h\}$.

Let $\mathcal{A}$ be the set of vectors $\mathbf{a} \in (\mathbb{F}_l^*)^r$ such that simultaneously

$$T_{(a_1,\ldots,a_s)}(l-1) \geq \frac{l^{-\delta}(l-1)2^s}{l+2^s} \tag{24.5}$$

and

$$T_{(a_{s+1},\ldots,a_r)}(h) \geq \frac{l^{-\delta}h2^t}{l+2^t}. \tag{24.6}$$

Thus the subset products of the last t components of $\mathbf{a}$ generate sufficiently many elements of the set $\{1,\ldots,h\}$ and the first s components of $\mathbf{a}$ generate sufficiently many elements in the whole set $\mathbb{F}_l^*$. In particular, $T_{(a_1,\ldots,a_s)}(l-1) \geq h$.

From Lemma 4.6, which applies because of the bound (24.4), we derive

$$|\mathcal{A}| = (l-1)^r + O\left((l-1)^{r-\delta}\right).$$

We show that the bound of the theorem holds for any $\mathbf{a} \in \mathcal{A}$. Indeed, let us fix $\mathbf{a} \in \mathcal{A}$ and assume that

$$L_{\mathbf{a}} \leq T_{(a_{s+1},\ldots,a_r)}(h) - 1. \tag{24.7}$$

Then we can select $L_{\mathbf{a}}+1$ integers $1 \leq d_0 < \ldots < d_{L_{\mathbf{a}}} \leq h \leq l-1 < p$ which are represented as non-empty subset products of $\{a_{s+1},\ldots,a_r\}$. Let $1 \leq e_0,\ldots,e_{L_{\mathbf{a}}} \leq 2^t-1$ be the t-bit integers whose bit representations correspond to these products. That is,

$$d_i = \prod_{j=1}^{t} a_{s+j}^{e_{ij}},$$

where $e_i = e_{i1}\ldots e_{it}$ is the bit representation of e_i, $i = 0,\ldots,L_{\mathbf{a}}$.

Using the corresponding linear recurrence of order $L_{\mathbf{a}} \leq 2^n$, we can extend the sequence $f_{\mathbf{a}}(k)$ to all integers $k \geq 0$.

Let us consider $L_{\mathbf{a}}+1$ sequences

$$f_{\mathbf{a},i}(k) = f_{\mathbf{a}}(2^tk+e_i), \qquad i = 0,\ldots,L_{\mathbf{a}},$$

which are defined for $k = 0,\ldots,2^s-1$.

From Lemma 2.2 we see that there are some $c_0,\ldots,c_{L_{\mathbf{a}}} \in \mathbb{F}_p$ not all equal to zero such that

$$c_0f_{\mathbf{a}}(2^tk+e_0) + \ldots + c_{L_{\mathbf{a}}}f_{\mathbf{a}}(2^tk+e_{L_{\mathbf{a}}}) = 0, \qquad k = 0,\ldots,2^s-1.$$

Hence for at least $T_{(a_1,\ldots,a_s)}(l-1)+1$ elements $u \in \mathbb{F}_p$, the polynomial

$$F(X) = c_0X^{d_0} + c_1X^{d_1} + \ldots + c_LX^{d_L}$$

vanishes. Namely, $F(u) = 0$ for all

$$u = g^{a_1^{w_1}\ldots a_s^{w_s}}, \qquad (w_1,\ldots,w_s) \in \mathfrak{B}_s^*,$$

and for $u = 0$. Thus this is impossible because it follows from (24.5) that

$$\deg F \le h \le T_{(a_1,\dots,a_s)}(l-1).$$

Therefore, the assumption (24.7) is false and from (24.6) we derive

$$L_{\mathbf{a}} \ge T_{(a_{s+1},\dots,a_r)}(h) \ge \frac{l^{-\delta}h2^t}{l+2^t} \ge \frac{l^{1-2\delta}2^r}{(l+2^s)(l+2^t)} \gg \frac{l^{-2\delta}2^r}{l+2^t} \gg \frac{l^{1+\gamma-2\delta}}{l+l^\gamma}$$

which implies the desired result. □

We observe that the implicit constant in the above estimate can easily be evaluated.

Chapter 25

$1/M$ Generator

Let $g \geq 2$ be a fixed integer and let M be a positive integer whose g-ary expansion is of length at most L, that is, $0 \leq M < g^L$.

It is proved in [50] that given $k = 2L+3$ consecutive digits of the g-ary expansion of $1/M$ one can find M in polynomial time $L^{O(1)}$. The algorithm of [50] makes use of continued fractions and is very efficient indeed.

It has also been shown in [50] that, under Artin's conjecture, $k = L-1$ digits are not enough to determine M unambiguously.

Here we present a weaker but unconditional result from Chapter 11 of [306] which implies that any string of $k = \lfloor (1-\varepsilon)L \rfloor$ consecutive digits provides no or very little information about M.

Namely, using results of [308] one can show that for an arbitrary $\varepsilon > 0$ any string of $k \leq (1-\varepsilon)L$ consecutive digits appears in the g-ary expansion of $1/M$ for at least $C(g)g^{\varepsilon L/2}$ values of $M < g^L$. Here $C(g)$ is some constant depending only on g.

Thus this *unpredictability* bound $k = \lfloor (1-\varepsilon)L \rfloor$ is of the same order of magnitude as the *predictability* bound $k = 2L+3$ from [50]. Several other results about predictability of pseudorandom number generators can be found in [197, 287, 313, 319]. On the other hand, Chapter 11 of [306] provides some unpredictability results.

Theorem 25.1. *There is a constant $c(g) > 0$, depending only on g, such that the set*

$$\mathcal{M} = \{M = p^\alpha \mu \mid 1 \leq \mu \leq Q, \ (\mu, g) = 1\},$$

where

$$Q = \left\lfloor c(g) g^{\varepsilon L/2} \right\rfloor,$$

p is the smallest odd prime number with $\gcd(p, g) = 1$, *and p^α is the largest power of p that is less than g^L/Q, satisfies the following property. For every element*

$M \in \mathcal{M}$, the g-ary expansion of $1/M$ contains every string of $k = \lfloor (1-\varepsilon)L \rfloor$ consecutive g-digits.

Proof. We denote by $N_M(d_1, \dots, d_k)$ the number of appearances of a g-ary string $(d_1, \dots, d_k)$ on the period of the g-ary expansion of $1/M$ (it is known that for $\gcd(M, g) = 1$ the period of this expansion is the multiplicative order t_M of g modulo M). It is evident that $t_M \le \varphi(M) < M$.

We define

$$D_M = \max_k \max_{(d_1, \dots, d_k)} |N_M(d_1, \dots, d_k) - t_M/g^k|$$

over all g-ary strings

$$(d_1, \dots, d_k), \qquad 0 \le d_i \le g-1: \quad i = 1, \dots, k, \quad k = 1, 2, \dots .$$

Let $M = p_1^{\alpha_1} \dots p_r^{\alpha_r}$ be the prime number factorisation of M. Set $m = p_1 \dots p_r$. Assuming $\gcd(M, g) = 1$, we denote by $\tau_M = t_m$ the multiplicative order of g modulo m.

We need the bound

$$D_M < 2\tau_M$$

which is a relaxed version of Theorem 1 of [308].

Evidently, for any $M \in \mathcal{M}$, $t_M \ge t_{p^\alpha}$. It is known, see [308], that if we define

$$\beta = \operatorname{ord}_p(g^{p-1} - 1) < \frac{p \ln g}{\ln p},$$

then

$$t_{p^\alpha} = t_p p^{\alpha - \beta}, \qquad \alpha = \beta, \beta + 1, \dots .$$

Thus $t_M \ge C(g) g^L Q^{-1}$ where $C(g)$ is some effective constant depending only on g (since p can be estimated in terms of g). Also we have the bound

$$\tau_M = t_m < m \le pQ.$$

Therefore, if $k = \lfloor (1-\varepsilon)L \rfloor$, then for any $M \in \mathcal{M}$ we have

$$N_M(d_1, \dots, d_k) \ge t_M g^{-k} - D_M > C(g) g^L / Q g^k - 2pQ.$$

Putting $c(g) = (C(g)/2p)^{-1/2}$ we get that $N_M(d_1, \dots, d_k) > 0$ for any $M \in \mathcal{M}$ and any string $(d_1, \dots, d_k)$. □

Since the set $\mathcal{M}$ is exponentially large, this result means that even as many as $k = \lfloor (1-\varepsilon)L \rfloor$ digits give us very little information on the possible values of M.

Using Lemma 4.3, it has also been proved in Chapter 11 of [306] that for a smaller value of k, namely for $k = \lfloor (3/37 - \varepsilon)L \rfloor$, one can include almost all prime numbers $p \le g^L - 1$ in the corresponding set $\mathcal{M}$.

Theorem 25.2. *For any $\varepsilon > 0$, given a string of $k = \lfloor (3/37 - \varepsilon)L \rfloor$ consecutive g-ary digits, there are at least $(1 + o(1))\pi(g^L)$ prime numbers $p < g^L$ such that the g-ary expansion of $1/p$ contains this string.*

Proof. For a prime p with $\gcd(p, g) = 1$, and a sequence $d_1, \dots, d_k$ of g-ary digits,

$$0 \le d_i \le g - 1, \qquad i = 1, \dots, k,$$

we denote by $N_p(d_1, \dots, d_k)$ the number of appearances of the string $(d_1 \dots d_k)$ in the period of the g-ary expansion of $1/p$ (it is known that for $\gcd(g, p) = 1$ the period length of such an expansion is t_p where, as before, t_p is the multiplicative order of g modulo p).

We show that $N_p(d_1, \dots, d_k) > 0$ for any string $(d_1 \dots d_k)$ and almost all primes p.

To be more precise, denote by δ_j, $0 \le \delta_j \le g - 1$, the g-ary digits of $1/p$,

$$\frac{1}{p} = \sum_{j=1}^{\infty} \delta_j g^{-j}.$$

We observe that, for any integer $j \ge 0$ and any g-ary string $(d_1 \dots d_k)$ we have $\delta_{j+i} = d_i$, $i = 1, \dots, k$, if and only if the fractional part of g^j/p satisfies the inequalities

$$\frac{D}{g^k} \le \left\{ \frac{g^j}{p} \right\} < \frac{D+1}{g^k}, \tag{25.1}$$

where $D = d_1 g^{k-1} + d_2 g^{k-2} + \ldots + d_k$.

Jointly, the inequalities (25.1) are equivalent to the congruence $g^j \equiv y \pmod p$ with some y from the interval

$$\frac{Dp}{g^k} \le y < \frac{(D+1)p}{g^k}.$$

Let

$$r = \left\lfloor \frac{(2D+1)p}{2g^k} \right\rfloor, \qquad h = \left\lfloor \frac{p}{2g^k} \right\rfloor.$$

Then, it is enough to show that the following congruence

$$t \equiv r + y \pmod p, \qquad t \in \mathcal{G},\ y = 0, \dots, h-1,$$

has at least one solution, where $\mathcal{G}$ is the subgroup of $\mathbb{F}_p^*$ generated by g.

As we have remarked for any any function $\varepsilon(p) \to 0$, for all primes $p < g^L$ except possibly $o(\pi(g^L))$, the multiplicative order of g modulo p is at least $p^{1/2+\varepsilon(p)}$, see [172]. One also verifies that $h \gg p^{34/37+\varepsilon}$. We can now apply Lemma 4.3 to such primes from which the desired result follows. □

As we have mentioned, the function $N_M(d_1, \dots, d_k)$ has been introduced and treated in [308] with help of the bound of exponential sums given by Lemma 3.13. For our application this bound is not sharp enough, but the bound of Lemma 3.15 can already produce some non-trivial results, although for smaller values k than in Theorem 25.2. The extension of Lemma 3.15 to smaller values of T, given in [304], can be used as well.

It has also been indicated in Chapter 11 of [306] how to improve the constant $3/37$ for a set of primes of positive density. However details have not been worked out yet.

Chapter 26

Inversive, Polynomial and Quadratic Exponential Generators

Here we consider several recently proposed generators of pseudorandom numbers, that, although mainly used for Monte Carlo type applications [163, 164, 165, 183, 426, 427, 428, 435, 438], may turn out to be useful for cryptography as well. Another reason for including these results in this book is that the underlying method has been a prototype of the method of estimation of exponential sums of Chapter 23, which are associated with the power generator. All the aforementioned generators are based on iterations of some function with well studied algebraic structure (such as a polynomial or a rational function).

Let ψ be the permutation of $\mathbb{F}_q$ defined by

$$\psi(w) = \begin{cases} aw^{-1} + b, & \text{if } w \neq 0, \\ b, & \text{if } w = 0. \end{cases} \tag{26.1}$$

We define the sequence (u_n) of elements of $\mathbb{F}_q$ by the recurrence relation

$$u_{n+1} = \psi(u_n), \qquad n = 0, 1, \ldots, \tag{26.2}$$

where u_0 is the *initial value*.

If $q = p$ is a prime number then as usual we identify $\mathbb{F}_p$ with the set $\{0, 1, \ldots, p-1\}$ thus the elements of the sequence (u_n) are integers in the range $[0, p-1]$. In this case the numbers $u_0/p,\ u_1/p, \ldots$ in the interval $[0, 1)$ form a sequence of *inversive congruential pseudorandom numbers*.

It is obvious that the sequence (26.2) is purely periodic with some period $t \leq q$.

The inversive congruential generator provides a very attractive alternative to linear congruential generators, and has been extensively studied in the literature. In particular, several results about the period of this generator have been obtained. For example, it is known when such sequences achieve the largest possible period, which is obviously $t = q$, see [183]. For such sequences of period $t = q$, a number of results about the period, distribution and statistical almost-independence of their terms over the full period have been established, starting with the paper [424]. Many of these results are essentially the best possible. We refer to [163, 164, 165, 183, 426, 427, 428, 435, 438, 456] for more details and references to original papers.

In the case where $q = p$ is prime, the first non-trivial results about the distribution of the above fractions in parts $N < t$ of the period (including the case of periods $t < p$ which is of interest as well) have recently been obtained in [246, 434]. Here we present the results of [246, 434] as well as some similar results obtained in the series of papers [128, 234, 245, 246, 431, 432, 433, 434] for other generators, for example for generators of the similar form with ψ being a small degree polynomial over $\mathbb{F}_p$. The same method has also given new results about the distribution of the inversive congruential generator in very small parts of the period "on average" over all initial values, see [436]. In [429, 440, 441, 442] this method has also been applied to several other pseudorandom numbers generated by similar methods, see also the surveys [435, 438]. Finally, in [437], the same method has been used to estimate sums of multiplicative characters with elements of the sequence (26.2), and thus to study the distribution of power residues and primitive roots.

Theorem 26.1. *For any integer $s \geq 1$, the discrepancy $D_s(N)$ of the s-tuples*

$$\left(\frac{u_n}{p}, \ldots, \frac{u_{n+s-1}}{p}\right), \qquad 0 \leq n \leq N-1,$$

where the sequence (u_n) is given by (26.2) and $q = p$ is prime, satisfies

$$D_s(N) = O(N^{-1/2} p^{1/4} \log^s p), \qquad 1 \leq N \leq t,$$

where the implied constant depends only on s.

Proof. According to a standard principle, we can estimate the discrepancy $D_s(N)$ by bounding the corresponding exponential sums. For an integer vector $\mathbf{a} = (a_0, \ldots, a_{s-1}) \in \mathbb{Z}^s$ we define the exponential sum

$$S_{\mathbf{a}}(N) = \sum_{n=0}^{N-1} \mathbf{e}_p\left(\sum_{j=0}^{s-1} a_j u_{n+j}\right)$$

and show that for any vector $\mathbf{a}$ with $\gcd(a_0, \ldots, a_{s-1}, p) = 1$ we have

$$|S_{\mathbf{a}}(N)| = O\left(N^{1/2} p^{1/4}\right), \qquad 1 \leq N \leq t, \tag{26.3}$$

where the implied constant depends only on s.

For any integer m let ψ^m denote the mth power of the permutation ψ given by (26.1) in the symmetric group on p symbols. Then $u_n = \psi^n(u_0)$ for all integers $n \geq 0$, and we use this identity to define u_n for all negative integers n.

It is easy to see that for any integer k we have

$$\left| S_{\mathbf{a}}(N) - \sum_{n=0}^{N-1} \mathbf{e}_p \left(\sum_{j=0}^{s-1} a_j u_{n+k+j} \right) \right| \leq 2|k|. \tag{26.4}$$

For an integer $K \geq 1$ put

$$\mathcal{R}(K) = \begin{cases} \{k \in \mathbb{Z} \mid -(K-1)/2 \leq k \leq (K-1)/2\}, & \text{if } K \text{ is odd,} \\ \{k \in \mathbb{Z} \mid -K/2+1 \leq k \leq K/2\}, & \text{if } K \text{ is even,} \end{cases}$$

thus,

$$\sum_{k \in \mathcal{R}(K)} |k| \leq K^2/4.$$

Therefore, if we use (26.4) for all $k \in \mathcal{R}(K)$, then we get

$$K|S_{\mathbf{a}}(N)| \leq W + K^2/2,$$

where

$$\begin{aligned} W &= \left| \sum_{n=0}^{N-1} \sum_{k \in \mathcal{R}(K)} \mathbf{e}_p \left(\sum_{j=0}^{s-1} a_j u_{n+k+j} \right) \right| \\ &\leq \sum_{n=0}^{N-1} \left| \sum_{k \in \mathcal{R}(K)} \mathbf{e}_p \left(\sum_{j=0}^{s-1} a_j u_{n+k+j} \right) \right| \\ &= \sum_{n=0}^{N-1} \left| \sum_{k \in \mathcal{R}(K)} \mathbf{e}_p \left(\sum_{j=0}^{s-1} a_j \psi^{k+j}(u_n) \right) \right|. \end{aligned}$$

By the Cauchy inequality we obtain

$$\begin{aligned} W^2 &\leq N \sum_{n=0}^{N-1} \left| \sum_{k \in \mathcal{R}(K)} \mathbf{e}_p \left(\sum_{j=0}^{s-1} a_j \psi^{k+j}(u_n) \right) \right|^2 \\ &\leq N \sum_{w \in \mathbb{F}_p} \left| \sum_{k \in \mathcal{R}(K)} \mathbf{e}_p \left(\sum_{j=0}^{s-1} a_j \psi^{k+j}(w) \right) \right|^2 \\ &= N \sum_{k,l \in \mathcal{R}(K)} \sum_{w \in \mathbb{F}_p} \mathbf{e}_p \left(\sum_{j=0}^{s-1} a_j \left(\psi^{k+j}(w) - \psi^{l+j}(w) \right) \right). \end{aligned}$$

Taking into account that ψ is a permutation, we conclude that the inner sum depends only on $m = k - l$. If $k = l$, then the inner sum equals p. Therefore,

$$W^2 \le KNp + 2N \sum_{m=1}^{K-1} (K-m) \left| \sum_{w \in \mathbb{F}_p} \mathbf{e}_p \left(\sum_{j=0}^{s-1} a_j \left(\psi^{m+j}(w) - \psi^j(w) \right) \right) \right|.$$

Without loss of generality we can assume that $a_0 \not\equiv 0 \pmod p$.

For $m \ge 0$ denote by $\mathcal{F}_m$ the set of poles of the rational functions H_{m+j}, $0 \le j \le s-1$, given by

$$H_0(X) = X, \qquad H_i(X) = H_{i-1}\left(aX^{-1} + b\right), \quad i = 1, 2, \dots .$$

It is obvious that this sequence is purely periodic with period $T \le t$. Thus $|\mathcal{F}_m| \le s$. We also put

$$\mathcal{E}_m = \cup_{i=0}^{m} \mathcal{F}_i,$$

thus $\mathcal{E}_m$ consists of the poles of the rational functions H_i, $0 \le i \le m + s - 1$.

It follows by straightforward induction that for $0 \le m \le t-1$ and $0 \le j \le s-1$ we have

$$\psi^{m+j}(w) = H_{m+j}(w)$$

for all $w \in \mathbb{F}_p \backslash \mathcal{E}_m$. Assuming that $1 \le m \le K-1$ and $K \le t - s$, we obtain

$$\left| \sum_{w \in \mathbb{F}_p} \mathbf{e}_p \left(\sum_{j=0}^{s-1} a_j \left(\psi^{m+j}(w) - \psi^j(w) \right) \right) - \sum_{\substack{w \in \mathbb{F}_p \\ w \notin \mathcal{F}_m \cup \mathcal{F}_0}} \mathbf{e}_p\left(G_{\mathbf{a}}(w)\right) \right| \le 2|\mathcal{E}_m| \le 2(m+s),$$

where

$$G_{\mathbf{a}}(X) = \sum_{j=0}^{s-1} a_j \left(H_{m+j}(X) - H_j(X) \right).$$

Because $a_0 \not\equiv 0 \pmod p$, from Lemma 2.9 we conclude that the rational function $G_{\mathbf{a}}(X)$ is not a constant. We claim that if $4s < p$ (otherwise the result of the theorem is trivial), then the function $G_{\mathbf{a}}(X)$ is not constant on $\mathbb{F}_p$. Indeed, otherwise we see that for some $d \in \mathbb{F}_p$, $G_{\mathbf{a}}(w) = d$ for each $w \in \mathbb{F}_p$ which is not a pole of this function. Thus $G_{\mathbf{a}}(w) = d$ for at least $p - 2s$ values of $w \in \mathbb{F}_p$. The degrees of the numerator and the denominator of $G_{\mathbf{a}}$ do not exceed $2s$. Therefore, if $2s < p - 2s$, then this is not possible. This means that we can apply Lemma 3.2 to the last exponential sum, getting the upper bound $4sp^{1/2}$. Hence,

$$W^2 \le KNp + 4sK^2Np^{1/2} + \frac{2}{3}K^3N + O(K^2N).$$

Select $K = \lceil (3p/2)^{1/2} \rceil$. If $K > t - s$, then the bound of the theorem is trivial. Otherwise, the above considerations apply and after simple calculations we obtain the bound (26.3).

Now the desired result follows from Lemma 4.1 and the bound (26.3). □

Theorem 26.1 yields a non-trivial discrepancy bound only in the case where N, and therefore t, is at least of the order of magnitude $p^{1/2} \log^{2s+\varepsilon} p$ for some $\varepsilon > 0$.

It has been shown in [436] that for almost all initial values for every N, $1 \le N \le t$, the discrepancy is bounded by $N^{-1/2}(\log N)^{s+1} \log T$ if $N \le p^{1/2}$, and by $p^{-1/4}(\log N)^{s+1} \log T$ if $N > p^{1/2}$, where $T \le p$ is the period of the sequence of the polynomials $H_i(X)$ defined in the proof of Theorem 26.1, see also Lemma 2.9.

Theorem 26.2. *The linear complexity profile $L(N)$ of a sequence (u_n) produced by the inversive generator (26.2), which is purely periodic with period t, satisfies*

$$L(N) \ge \min \{ \lceil N/3 \rceil, \lceil t/2 \rceil \}.$$

Proof. Let k be the least positive integer with

$$u_{n+k} = a_{k-1} u_{n+k-1} + \ldots + a_0 u_n, \qquad 0 \le n \le N - k - 1,$$

and l be the smallest integer with $a_l \ne 0$. Then

$$-\psi^{k-l}(u_{n+l}) + a_{k-1}\psi^{k-l-1}(u_{n+l}) + \ldots + a_l \psi^0(u_{n+l}) = 0, \qquad (26.5)$$

for $0 \le n \le N - k - 1$.

As in the proof of Theorem 26.1 we consider the following sequence of rational functions over $\mathbb{F}_q$:

$$H_0(X) = X, \qquad H_i(X) = H_{i-1}(aX^{-1} + b), \qquad i = 1, 2, \ldots .$$

It is obvious that this sequence is purely periodic. Denote by T the smallest period. Obviously $T \ge t$. It has been shown in Lemma 2.9 that if for some $b_0, \ldots, b_I \in \mathbb{F}_q$ we have an identity

$$H_0(X) = b_0 + \sum_{i=1}^{I} b_i H_i(X),$$

then $I \ge T$ (the proof is given in the case of prime fields but it can be extended to any field without any changes). For $j \ge 0$ let denote by E_j the set of poles of the rational functions $H_0, \ldots, H_j$. Thus $|E_j| \le j$. By induction we have

$$\psi^j(x) = H_j(x), \qquad x \in \mathbb{F}_q \setminus E_j. \qquad (26.6)$$

Hence, we have either $k \ge k - l \ge T \ge t$ or the polynomial

$$H(X) = -H_{k-l}(X) + a_{k-1} H_{k-l-1}(X) + \ldots + a_l H_0(X)$$

does not vanish identical to zero. In the latter case we see that $H_i(X) = f_i(X)/g_i(X)$ are nonconstant rational functions, where $f_i(X), g_i(X) \in \mathbb{F}_q[X]$ with $\max\{\deg f_i, \deg g_i\} = 1$. Hence, $H(X) = F(X)/G(X)$ with $F(X), G(X) \in \mathbb{F}_q[X]$ and $\deg F \le k - l \le k$. On the other hand, by (26.5) and (26.6) the polynomial $F(X)$ has at least

$$\min\{N-k,\, t\} - (k-j) \ge \min\{N-2k,\, t-k\}$$

zeros and thus

$$\deg F \ge \min\{N-2k,\, t-k\}.$$

Hence, we have $k \ge \min\{N-2k,\, t-k\}$ from which the result follows. □

In [247] an extension of Lemma 2.9 has been used to obtain the lower bound

$$NL_m(N) \ge \min\left\{\left\lfloor\frac{N-1}{m+2}\right\rfloor, \left\lfloor\frac{t-1}{m+1}\right\rfloor\right\}$$

on the non-linear complexity profile of the inversive generator (26.2).

Now we show how to apply similar arguments to studying the distribution and linear complexity of some similar sequences.

Given a polynomial $f(X) \in \mathbb{F}_q[X]$ we define the sequence (u_n) of elements of $\mathbb{F}_q$ by the recurrence relation

$$u_{n+1} = f(u_n), \qquad n = 0, 1, \ldots, \tag{26.7}$$

where u_0 is the *initial value*.

If $q = p$ is prime, then one consider the numbers $u_0/p, u_1/p, \ldots$ in the interval $[0,1)$ which are known as *polynomial congruential pseudorandom numbers*.

It is obvious that the sequence (u_n) given by (26.7) eventually becomes periodic with some period $t \le q$. We assume that this sequence is *purely periodic*, that is, that $u_n = u_{n+t}$ beginning with $n = 0$, otherwise we consider a shift of the original sequence.

Theorem 26.3. *If the sequence (u_n), given by (26.7) with a polynomial $f(X) \in \mathbb{F}_p[X]$ of degree $d \ge 2$, where $q = p$ is prime, is purely periodic with period t, then for any integer $s \ge 1$ the discrepancy $D_s(N)$ of the s-tuples*

$$\left(\frac{u_n}{p}, \ldots, \frac{u_{n+s-1}}{p}\right), \qquad 0 \le n \le N-1,$$

satisfies the bound

$$D_s(N) = O\left(N^{-1/2}p^{1/2}\log^{-1/2}p\ \log^s\log p\right), \qquad 1 \le N \le t,$$

where the implied constant depends only on d and s.

Proof. For an integer vector $\mathbf{a} = (a_0, \dots, a_{s-1}) \in \mathbb{Z}^s$ we introduce the exponential sum

$$T_{\mathbf{a}}(N) = \sum_{n=0}^{N-1} \mathbf{e}_p \left(\sum_{j=0}^{s-1} a_j u_{n+j} \right)$$

and show that for any vector $\mathbf{a}$ with $\gcd(a_0, \dots, a_{s-1}, p) = 1$ we have

$$|T_{\mathbf{a}}(N)| = O\left(N^{1/2} p^{1/2} \log^{-1/2} p\right), \qquad 1 \le N \le t, \tag{26.8}$$

where the implied constant depends only on d and s.

Select any $\mathbf{a} = (a_0, \dots, a_{s-1}) \in \mathbb{Z}^s$ with $\gcd(a_0, \dots, a_{s-1}, p) = 1$. It is obvious that for any integer $k \ge 0$ we have

$$\left| T_{\mathbf{a}}(N) - \sum_{n=0}^{N-1} \mathbf{e}_p \left(\sum_{j=0}^{s-1} a_j u_{n+k+j} \right) \right| \le 2k.$$

Therefore, for any integer $K \ge 1$,

$$K |T_{\mathbf{a}}(N)| \le W + K^2,$$

where

$$W = \left| \sum_{n=0}^{N-1} \sum_{k=0}^{K-1} \mathbf{e}_p \left(\sum_{j=0}^{s-1} a_j u_{n+k+j} \right) \right| \le \sum_{n=0}^{N-1} \left| \sum_{k=0}^{K-1} \mathbf{e}_p \left(\sum_{j=0}^{s-1} a_j u_{n+k+j} \right) \right|.$$

Define the sequence of polynomials $f_k(X) \in \mathbb{F}_p[X]$ by the recurrence relation

$$f_k(X) = f\left(f_{k-1}(X)\right), \qquad k = 1, 2, \dots,$$

where $f_0(X) = X$. It is clear that $\deg f_k = d^k$ and that $u_{n+k} = f_k(u_n)$. Accordingly, we obtain

$$\begin{aligned} W^2 &\le N \sum_{n=0}^{N-1} \left| \sum_{k=0}^{K-1} \mathbf{e}_p \left(\sum_{j=0}^{s-1} a_j f_{k+j}(u_n) \right) \right|^2 \\ &\le N \sum_{x=1}^{p} \left| \sum_{k=0}^{K-1} \mathbf{e}_p \left(\sum_{j=0}^{s-1} a_j f_{k+j}(x) \right) \right|^2 \\ &= N \sum_{k=0}^{K-1} \sum_{l=0}^{K-1} \sum_{x=1}^{p} \mathbf{e}_p \left(\sum_{j=0}^{s-1} a_j \left(f_{k+j}(x) - f_{l+j}(x)\right) \right). \end{aligned}$$

If $k = l$, then the inner sum is trivially equal to p. There are K such sums. Otherwise we can apply Lemma 3.1 to the inner sum, getting the upper bound $d^{K+s-2}p^{1/2}$ for at most K^2 sums. Hence,

$$W^2 \leq KNp + d^{K+s-2}K^2Np^{1/2}.$$

Select

$$K = \left\lceil 0.4\frac{\log p}{\log d}\right\rceil.$$

Then it is easy to see that the first term dominates the second one, and after simple calculations we obtain the bound (26.8). Now the desired result follows from Lemma 4.1 and the bound (26.8). □

The result of Theorem 26.3 is non-trivial only for sufficiently large values of N, namely for $t \geq N \geq p\log^{-1+\varepsilon} p$ with some fixed $\varepsilon > 0$, and it would be very important to extend the range of N for which $\Delta_s(N) = o(1)$.

It is easy to see why the bound of Theorem 26.1 is stronger than that of Theorem 26.3. The reason is that rational functions $(AX + B)/(CX + D)$ preserve their form under iterations, while the degree of iterations of non-linear polynomials grows exponentially.

The same method has also been applied to some other sequences of pseudorandom numbers generated by similar relations with multivariate polynomials, that is of the form

$$u_{n+m} = f\left(u_{n+m-1}, \dots, u_n\right), \qquad n = 0, 1, \dots,$$

where $f(X_1, \dots, X_m) \in \mathbb{F}_q[X_1, \dots, X_m]$, see [234, 245]. Generalisations to prime power moduli are known as well [432]. These and several other results and open problems can be found in [128, 234, 245, 246, 429, 431, 432, 433, 434, 435, 436, 438, 440, 441, 442].

Now we turn to studying some lattice properties of pseudorandom numbers given by (26.7).

Let $q = p$ be prime. Denote by h the unique polynomial over $\mathbb{F}_p$ of degree at most $t-1$ which interpolates a sequence (u_n) satisfying (26.7) at the points $n = 1, \dots, t$, where t is the period of this sequence, that is,

$$u_n = h(n), \qquad n = 1, \dots, t. \tag{26.9}$$

It is known, see Section 8.1 of [426], that the degree of h is a very important characteristic of this sequence of pseudorandom numbers.

Here we show that for any non-linear polynomial f of small degree the degree of h is large. We remark that only this case of polynomials f of small degree is of practical interest because for polynomials of large degree the generator becomes computationally less efficient (even if one uses sparse polynomials).

Theorem 26.4. *Let $h(X) \in \mathbb{F}_p[X]$ be a polynomial of degree $\deg h = D$ such that (26.9) holds for a sequence (u_n) which satisfies (26.7) with a polynomial $f(X) \in \mathbb{F}_p[X]$ of degree $d \geq 2$ and is purely periodic with period t. Then*

$$dD \geq \begin{cases} t-1, & \text{if } t < p-1; \\ p, & \text{if } t = p. \end{cases}$$

Proof. If $t = 1$ there is nothing to prove. Otherwise we see that h is a non-constant polynomial. We have $f(h(n-1)) = h(n)$ for $n = 1, \ldots, t-1$. Therefore the polynomial $F(X) = f(h(X-1)) - h(X)$ has at least $t-1$ zeros. It is easy to see that $F(X)$ is a non-zero polynomial of degree precisely dD (since $d \geq 2$) which has at least $t-1$ zeros. Therefore $dD \geq t-1$. Moreover, it is easy to see that if $t = p$, then we have $f(h(n-1)) = h(n)$ for $n = 1, \ldots, p$ and thus $F(X)$ has at least p zeros in this case. □

From Theorem 26.4 we derive that if the corresponding sequence of pseudorandom numbers is of largest possible period $t = p$, then it passes the so called s-dimensional lattice test for rather large values of s.

We recall that a periodic sequence (w_n) over a field $\mathbb{F}$ with period t is said to pass the *s-dimensional lattice test* if the vectors

$$(w_n - w_0, \ldots, w_{n+s-1} - w_{s-1}), \qquad n = 0, \ldots, t-1,$$

span the vector space $\mathbb{F}^s$, see [368, 426, 444].

It has been shown in Theorem 8.2 of [426] that if $t = p$, then the corresponding sequence passes the s-dimensional lattice test for all positive $s \leq D$; thus, as we see from Theorem 26.4, for all positive $s \leq \lceil p/d \rceil$. Several more results in this direction have recently been given in [444].

The lattice structure of pseudorandom numbers obtained from polynomial generators over arbitrary finite fields have been studied in [443].

In [247] one can find some lower bounds on the non-linear complexity profile of the polynomial generator (26.7) but they are rather weak. However, even for the linear complexity profile no better general bounds are known. Namely, the paper [247] establishes the bound

$$NL_{d-1}(N) \geq \min\{\lceil \log_d(N - \lfloor \log_d N \rfloor) \rceil, \lceil \log_d t \rceil\}$$

on the non-linear complexity profile $NL_{d-1}(N)$ of order $d-1$ of a sequence (u_n) produced by the polynomial generator (26.7), with a polynomial of degree $d \geq 2$ which is purely periodic with period t. Even for $NL_1(N)$ no better bounds are known. On the other hand, it is clear that $NL_d(N) = 1$. Although, as we have just mentioned, no stronger general bounds are known, in the case of the largest possible period $t = p$. Theorem 3 of [383], together with Theorem 5 of [431], implies the following bound on the linear complexity profile

$$L(N) \geq \min\left\{N - p + 1, \left\lceil \frac{p}{d} \right\rceil\right\}.$$

Finally, we introduce one more nonlinear pseudorandom number generator which we call *quadratic exponential generator*. This generator seems to be new and has not been considered in the literature, except for [247] where it has been introduced. We believe it deserves a more detailed study. Namely, given an element $\vartheta \in \mathbb{F}_q^*$, we consider the sequence (u_n) where

$$u_n = \vartheta^{n^2}, \qquad n = 0, 1, \ldots . \tag{26.10}$$

The period of this sequence is at least $\tau/2$, where τ is the multiplicative order of ϑ.

In fact the expression (26.10) is only convenient for studying the sequence $\mathcal{U}$ while for the actual generation of its elements one should probably use the recurrence formulas

$$w_n = \lambda w_{n-1}, \qquad u_n = u_{n-1} w_{n-1}, \qquad n = 1, 2, \ldots ,$$

where $\lambda = \vartheta^2$, $u_0 = 1$, $w_0 = \vartheta$. If $q = p$ is a prime number and λ is a very small number (say $\vartheta = 2$ and $\lambda = 4$) then each step of computation of w_n is computationally inexpensive and each step of computation of u_n costs only one multiplication. Thus overall this sequence could be faster to generate than the sequences produced by the inversive generator (26.2).

Also the bound of exponential sums from [187] shows that the sequence $\mathcal{U}$ is asymptotically uniformly distributed. Indeed, combining Lemma 3.24 with Lemma 4.1, we derive:

Theorem 26.5. *The discrepancy D of the fractional parts $\{u_n/p\}$, $n = 1, \ldots, t$ for a sequence (u_n) produced by the quadratic exponential generator (26.10), where $q = p$ is prime, which is purely periodic with period t, satisfies the inequality*

$$D = O\left(t^{-1/4} p^{1/8+\varepsilon}\right).$$

For the same generator defined modulo a composite integer one can use the corresponding results from [189].

The paper [247] also provides lower bounds on the linear complexity profile of this sequence.

Theorem 26.6. *The linear complexity profile $L(N)$ of a sequence (u_n) produced by the quadratic exponential generator (26.10), which is purely periodic with period t, satisfies the inequality*

$$L(N) \geq \left\lceil \frac{\min\{N, t\}}{2} \right\rceil.$$

Proof. Let k be the least positive integer with

$$u_{n+k} = a_{k-1}u_{n+k-1} + \ldots + a_0 u_n, \qquad 0 \le n \le N-k-1,$$

and verify that

$$u_{n+l} = \vartheta^{l^2} u_n \vartheta^{2nl}, \qquad l, n \ge 0. \tag{26.11}$$

Then with $a_k = -1$ and $b_l = \vartheta^{l^2} a_l$ for $0 \le l \le k$ we have

$$b_k \vartheta^{2nk} + \ldots + b_0 = 0, \qquad 0 \le n \le N-k-1.$$

Since for even values of τ the elements $1, \vartheta^2, \vartheta^4, \ldots, \vartheta^{\tau-2}$ and for odd values of τ the elements $1, \vartheta^2, \vartheta^4, \ldots, \vartheta^{2\tau-2}$ are distinct, the polynomial $F(X) = b_k X^k + \ldots + b_0$ of degree k has at least $\min\{N-k, \tau/2\}$ zeros if τ is even and at least $\min\{N-k, \tau\}$ zeros if τ is odd. Since $t \le \tau$ we get the result. □

One can also find in [247] some lower bounds on the non-linear complexity profile of the quadratic exponential generator (26.10). It is interesting to note that this bound of non-linear complexity profile is based on Theorem 12.5.

Theorem 26.7. *The nonlinear complexity profile $NL_m(N)$ of order $m \ge 1$ of a sequence (u_n) produced by the quadratic exponential generator (26.10), which is purely periodic with period t, satisfies the inequality*

$$NL_m(N) \ge C \frac{\min\{N^{1/2}, t^{1/2}\}\, t}{mq}$$

with an absolute effectively computable constant $C > 0$.

Proof. Let k be the least positive integer with

$$u_{n+k} = \Psi(u_{n+k-1}, \ldots, u_n), \qquad 0 \le n \le N-k-1,$$

with a polynomial $\Psi(\lambda_1, \ldots, \lambda_k)$ over $\mathbb{F}_q$ of total degree at most m. By (26.11) we have

$$\vartheta^{k^2} u_n \vartheta^{2nk} = \Psi(\vartheta^{(k-1)^2} u_n \vartheta^{2n(k-1)}, \ldots, u_n), \qquad 0 \le n \le N-k-1.$$

Hence the polynomial

$$F(X,Y) = -\vartheta^{k^2} Y X^{2k} + \Psi(\vartheta^{(k-1)^2} Y X^{2(k-1)}, \ldots, Y)$$

satisfies

$$F(\vartheta^n, \vartheta^{n^2}) = 0, \qquad 0 \le n \le \min\{N-k-1, t-1\}.$$

If $\deg \Psi \ge 2$, then the polynomial $F(X,Y)$ is not identically zero since

$$\deg_Y F = \deg \Psi = m.$$

If $\deg \Psi = 1$, then the polynomial $F(X, Y)$ is not identically zero since $\deg_X F = 2k$.

As before we denote by τ the multiplicative order of ϑ. Let $r = (q-1)/\tau$. Then $\vartheta = g^r$ for some primitive element g of $\mathbb{F}_q$ and we obtain

$$G(g^n, g^{n^2}) = 0, \qquad 0 \le n \le \min\{N - k - 1, t - 1\},$$

where $G(X, Y) = F(X^r, Y^r)$. Now Theorem 12.5 implies

$$\deg G \ge c \min\{N - k, t\}^{1/2}$$

with an absolute effectively computable constant $c > 0$. Since

$$\deg G \le r \max\{(2k-1)m, 2k + 1\}$$

and $t \ge (q-1)/2$, the result follows. □

Thus, if ϑ is a primitive root, then $t \ge (q-1)/2$ and we obtain

$$NL_m(N) \gg \frac{\min\left\{N^{1/2}, t^{1/2}\right\}}{m}.$$

Several results about the linear complexity profile of sequences of the values of polynomials and rational functions in finite fields are given in [382, 383, 385].

Chapter 27

Subset Sum Generators

Let $(u(x))$ be a linear recurrence sequence of order r over $\mathbb{F}_2$.

For an integer $m \geq 1$ one can consider the following *subset sum* generator of pseudorandom numbers. Given an r-dimensional vector $\mathbf{z} = (z_1, \ldots, z_r) \in \mathbb{Z}_m^r$ of weights one can consider the sequence

$$V_{\mathbf{z}}(n) = \left\lfloor \sum_{j=1}^{r} u(n+j-1) z_j \right\rfloor_m, \qquad n = 1, 2, \ldots$$

of elements of $\mathbb{Z}_m$.

This generator, which is also known as *knapsack generator*, has been introduced in [485] and studied in [483], see also Section 6.3.2 of [389] and Section 3.7.9 of [484]. For cryptographic applications, it is typically recommended to use a linear recurrence sequence of maximal period $\tau = 2^r - 1$ and also the modulus $m = 2^r$, but here we consider more general settings.

We remark that in the special case of weights $z_j = 2^{r-j+1}$, $j = 1, \ldots, r$, and the modulus $m = 2^r$ this generator is well known in the theory of Quasi Monte Carlo methods. An exhaustive survey of known results about the distribution of this and more general generators can be found in Chapter 9 of [426]. Although, for this (deterministic) choice of weights some uniformity of distribution results are known, they are weaker than our results which, however, apply only to randomised choice of weights. It is also clear that this choice of weights corresponds to very easy instances of the knapsack problem and thus is probably not suitable for cryptographic applications.

We denote by $D_{\mathbf{z}}(N)$ the discrepancy of the sequence $V_{\mathbf{z}}(n)/m$, $n = 1, \ldots, N$.

Theorem 27.1. *Let the linear recurrence sequence $(u(x))$ be purely periodic with period τ. Then for any $\delta > 0$ for all $\mathbf{z} \in \mathbb{Z}_m^r$ except maybe $O(\delta m^r)$ of them, the bound*

$$D_{\mathbf{z}}(N) \ll N^{-1/2} \log m \log^2 \tau$$

holds for all $1 \le N \le \tau$.

Proof. We denote by $\mathbf{u}_n$ the following binary vector

$$\mathbf{u}_n = (u_n, u_{n+1}, \dots, u_{n+r-1}) \in \mathfrak{B}_r.$$

As in Chapter 21, for two r-dimensional vectors $\mathbf{z} = (z_1, \dots, z_r) \in \mathbb{Z}_m^r$ and $\mathbf{u} = (u_1, \dots, u_r) \in \mathfrak{B}_r$ we define $s_{\mathbf{z}}(\mathbf{u})$ by (21.2). Then $V_{\mathbf{z}}(n) = \lfloor s_{\mathbf{z}}(\mathbf{u}_n) \rfloor_m$. From Lemma 4.1 we derive

$$D_{\mathbf{z}}(N) \ll \frac{1}{m} + \frac{1}{N} \sum_{0 \le |a| < m} \frac{1}{|a|} \left| \sum_{n=1}^{N} \mathbf{e}_m \left(a s_{\mathbf{z}}(\mathbf{u}_n) \right) \right|.$$

Let $N_\nu = 2^\nu$, $\nu = 0, 1, \dots$. Define k by the inequality $N_{k-1} < N \le N_k$. Then from (3.2) we derive

$$\sum_{n=1}^{N} \mathbf{e}_m \left(a s_{\mathbf{z}}(\mathbf{u}_n) \right) = \frac{1}{N_k} \sum_{n=1}^{N_k} \sum_{h=1}^{N} \sum_{c=0}^{N_k} \mathbf{e}_m \left(a s_{\mathbf{z}}(\mathbf{u}_n) \right) \mathbf{e}_{N_k} \left(c(n-h) \right).$$

Hence,

$$D_{\mathbf{z}}(N) \ll \frac{1}{N N_k} \Delta_{\mathbf{z}}(k) + \frac{1}{m} \tag{27.1}$$

where

$$\Delta_{\mathbf{z}}(k) = \sum_{0 < |a| < m} \frac{1}{|a|} \sum_{c=0}^{N_k} \left| \sum_{h=1}^{N} \mathbf{e}_{N_k}(-ch) \right| \left| \sum_{n=1}^{N_k} \mathbf{e}_m \left(a s_{\mathbf{z}}(\mathbf{u}_n) \right) \mathbf{e}_{N_k}(cn) \right|.$$

Applying the Cauchy inequality we derive

$$\begin{aligned} &\left(\sum_{\mathbf{z} \in \mathbb{Z}_m^r} \left| \sum_{n=1}^{N_k} \mathbf{e}_m \left(a s_{\mathbf{z}}(\mathbf{u}_n) \right) \mathbf{e}_{N_k}(cn) \right| \right)^2 \\ &\quad \le m^r \sum_{\mathbf{z} \in \mathbb{Z}_m^r} \left| \sum_{n=1}^{N_k} \mathbf{e}_m \left(a s_{\mathbf{z}}(\mathbf{u}_n) \right) \mathbf{e}_{N_k}(cn) \right|^2 \\ &\quad = m^r \sum_{n,l=1}^{N_k} \mathbf{e}_{N_k} \left(c(n-l) \right) \sum_{\mathbf{z} \in \mathbb{Z}_m^r} \mathbf{e}_m \left(a \left(s_{\mathbf{z}}(\mathbf{u}_n) - s_{\mathbf{z}}(\mathbf{u}_l) \right) \right). \end{aligned}$$

As in the proof of Theorem 22.1, we see that the inner sum vanishes unless $\mathbf{u}_n = \mathbf{u}_l$. But this is possible only for $n \equiv l \pmod t$ in which case the inner sum is equal to m^r. There are at most $N_k(\lfloor N_k/t \rfloor + 1)$ integers $1 \le n, l \le N_k$ with $n \equiv l \pmod t$. Because $N_k = 2N_{k-1} < 2N \le 2\tau$ we see that $N_k(\lfloor N_k/\tau \rfloor + 1) \le 2N_k$ Thus the whole sum is at most $2m^{2r}N_k$. Therefore

$$\sum_{\mathbf{z}\in \mathbb{Z}_m^r} \left| \sum_{n=1}^{N_k} \mathbf{e}_m\left(as_{\mathbf{z}}(\mathbf{u}_n)\right) \mathbf{e}_{N_k}(cn) \right| \le 2^{1/2} m^r N_k^{1/2}.$$

Thus, by (3.3),

$$\begin{aligned}\sum_{\mathbf{z}\in \mathbb{Z}_m^r} \Delta_{\mathbf{z}}(k) &\le m^r N_k^{1/2} \sum_{0<|a|<m} \frac{1}{|a|} \sum_{c=0}^{N_k} \left| \sum_{h=1}^{N} \mathbf{e}_{N_k}(-ch) \right| \\ &\ll m^r N_k^{3/2} \log m \log N_k \ll m^r N_k^{3/2} \log m \log \tau.\end{aligned}$$

This implies that for any k the number of vectors $\mathbf{z} \in \mathbb{Z}_m^r$ with

$$\Delta_{\mathbf{z}}(k) \ge \delta^{-1} N_k^{3/2} \log m \log^2 \tau$$

is at most $O(\delta^{-1} m^r \log^{-1} \tau)$. Therefore, the number of vectors $\mathbf{z} \in \mathbb{Z}_m^r$ with

$$\Delta_{\mathbf{z}}(k) \ge \delta^{-1} N_k^{3/2} \log m \log^2 \tau$$

for at least one $k = 1, \dots, \lceil \log \tau \rceil$ is at most $O(\delta m^r)$. For other $\mathbf{z} \in \mathbb{Z}_m^r$, from (27.1), we obtain

$$D_{\mathbf{z}}(N) \ll \frac{1}{NN_k} \Delta_{\mathbf{z}}(k) + \frac{1}{m} \le \delta^{-1} N^{-1} N_k^{1/2} \log m \log^2 \tau.$$

Taking into account the inequality $N^{-1}N_k^{1/2} \le 2N^{-1/2}$, we obtain the desired result. □

For linear recurrence sequences $(u(x))$ with irreducible characteristic polynomial, similar results for the multidimensional discrepancy have been obtained in [129].

A very different construction of pseudorandom numbers is given by the *iterative subset sum* generator of pseudorandom numbers which has been introduced in [280]. In this construction the function which is iterated is more of combinatorial nature rather than of an algebraic origin as in (26.2) and (26.7).

Let $t \ge 2^r$. For an integer $w \in [0, t-1]$ we also define the function

$$S_{\mathbf{z},t}(w) = \lfloor s_{\mathbf{z}}(\mathbf{u}) \rfloor_t$$

where $\mathbf{u}$ is the vector consisting of the r rightmost bits of w and as before, for two r-dimensional vectors $\mathbf{z} = (z_1, \dots, z_r) \in \mathbb{Z}_t^r$ and $\mathbf{u} = (u_1, \dots, u_r) \in \mathfrak{B}_r$ we define $s_{\mathbf{z}}(\mathbf{u})$ by (21.2).

For a fixed vector $\mathbf{z}$ and a given integer $w_0 \in [0, t-1]$ one can consider the sequence

$$W_{\mathbf{z}}(0) = w_0, \quad W_{\mathbf{z}}(n+1) = S_{\mathbf{z},t}\left(W_{\mathbf{z}}(n)\right), \qquad n = 0, 1, \ldots . \tag{27.2}$$

In the special case of $t = 2^k$, $k \geq r$, this construction has been shown to have some important cryptographic applications, see [280]. It has also been shown in [280] that this sequence admits a very efficient generation algorithm.

We have already mentioned in Chapter 21 the results of [45] about fixed points of the RSA mapping and the results of [92, 116, 565] about fixed points of the discrete logarithm. Here we show that using Theorem 22.1 one can estimate the average number

$$M_{r,t} = \frac{1}{t^r} \sum_{\mathbf{z} \in \mathbb{Z}_t^r} |\{w \in [0, t-1] \mid w = S_{\mathbf{z},t}(w)\}|$$

of fixed points of the family of the functions $S_{\mathbf{z},t}(w)$.

Theorem 27.2. *The bound*

$$M_{r,t} \leq (2t)^{1/2} + 2$$

holds.

Proof. If $w = S_{\mathbf{z},t}(w)$ and $w = u_m \ldots u_1$ is the bit representation of w with $m = \lceil \log t \rceil$, then we have

$$\sum_{j=1}^{m} u_j 2^j \equiv \sum_{j=1}^{r} u_j z_j \pmod{t}$$

or

$$\sum_{j=1}^{r} u_j \left(z_j - 2^j\right) \equiv \sum_{j=r+1}^{m} u_j 2^j \pmod{t}.$$

Let C be the set of all possible values taken by the right-hand side of the last congruence when w runs through the interval $[0, t-1]$. Then, in the notation of Theorem 22.1

$$M_{r,t} = \frac{1}{t^r} \sum_{\mathbf{z} \in \mathbb{Z}_t^r} \sum_{c \in C} N_{\mathbf{z}}\left(\mathfrak{B}_r, c\right).$$

Thus, from Theorem 22.1 and the Cauchy inequality we derive

$$\begin{aligned} M_{r,t} &= \frac{1}{t^r} \sum_{\mathbf{z} \in \mathbb{Z}_t^r} \sum_{c \in C} \left(N_{\mathbf{z}}\left(\mathfrak{B}_r, c\right) - 2^r/t\right) + |C| 2^r/t \\ &\leq \left(\frac{|C|}{t^r} \sum_{\mathbf{z} \in \mathbb{Z}_t^r} \sum_{c \in C} \left(N_{\mathbf{z}}\left(\mathfrak{B}_r, c\right) - 2^r/t\right)^2 \right)^{1/2} + |C| 2^r/t \\ &\leq |C|^{1/2} 2^{r/2} + |C| 2^r/t. \end{aligned}$$

From the inequality $|\mathcal{C}| \le 2^{m-r} \le t2^{-r+1}$ we obtain the desired result. □

The function $S_{\mathbf{z},t}(w)$ has also been proposed in [280] to be used as a hash function. So our estimate gives some information on the concealing properties of this function. One can also derive an exact formula for the average *collision probability* $P_{r,t}(\mathcal{M})$ of this function applied to an arbitrary set of binary messages $\mathcal{M} \subseteq \mathfrak{B}_r$ (where as usual we identify a binary vector from $\mathfrak{B}_r$ and the corresponding r-bit integer whose binary expansion is given by this vector). That is, $P_{r,t}(\mathcal{M})$ is the average value over $z \in \mathbb{Z}$ of the probability $S_{\mathbf{z},t}(m_1) = S_{\mathbf{z},t}(m_2)$ for a randomly chosen pair $(m_1, m_2) \in \mathcal{M}^2$ with $m_1 \ne m_2$ (chosen uniformly at random). Indeed

$$\begin{aligned} P_{r,t}(\mathcal{M}) &= \frac{1}{t^r|\mathcal{M}|\left(|\mathcal{M}|-1\right)} \sum_{\mathbf{z}\in\mathbb{Z}_t^r} \left(\sum_{c\in\mathbb{Z}_t} N_{\mathbf{z}}^2\left(\mathcal{M},c\right) - |\mathcal{M}| \right) \\ &= \frac{1}{t^r|\mathcal{M}|\left(|\mathcal{M}|-1\right)} \sum_{\mathbf{z}\in\mathbb{Z}_t^r} \sum_{c\in\mathbb{Z}_t} N_{\mathbf{z}}^2\left(\mathcal{M},c\right) - \frac{1}{|\mathcal{M}|-1}. \end{aligned}$$

Hence, using the identity

$$\frac{1}{t^r} \sum_{\mathbf{z}\in\mathbb{Z}_t^r} \sum_{c\in\mathbb{Z}_t} N_{\mathbf{z}}\left(\mathcal{M},c\right) = |\mathcal{M}|$$

and Theorem 22.1, we derive

$$\begin{aligned} \frac{t-1}{t}|\mathcal{M}| &= \frac{1}{t^r} \sum_{\mathbf{z}\in\mathbb{Z}_t^r} \sum_{c\in\mathbb{Z}_t} \left(N_{\mathbf{z}}\left(\mathcal{M},c\right) - \frac{|\mathcal{M}|}{t} \right)^2 \\ &= \frac{1}{t^r} \sum_{\mathbf{z}\in\mathbb{Z}_t^r} \sum_{c\in\mathbb{Z}_t} N_{\mathbf{z}}\left(\mathcal{M},c\right)^2 - 2\frac{|\mathcal{M}|}{t^{r+1}} \sum_{\mathbf{z}\in\mathbb{Z}_t^r} \sum_{c\in\mathbb{Z}_t} N_{\mathbf{z}}\left(\mathcal{M},c\right) + \frac{|\mathcal{M}|^2}{t} \\ &= \frac{1}{t^r} \sum_{\mathbf{z}\in\mathbb{Z}_t^r} \sum_{c\in\mathbb{Z}_t} N_{\mathbf{z}}\left(\mathcal{M},c\right)^2 - \frac{|\mathcal{M}|^2}{t}. \end{aligned}$$

Therefore

$$P_{r,t}(\mathcal{M}) = \frac{1}{t}.$$

It would be interesting to study other properties of the iterative *subset sum* generator of pseudorandom numbers but unfortunately the methods presented here do not seem to apply.

Part VI

Other Applications

Chapter 28

Square-Freeness Testing and Other Number-Theoretic Problems

We extend the area of applications of our methods to lower bounds on the circuit and decision tree complexity of Boolean functions related to some number-theoretic problems. In particular, we show that deciding whether a given integer is *square-free* and testing *co-primality* of two integers by unbounded fan-in circuits of bounded depth requires superpolynomial size, see [9, 38, 39, 40, 41, 460]. This method can be applied to other number-theoretic problems related to arithmetical properties of integers. Unfortunately this approach does not seem to apply to the complexity of the *primality testing* problem, for which an alternative approach has been developed in [9], which in fact implies stronger but less explicit results (which apply to square-freeness and co-primality testing as well).

The above problems, that is, square-freeness, co-primality and primality testing, have been extensively studied in the literature and many very efficient algorithms (sequential and parallel) are known [16, 127, 136, 203]. The complexity of checking of an oracle which presumably computes the greatest common divisor has been studied in [2]. Nevertheless very little is known about lower bounds on their complexity.

In the notation of Chapter 11, the work [380] provides the lower bound $\Omega(\log^{1/2} r)$ for the sequential arithmetic RAM complexity (with the operation set $\{+, -, \times, /, \lfloor\cdot\rfloor\}$, see Chapter 11) of the primality testing, that is, for the function

$$f(x) = \begin{cases} 0, & \text{if } x \text{ is not prime,} \\ 1, & \text{if } x \text{ is prime,} \end{cases}$$

for all r-bit integers. This result has been derived from a lower bound on the $M_f(r)$-invariant of the above function.

For computation of the greatest common divisor of two integers, a similar lower bound $\Omega(\log^{1/2} r)$ is obtained in [365].

Studying the communication complexity of computation of the greatest common divisor is of interest as well; see [454].

We first consider a Boolean function B which decides whether a given $(r+1)$-bit odd integer is square-free, that is,

$$B(u_1, \dots, u_r) = \begin{cases} 0, & \text{if } 2x+1 \text{ is not square-free,} \\ 1, & \text{if } 2x+1 \text{ is square-free,} \end{cases} \tag{28.1}$$

where $x = u_1 \dots u_r 1$ is the bit representation of x, $0 \le x \le 2^r - 1$ (if necessary we add several leading zeros).

Theorem 28.1. *Assume that there is a circuit $C \in \mathrm{UBC}(d, S)$ such that, given the bit representation $x = u_1 \dots u_r$ of x, $0 \le x \le 2^r - 1$, it computes the Boolean function B given by (28.1),*

$$C(u_1, \dots, u_r) = B(u_1, \dots, u_r)$$

Then the bound

$$d \log\log S \ge 0.5 \log r + O(\log\log r)$$

holds.

Proof. Put

$$w = \left\lfloor \left(\frac{2r}{3 \log r} \right)^{1/2} \right\rfloor - 1.$$

It follows from Lemma 5.5 that, for sufficiently large r, the Boolean function B is δ-approximated at level $k = r - w$ with

$$\widehat{B}(0) = \left(1 - \frac{16}{\pi^2} + o(1)\right) 2^r \qquad \text{and} \qquad \delta = O\left(2^{-r/3(w+1)}\right).$$

An application of Lemma 7.2 and the Parseval identity (7.1) yields

$$\begin{aligned} S 2^{-0.05 w^{1/d} + 1} &\ge 2^{-2r} \sum_{\substack{a \in \mathfrak{B}_r \\ \|a\| > w}} |\widehat{B}(a)|^2 \\ &= 2^{-2r} \sum_{a \in \mathfrak{B}_r} |\widehat{B}(a)|^2 - 2^{-2r} \widehat{B}(0)^2 - 2^{-2r} \sum_{\substack{a \in \mathfrak{B}_r \\ 1 \le \|a\| \le w}} |\widehat{B}(a)|^2 \\ &= 1 - 2^{-2r} \widehat{B}(0)^2 - 2^{-2r} \sum_{\substack{a \in \mathfrak{B}_r \\ 1 \le \|a\| \le w}} |\widehat{B}(a)|^2 . \end{aligned}$$

From Lemma 7.4 we derive that

$$\sum_{\substack{a\in\mathfrak{B}_r\\ 1\le\|a\|\le w}} |\widehat{B}(a)|^2 \le \delta^2 2^{2r} \sum_{j=1}^{w} \binom{r}{j} \le \delta^2\, 2^{2r+w\log r},$$

where we have applied the inequality

$$\sum_{j=1}^{w} \binom{r}{j} \le r^w .$$

Therefore,

$$S2^{-0.05w^{1/d}+1} \ge \left(1 - 2^{-2r}\widehat{B}(0)^2 - \delta^2 2^{w\log r}\right).$$

Since $-2r/3(w+1) + w\log r \le -\log r$ we have $\delta^2 2^{w\log r} = O\left(r^{-1}\right)$ and the desired statement follows. □

One can also establish a result for the decision tree complexity of the Boolean function (28.1).

Theorem 28.2. *For the Boolean function B given by (28.1) the bound*

$$\log \mathrm{DT}\,(B) \ge \left(\left(\frac{2}{3}\right)^{1/2} + o(1)\right) r^{1/2} \log^{-1/2} r$$

holds.

Proof. Put

$$w = \left\lfloor \left(\frac{2r}{3\log r}\right)^{1/2} \right\rfloor - 1.$$

As in the proof of Theorem 28.1 we remark that it follows from Lemma 5.5 that, for sufficiently large r, the Boolean function B is δ-approximated at level $k = r - w$ with

$$\widehat{B}(0) = \left(1 - 16/\pi^2 + o(1)\right) 2^r \qquad \text{and} \qquad \delta = O\left(2^{-r/3(w+1)}\right).$$

An application of Lemma 7.3 yields

$$\mathrm{DT}\,(B) \ge 2^{-2r} \sum_{a\in\mathfrak{B}_r} 2^{\|a\|} \widehat{B}(a)^2 \ge 2^{-2r+w+1} \sum_{\substack{a\in\mathfrak{B}_r\\ \|a\|>w}} \widehat{B}(a)^2.$$

As in the proof of Theorem 28.1, from Lemma 7.4 we obtain

$$2^{-2r} \sum_{\substack{a\in\mathfrak{B}_r\\ \|a\|>w}} \widehat{B}(a)^2 \gg 1 - 2^{-2r}\widehat{B}(0)^2 - \delta^2 2^{w\log r} \gg 1.$$

Hence $\mathrm{DT}\,(B) \gg 2^w$, and the desired statement follows. □

We are also able to derive an asymptotic formula for the average sensitivity of testing square-free numbers.

Theorem 28.3. *For the Boolean function B given by (28.1) the bound*

$$\sigma_{av}(B) = 2\gamma r + o(r) \sim 0.330\ldots r$$

holds, where γ is defined by (5.10).

Proof. It is easy to see that, in notations of Lemma 5.7,

$$\sigma_{av}(B) = 2^{-r} \sum_{i=1}^{r} \left(N_i(r) + M_i(r)\right) .$$

Applying the bounds of Lemma 5.7 we obtain the desired estimate. □

Combining Theorem 28.3 and Lemma 7.5 one can obtain lower bounds of degree of real polynomial representations of the Boolean function (28.1):

$$\Delta(B) \geq 2\gamma r + o(r) \sim 0.330\ldots r$$

and

$$\delta(B) \geq \frac{\gamma^{1/2}}{3^{1/2}} r^{1/2} + o(r^{1/2}) \sim 0.234\ldots r^{1/2}.$$

These and some other applications of Theorem 28.3, such as lower bounds on the formula size and the depth of the decision tree for this function can be found in [39].

Some, rather weak, analogues of Theorems 10.1 and 11.7 for the Boolean function (28.1) have been obtained [460]; they apply to more general representations, though.

The same technique can be applied to obtain similar lower bounds for the Boolean function B of $r = 2m$ variables which decides whether two given $(m+1)$-bit odd integers are co-prime, that is, for which

$$B(u_1, \ldots, u_m, v_1, \ldots, v_m) = \begin{cases} 0, & \text{if } \gcd(2x+1, 2y+1) = 1, \\ 1, & \text{if } \gcd(2x+1, 2y+1) > 1, \end{cases} \qquad (28.2)$$

where $x = u_1 \ldots u_m$ and $y = v_1 \ldots v_m$ are the bit representations of x and y, $0 \leq x, y \leq 2^m - 1$ (if necessary we add several leading zeros).

Using Lemma 5.6 quite similarly one can obtain the following two results.

Theorem 28.4. *Assume that there is a circuit $C \in \mathrm{UBC}(d, S)$ such that, given the bit representations $x = u_1 \ldots u_m$ and $y = v_1 \ldots v_m$ of x and y, $0 \leq x, y \leq 2^m - 1$, it computes the Boolean function B given by (28.2)*

$$C(u_1, \ldots, u_m, v_1, \ldots, v_m) = B(u_1, \ldots, u_m, v_1, \ldots, v_m).$$

Then the bound

$$d \log\log S \geq 0.5 \log r + O(\log\log r)$$

holds, where $r = 2m$.

Theorem 28.5. *For the Boolean function* B *given by (28.2) the bound*

$$\log \mathrm{DT}(B) \geq \left(3^{-1/2} + o(1)\right) r^{1/2} \log^{-1/2} r$$

holds, where $r = 2m$.

Unfortunately the above do not apply to the primality testing problem, that is, to the Boolean function

$$B(u_1, \ldots, u_r) = \begin{cases} 0, & \text{if } x \text{ is not prime,} \\ 1, & \text{if } x \text{ is prime,} \end{cases} \tag{28.3}$$

where $x = u_1 \ldots u_r$ is the bit representation of x, $0 \leq x \leq 2^r - 1$.

For this function a different method has been developed in [9], which also applies to the Boolean functions (28.1) and (28.2) as well as to the Boolean function

$$B(u_1, \ldots, u_r) = \begin{cases} 0, & \text{if } \nu(x) \text{ is even,} \\ 1, & \text{if } \nu(x) \text{ is odd,} \end{cases}$$

where $x = u_1 \ldots u_r$ is the bit representation of x, $0 \leq x \leq 2^r - 1$ (we put $\nu(0) = \nu(1) = 0$).

In fact this method produces even stronger results which apply to circuits of the class $\mathrm{UBC}(d, S)$ enhanced by some other type of gate. On the other hand, the results are less explicit because they are based on a chain of reductions to some other problems which are known not to belong to that class.

We also remark that the sensitivity of the Boolean function (28.3) can be estimated via very elementary considerations. Indeed, let us select an r bit prime $p \equiv 1 \pmod 3$ and $p \equiv -1 \pmod 5$. We see that switching its ith rightmost bit for i of the form $i = 4k + 1$, we obtain a number which is

- divisible by 3 if this bit is 1 because

$$p - 2^{i-1} \equiv p - 2^{4k} \equiv 0 \pmod 3;$$

- divisible by 5 if this bit is 0 because

$$p + 2^{i-1} \equiv p + 2^{4k} \equiv 0 \pmod 5.$$

Therefore,

$$\sigma(B) \geq 0.25r + O(1)$$

and we obtain the lower bound $0.5 \log r + O(1)$ on the CREW PRAM complexity of primality testing of r-bit integers.

We remark that each of the above problems, that is, square-freeness, co-primality and primality testing, is no harder than the *integer factorisation* problem, thus any lower bound on one of those problem implies a lower bound for the integer factorisation as well.

Several very interesting results on arithmetic complexity of integer factorisation have been obtained in [60, 61, 66, 349].

Some efficient bounded and unbounded fan-in *probabilistic* Boolean circuits for integer factorisation are constructed in [537].

Studying the complexity of other problems related to deciding certain natural arithmetic properties of integers is of ultimate interest as well. Such problems have never been studied systematically although several interesting results have recently been obtained.

For example, the complexity of deciding if a given integer x is a *perfect power*, that is $x = k^l$ for some integers $k, l \geq 2$, has been studied in several papers [17, 42]. The question is of independent interest and is also a part of some modern integer factorisation algorithms; see a discussion in [42]. For the function $f(x)$, deciding if an r-bit integer x is a perfect square, the upper bound $O(\log^{1/2} r)$ is obtained in [81] for the computational tree complexity and for the sequential arithmetic RAM complexity with the operation set $\{+, -, \times, /, \lfloor \cdot \rfloor\}$. This upper bound matches the lower bound $\Omega(\log^{1/2} r)$ from [366].

A very interesting alternative approach to obtaining lower bounds on the complexity of primality has recently been proposed in [562]. This approach is based on a very interesting combination of combinatorial and analytic number theory (and of course on some complexity theory tools).

In [10], some of the results presented in this chapter have been extended to similar problems with polynomials over $\mathbb{F}_2$. In particular, some analogues of these results are obtained for the complexity of irreducibility, square-freeness and co-primality testing of polynomials over $\mathbb{F}_2$.

Chapter 29

Trade-off Between the Boolean and Arithmetic Depths of Modulo p Functions

For a polynomial $f(X) \in \mathbb{Z}[X]$ we consider Boolean functions producing the second leftmost bit of $\lfloor f(x) \rfloor_p$ from the bit representation of x and obtain a lower bound on their sensitivity. Then a similar but a weaker bound is obtained for the sensitivity of Boolean functions producing the second leftmost bit of rational functions modulo p.

We apply these results to show that either the arithmetic or the Boolean depth (or both) of any non-linear function modulo p is large enough.

As before, we assume that all arguments x contain the same number r of bits (adding several leading zeros if necessary), where $r = \lfloor \log p \rfloor$. Thus, as before, each such function is defined on a portion $1 \le x \le 2^r - 1 \le p-1$ of the complete residue system modulo p. Moreover, we assume that the values of functions all contain the same number $r+1$ of bits. Certainly the leftmost bit could be zero for almost all values. For example, this is true if $p = 2^{2^n} + 1$ is a Fermat number. This is why the second leftmost bit is much more convenient to work with.

Theorem 29.1. *Suppose that $f(X) \in \mathbb{Z}[X]$ is a polynomial of degree $n \ge 3$ with a non-vanishing modulo p leading coefficient. Let $r = \lfloor \log p \rfloor$ and let a Boolean function $B(U_1, \ldots, U_r)$ be such that for any x, $0 \le x \le 2^r - 1$, $B(u_1, \ldots, u_r)$ equals the second leftmost bit of $\lfloor f(x) \rfloor_p$, where $x = u_1 \ldots u_r$ is the bit representation of x. For any $\varepsilon > 0$ there is a constant $c(\varepsilon) > 0$ depending only on ε such that if*

$$n \le 2^{(1-\varepsilon)r/2},$$

then the bound

$$\sigma(B) \geq c(\varepsilon) r^{1/2}$$

holds.

Proof. We define

$$k = \left\lfloor r/2 - \log n - r^{1/2} \log^2 r \right\rfloor, \quad s = \left\lfloor k^{1/2} \right\rfloor, \quad m = \left\lfloor \frac{s}{\log 9C} \right\rfloor,$$

where C is the constant of Lemma 4.1.

We remark that

$$s(m+1) \leq k \leq r$$

for p large enough and that

$$m \geq c(\varepsilon) r^{1/2}$$

for some constant $c(\varepsilon) > 0$ depending only on ε.

We claim that it is enough to prove that there exists an x, $0 \leq x \leq 2^{r-k} - 1$, such that the fractional parts

$$\left\{ \frac{f(2^k x)}{p} \right\} < \frac{2^{r-2}}{p},$$

and

$$\frac{2^{r-2}}{p} \leq \left\{ \frac{f(2^k x + 2^{si})}{p} \right\} < \frac{2^{r-1}}{p}, \qquad i = 1, \ldots, m.$$

Indeed from these inequalities we conclude that the second leftmost bit of $\lfloor f(2^k x) \rfloor_p$ is 0 and the second leftmost bit of $\lfloor f(2^k x + 2^{si}) \rfloor_p$ is 1 for $i = 1, \ldots, m$. Hence the existence of such x implies that $\sigma(B) \geq m$, which yields the required estimate.

To prove that such an x exists we show that the discrepancy Δ of the sequence

$$\left(\left\{ \frac{f(2^k x)}{p} \right\}, \left\{ \frac{f(2^k x + 2^s)}{p} \right\}, \ldots, \left\{ \frac{f(2^k x + 2^{sm})}{p} \right\} \right)_{x=0}^{2^{r-k}-1}$$

of 2^{r-k} points of the $(m+1)$-dimensional unit cube satisfies the inequality

$$\Delta < \left(\frac{2^{r-2}}{p} \right)^{m+1}.$$

Let us put $L = 2^s - 1$. From Lemmas 2.10, 3.6 and 4.1 we derive that the discrepancy of this sequence satisfies

$$\begin{aligned}\Delta &\le C^{m+1}2^{-s+O(1)} + n2^{k-r+O(m)}p^{1/2}\log p \sum_{\substack{\mathbf{a}\in\mathbb{Z}^{m+1}\\0<|\mathbf{a}|\le L}} \frac{1}{r(\mathbf{a})}\\ &\le 9^{-m+O(1)} + n2^{k-r/2+O(m)}r^{m+2}\\ &= 9^{-m+O(1)} + 2^{-r^{1/2}\log^2 r+O(m\log r)} \le 8^{-m-1}\\ &< \left(\frac{2^{r-2}}{p}\right)^{m+1},\end{aligned}$$

provided that p is large enough. Therefore the claimed x exists and the desired result follows. □

The previous results cannot be applied to quadratic polynomials and it is still unclear how to consider that case. However, the special case of the polynomial $f(X) = X^2$ can be dealt with.

Theorem 29.2. *Let a Boolean function $B(U_1, \dots, U_r)$ be such that for any x, $0 \le x \le 2^r - 1$, where $r = \lfloor \log p \rfloor$, $B(u_1, \dots, u_r)$ equals the second leftmost bit of $\lfloor x^2 \rfloor_p$, where $x = u_1 \dots u_r$ is the bit representation of x. Then the bound*

$$\sigma(B) \ge \left(2^{-3/2} + o(1)\right) r^{1/2} \log^{-1/2} r$$

holds.

Proof. We assume that p is large enough and put

$$m = \left\lfloor r/2 - r^{2/3} \right\rfloor, \quad s = \left\lfloor 0.5m^{1/2}\log^{1/2} m \right\rfloor, \quad k = \left\lfloor \frac{s}{\log s + \log^{1/2} s} \right\rfloor.$$

In particular, $sk < m$. We also put $\lambda = 2^s$.

It is enough to show that there is an $1 \le x \le 2^{r-m} - 1$ for which

$$\left\{ \frac{(2^m x)^2}{p} \right\} < \frac{2^{r-2}}{p},$$

but for the $k-1$ flips we have

$$\frac{2^{r-2}}{p} \le \left\{ \frac{(2^m x + \lambda^i)^2}{p} \right\} < \frac{2^{r-1}}{p}, \qquad i = 1, \dots, k-1,$$

because then $\sigma(B) \ge k - 1$.

Assume that such an x does not exist.

Put $H = 2^{r-m} - 1$, $e_1 = 0$ and $e_i = \lambda^{(i-1)}$, $i = 2, \dots, k$. It follows from our assumption that the sequence of fractional parts

$$\Gamma = \left(\left\{ \frac{(2^m x + e_1)^2}{p} \right\}, \dots, \left\{ \frac{(2^m x + e_k)^2}{p} \right\} \right)_{x=1}^{H}$$

does not hit the k-dimensional box

$$B = \left[0, \frac{2^{r-2}}{p}\right) \times \left[\frac{2^{r-2}}{p}, \frac{2^{r-1}}{p}\right)^{k-1}.$$

Thus the discrepancy of the sequence is at least

$$D_\Gamma \geq |B| = (2^{r-2}/p)^k \geq 2^{-3k}. \tag{29.1}$$

Let $L = \lambda - 1$. It is easy to verify that for p large enough

$$2^{m+1} L \lambda^k \leq 2^{m+1} 2^{s(k+1)} \leq 2^{2m+1} < p.$$

Thus all the conditions of Lemma 3.12 are satisfied (with $N = k$). Therefore, if $0 \leq |\mathbf{a}| \leq L$, then

$$\left| \sum_{x=1}^{H} \mathbf{e}_p \left(\sum_{i=1}^{k} a_i (2^m x + e_i)^2 \right) \right| \ll \max\{p^{1/2} \log p,\, p2^{-m-s}\}. \tag{29.2}$$

For p large enough, we have

$$p^{1/2} \log p \leq 2^{r/2 + \log(r+1) + 1} \leq p2^{-m-s-1}.$$

Hence, taking into account that $L \geq 1$, we obtain

$$\max\{p^{1/2} \log p,\, p2^{-m-s-1}\} = p2^{-m-s-1} \leq 2^{r-m-s} = (H+1)2^{-s}. \tag{29.3}$$

Substituting (29.2) and (29.3) in Lemma 4.1 and using (29.1), we derive the inequality

$$2^{-3k} \ll C^{k+1} \left(\frac{1}{L} + \frac{1}{2^s} \sum_{0 \leq |\mathbf{a}| \leq L} \frac{1}{r(\mathbf{a})} \right) \ll C^{k+1} 2^{-s} \left(1 + \sum_{0 \leq |\mathbf{a}| \leq L} \frac{1}{r(\mathbf{a})} \right).$$

Taking into account that

$$\sum_{0 \leq |\mathbf{a}| \leq L} \frac{1}{r(\mathbf{a})} \leq (c \log L)^k \leq (cs)^k,$$

where $c > 0$ is an absolute constant, we obtain

$$2^{-3k} \le 2^{-s+k\log s+O(k)}$$

or

$$k \ge \frac{s}{\log s + O(1)}$$

which contradicts the choice of k. Therefore our assumption was incorrect and the desired result follows. □

We now consider rational functions modulo p.

Theorem 29.3. *Suppose that $h(X) = f(X)/g(X) \in \mathbb{Z}[X]$ is a rational function which is not a polynomial modulo p,*

$$n = \max\{\deg f, \deg g\} \le 0.6r^{1/2},$$

where $r = \lfloor \log p \rfloor$. Let a Boolean function $B(U_1, \dots, U_r)$ be such that for any x, $0 \le x \le 2^r - 1$, with $g(x) \not\equiv 0 \pmod p$, $B(u_1, \dots, u_r)$ equals the second leftmost bit of $\lfloor h(x) \rfloor_p$, where $x = u_1 \dots u_r$ is the bit representation of x. Then the bound

$$\sigma(B) \ge \frac{1}{16}r + o(r)$$

holds.

Proof. Without loss of generality we assume that p is large enough. We proceed as in the proof of Theorem 29.1.

We define

$$k = \lfloor r/4 \rfloor, \qquad m = \left\lfloor \frac{r}{16} - r^{1/2} \right\rfloor.$$

It is easy to verify that

$$m \le k - n(n-1)/2,$$

thus we can select the first m elements $s_1, \dots, s_m$ of the sequence defined in Lemma 2.11 and put

$$e_0 = 0, \qquad e_i = 2^{s_i}, \quad i = 1, \dots, m.$$

For $g(x) \not\equiv 0 \pmod p$ we denote by $h(x)$ the smallest non-negative residue of the fraction $f(x)/g(x)$ taken modulo p.

Denote by $\mathcal{X}$ the set of x, $0 \le x \le 2^{r-k} - 1$, for which

$$\prod_{i=0}^{m} g(2^k x + e_i) \not\equiv 0 \pmod p.$$

Obviously $2^{r-k} - (m+1)n \le |\mathcal{X}| \le 2^{r-k}$.

We claim that it is enough to prove that there exists an $x \in \mathcal{X}$ such that

$$\left\{ \frac{h(2^k x)}{p} \right\} < \frac{2^{r-2}}{p},$$

and

$$\frac{2^{r-2}}{p} \le \left\{ \frac{h(2^k x + e_i)}{p} \right\} < \frac{2^{r-1}}{p}, \qquad i = 1, \dots, m.$$

Indeed from these inequalities we see that the second leftmost bit of $h(2^k x)$ modulo p is 0 and the second leftmost bit of $h(2^k x + e_i)$ modulo p is 1 for $i = 1, \dots, m$. Hence the existence of such an x implies that $\sigma(B) \ge m$, which yields the required estimate.

To prove that such x exists we show that the system of congruences

$$h(2^k x + e_i) \equiv b_i + y_i - z_i \pmod p, \qquad i = 0, \dots, m,$$

where $b_0 = 2^{r-3}$, $b_i = 2^{r-2} + 2^{r-3}$, $i = 1, \dots, m$, has a solution with

$$x \in \mathcal{X}, \qquad 0 \le y_i, z_i \le 2^{r-3} - 1, \quad i = 0, \dots, m.$$

For the number T of such solutions we have

$$T = \frac{1}{p^{m+1}} \sum_{x \in \mathcal{X}} \sum_{\substack{y_0, \dots, y_m = 0 \\ z_0, \dots, z_m = 0}}^{2^{r-3}-1} \sum_{|\mathbf{a}| \le (p-1)/2} \mathbf{e}_p \left(\sum_{i=0}^{m} a_i \left(h(2^k x + e_i) - b_i - y_i + z_i \right) \right),$$

where $\mathbf{a} = (a_0, \dots, a_m) \in \mathbb{Z}^{m+1}$ runs through all $(m+1)$-dimensional integer vectors with $|\mathbf{a}| \le (p-1)/2$ (thus $\mathbf{a}$ takes p^{m+1} distinct values) and $|\mathbf{a}|$ is defined by (1.8) with $N = m+1$.

Making the summation over $\mathbf{a}$ external and separating the term corresponding to the zero vector, we obtain

$$\left| T - |\mathcal{X}| 2^{2(m+1)(r-3)} p^{-m-1} \right|$$
$$\le \frac{1}{p^{m+1}} \sum_{0 < |\mathbf{a}| \le (p-1)/2} \left| \sum_{x \in \mathcal{X}} \mathbf{e}_p \left(\sum_{i=0}^{m} a_i h(2^k x + e_i) \right) \right| \prod_{i=0}^{m} \left| \sum_{y_i=0}^{2^{r-3}-1} \mathbf{e}_p (a_i y_i) \right|^2 .$$

From Lemma 3.6 and the choice of $s_1, \dots, s_m$ we obtain that

$$\left| \sum_{x \in \mathcal{X}} \mathbf{e}_p \left(\sum_{i=0}^{m} a_i h(2^k x + e_i) \right) \right| \le 2n(m+1) p^{1/2} \log p + n(m+1)$$

(the last term takes care of the at most $n(m+1)$ values which are not in $\mathcal{X}$ and are not poles of the rational function in the exponent). Therefore

$$\begin{aligned}
&|T - |\mathcal{X}|2^{2(m+1)(r-3)}p^{-m-1}| \\
&\quad \le \frac{4n(m+1)p^{1/2}\log p}{p^{m+1}} \sum_{0<|\mathbf{a}|\le(p-1)/2} \prod_{i=0}^{m} \left| \sum_{y_i=0}^{2^{r-3}-1} \mathbf{e}_p(a_i y_i) \right|^2 \\
&\quad \le \frac{4n(m+1)p^{1/2}\log p}{p^{m+1}} \prod_{i=0}^{m} \sum_{a_i=-(p-1)/2}^{(p-1)/2} \left| \sum_{y_i=0}^{2^{r-3}-1} \mathbf{e}_p(a_i y_i) \right|^2 \\
&\quad = \frac{4n(m+1)p^{1/2}\log p}{p^{m+1}} \left(\sum_{a=0}^{p-1} \left| \sum_{y=0}^{2^{r-3}-1} \mathbf{e}_p(ay) \right|^2 \right)^{m+1}.
\end{aligned}$$

Taking into account that

$$\sum_{a=0}^{p-1} \left| \sum_{y=0}^{2^{r-3}-1} \mathbf{e}_p(ay) \right|^2 = p2^{r-3},$$

we derive

$$|T - |\mathcal{X}|2^{2(m+1)(r-3)}p^{-m-1}| \le 4n(m+1)2^{(m+1)(r-3)}p^{1/2}\log p.$$

Hence

$$T \ge 2^{(m+1)(r-3)}Q,$$

where

$$\begin{aligned}
Q &= |\mathcal{X}|2^{(m+1)(r-3)}p^{-m-1} - 4n(m+1)p^{1/2}\log p \\
&= 2^{r-k+O(1)}\left(\frac{2^{r-3}}{p}\right)^{m+1} - 2^{r/2+O(\log r)} \\
&\ge 2^{r-k-4m+O(1)} - 2^{r/2+O(\log r)} \\
&= 2^{r/2+4r^{1/2}+O(1)} - 2^{r/2+O(\log r)} > 0,
\end{aligned}$$

provided that p is large enough.

Therefore the claimed x exists and the desired result follows. □

Certainly the bound is of the correct order because obviously $\sigma(B) \le r$.

In the next theorem we estimate the Fourier coefficients (1.7) of Boolean functions B giving the rightmost bit of rational functions modulo p.

Theorem 29.4. *Suppose that $h(X) = f(X)/g(X) \in \mathbb{Z}[X]$ is a rational function which is neither a constant nor a linear function modulo p,*

$$n = \max\{\deg f, \deg g\}.$$

Let a Boolean function $B(U_1, \ldots, U_r)$ be such that for any x, $0 \le x \le 2^r - 1$, where $r = \lfloor \log p \rfloor$, with $g(x) \not\equiv 0 \pmod p$, $B(u_1, \ldots, u_r)$ equals the rightmost bit of $\lfloor h(x) \rfloor_p$, where $x = u_1 \ldots u_r$ is the bit representation of x. Then the bound

$$\max_{a \in \mathfrak{B}_r} \left| \widehat{B}(a) \right| \le 5n^{1/4} p^{7/8} \log^{3/4} p$$

holds.

Proof. We put

$$k = \left\lceil 0.5 \log \left(20np^{3/2} \log^3 p \right) \right\rceil, \qquad N = 2^k, \qquad M = 2^{r-k}.$$

Then we obtain

$$\widehat{B}(a) = \sum_{y=0}^{N-1} \sum_{z=0}^{M-1} (-1)^{B(y+Nz)\langle y.b \rangle + \langle z.c \rangle},$$

where $b = (a_1, \ldots, a_k)$, $c = (a_{k+1}, \ldots, a_r)$, and accordingly,

$$\langle y, b \rangle = a_1 y_1 + \ldots + a_k y_k, \qquad \langle z, c \rangle = a_{k+1} z_1 + \ldots + a_r z_{r-k},$$

where $y = y_1 \ldots y_k$ and $z = z_1 \ldots z_{r-k}$ are the bit representations of y and z, respectively.

Therefore

$$\left| \widehat{B}(a) \right| \le \sum_{y=0}^{N-1} \left| \sum_{z=0}^{M-1} (-1)^{B(y+Nz) + \langle z.c \rangle} \right|,$$

hence we derive

$$\begin{aligned}
\left| \widehat{B}(a) \right|^2 &\le \left(\sum_{y=0}^{N-1} \left| \sum_{z=0}^{M-1} (-1)^{B(y+Nz)+\langle z.c \rangle} \right| \right)^2 \\
&\le N \sum_{y=0}^{N-1} \left| \sum_{z=0}^{M-1} (-1)^{B(y+Nz)+\langle z.c \rangle} \right|^2 \\
&= N \sum_{y=0}^{N-1} \sum_{z_1.z_2=0}^{M-1} (-1)^{B(y+Nz_1)+B(y+Nz_2)+\langle z_1.c \rangle + \langle z_2.c \rangle} \\
&\le N \sum_{z_1,z_2=0}^{M-1} \left| \sum_{y=0}^{N-1} (-1)^{B(y+Nz_1)+B(y+Nz_2)} \right|.
\end{aligned}$$

Now we claim that for $M(M-1)$ pairs (z_1, z_2) with $z_1 \neq z_2$ the bound

$$\left|\sum_{y=0}^{N-1} (-1)^{B(y+Nz_1)+B(y+Nz_2)}\right| \leq 20np^{1/2}\log^3 p \tag{29.4}$$

holds.

Let $\mathcal{Y}$ be the set of $y = 0, \dots, N-1$ with

$$g(y+Nz_1)g(y+Nz_2) \not\equiv 0 \pmod p.$$

In order to prove this bound we estimate the number $T(\vartheta_1, \vartheta_2)$ of values of $y \in \mathcal{Y}$ with

$$B(y+Nz_1) = \vartheta_1, \qquad B(y+Nz_2) = \vartheta_2 \tag{29.5}$$

for each pair $\vartheta_1, \vartheta_2 \in \{0,1\}$.

Put

$$b_1 = Nz_1, \quad b_2 = Nz_2, \quad K_1 = \left\lfloor \frac{p-1-\vartheta_1}{2} \right\rfloor, \quad K_2 = \left\lfloor \frac{p-1-\vartheta_2}{2} \right\rfloor.$$

Then the system of equations (29.5) is equivalent to the system of congruences

$$h(y+b_1) \equiv 2u_1 + \vartheta_1 \pmod p, \qquad h(y+b_2) \equiv 2u_2 + \vartheta_2 \pmod p,$$

where

$$y \in \mathcal{Y}, \qquad 0 \leq u_1 \leq K_1, \qquad 0 \leq u_2 \leq K_2.$$

From (3.2) we see that

$$\begin{aligned}
T(\vartheta_1, \vartheta_2) & \\
&= \frac{1}{p^2} \sum_{y \in \mathcal{Y}} \sum_{\lambda_1,\lambda_2=0}^{p-1} \\
&\quad \times \sum_{u_1=0}^{K_1} \sum_{u_2=0}^{K_2} \mathbf{e}_p\left(\lambda_1\left(h(y+b_1) - u_1\right) + \lambda_2\left(h(y+b_2) - u_2\right)\right) \\
&= \frac{1}{p^2} \sum_{\lambda_1,\lambda_2=0}^{p-1} \sum_{y \in \mathcal{Y}} \mathbf{e}_p\left(\lambda_1 h(y+b_1) + \lambda_2 h(y+b_2)\right) \\
&\quad \times \sum_{u_1=0}^{K_1} \mathbf{e}_p\left(-\lambda_1 u_1\right) \sum_{u_2=0}^{K_2} \mathbf{e}_p\left(-\lambda_2 u_2\right).
\end{aligned}$$

The term corresponding to $\lambda_1 = \lambda_2 = 0$ equals $|\mathcal{Y}|(K_1+1)(K_2+1)/p^2$.

For other values of λ_1, λ_2 we see that the linear combination

$$H_{\lambda_1,\lambda_2}(Y) = \lambda_1 h(Y + b_1) + \lambda_2 h(Y + b_2) = \frac{F_{\lambda_1,\lambda_2}(Y)}{G_{\lambda_1,\lambda_2}(Y)}$$

is a non-trivial (because $b_1 \not\equiv b_2 \pmod p$) rational function with

$$\deg F_{\lambda_1,\lambda_2}(Y) + \deg G_{\lambda_1,\lambda_2}(Y) \le 4n.$$

Therefore from Lemma 3.6, as before, we derive that

$$\left|\sum_{y\in\mathcal{Y}} \mathbf{e}_p\left(\lambda_1 h(y + b_1) + \lambda_2 h(y + b_2)\right)\right| \le 4np^{1/2}\log p + 2n$$

for such λ_1, λ_2.

Using this bound and (3.4), we obtain

$$\begin{aligned}
&\left|T(\vartheta_1, \vartheta_2) - \frac{|\mathcal{Y}|(K_1+1)(K_2+1)}{p^2}\right| \\
&\quad \le \frac{4.5np^{1/2}\log p}{p^2} \sum_{\lambda_1.\lambda_2=0}^{p-1} \left|\sum_{u_1=0}^{K_1} \mathbf{e}_p(-\lambda_1 u_1)\right| \left|\sum_{u_2=0}^{K_2} \mathbf{e}_p(-\lambda_2 u_2)\right| \\
&\quad = \frac{4.5np^{1/2}\log p}{p^2} \left(\sum_{\lambda_1=0}^{p-1} \left|\sum_{u_1=0}^{K_1} \mathbf{e}_p(-\lambda_1 u_1)\right|\right) \left(\sum_{\lambda_1=0}^{p-1} \left|\sum_{u_2=0}^{K_2} \mathbf{e}_p(-\lambda_2 u_2)\right|\right) \\
&\quad \le 4.5np^{1/2}\log^3 p.
\end{aligned}$$

After simple evaluation we obtain the inequality

$$\begin{aligned}
\left|\sum_{y\in\mathcal{Y}} (-1)^{B(y+Nz_1)+B(y+Nz_2)}\right| &= |T(0,0) + T(1,1) - T(1,0) - T(0,1)| \\
&\le |\mathcal{Y}|p^{-2} + 18np^{1/2}\log^3 p.
\end{aligned}$$

We also have

$$\left|\sum_{y\notin Y} (-1)^{B(y+Nz_1)+B(y+Nz_2)}\right| \le N - |\mathcal{Y}| \le 2n$$

and the bound (29.4) follows.

Applying the bound (29.4) for $M(M-1)$ pairs (z_1, z_2) with $z_1 \ne z_2$ and the trivial identity

$$\sum_{y=0}^{N-1} (-1)^{B(y+Nz_1)+B(y+Nz_2)} = N$$

for M pairs (z_1, z_2) with $z_1 = z_2$, we obtain

$$\left|\widehat{B}(a)\right|^2 \le N(20nM^2p^{1/2}\log^3 p + MN) \le 20np^{5/2}N^{-1}\log^3 p + pN.$$

Noting that

$$20np^{3/2}N^{-1}\log^3 p \le N \le 2\left(20np^{3/2}\log^3 p\right)^{1/2},$$

we obtain

$$\left|\widehat{B}(a)\right|^2 \le 4p\left(20np^{3/2}\log^3 p\right)^{1/2} \le 20n^{1/2}p^{7/4}\log^{3/2} p,$$

and the desired estimate follows. □

The most immediate application of this result is a linear lower bound on the average sensitivity.

Theorem 29.5. *Suppose that $h(X) = f(X)/g(X) \in \mathbb{Z}[X]$ is a rational function which is neither a constant nor a linear function modulo p and such that*

$$\max\{\deg f, \deg g\} \le p^{1/2-\delta}$$

for some $\delta > 0$. Let a Boolean function $B(U_1, \ldots, U_r)$ be such that for any x, $0 \le x \le 2^r - 1$, where $r = \lfloor \log p \rfloor$, with $g(x) \not\equiv 0 \pmod p$, $B(u_1, \ldots, u_r)$ equals the rightmost bit of $\lfloor h(x) \rfloor_p$, where $x = u_1 \ldots u_r$ is the bit representation of x. Then the bound

$$\sigma_{av}(B) \ge \vartheta r + o(r),$$

holds, where ϑ is the unique root of the equation

$$H(\vartheta) = \delta/2, \qquad 0 < \vartheta < 1/2,$$

with the binary entropy function $H(\gamma)$ given by (1.4).

Proof. The relation between the Fourier coefficients and the average sensitivity of a Boolean function is given by the following identity,

$$\sigma_{av}(B) = 2^{-2r}\sum_{a \in \mathfrak{B}_r} \|a\| \left|\widehat{B}(a)\right|^2,$$

where $\|a\|$ is the Hamming weight of a, which can be found in [33, 36].

Therefore, for any $w \le r$, from the Parseval identity (7.1), we have

$$\begin{aligned}\sigma_{av}(B) &\ge 2^{-2r} \sum_{\substack{a \in \mathfrak{B}_r \\ \|a\| < w}} \|a\| \left|\widehat{B}(a)\right|^2 + 2^{-2r} w \sum_{\substack{a \in \mathfrak{B}_r \\ \|a\| \ge w}} \left|\widehat{B}(a)\right|^2 \\ &= 2^{-2r} \sum_{\substack{a \in \mathfrak{B}_r \\ \|a\| < w}} \|a\| \left|\widehat{B}(a)\right|^2 + w \left(1 - 2^{-2r} \sum_{\substack{a \in \mathfrak{B}_r \\ \|a\| < w}} \left|\widehat{B}(a)\right|^2\right) \\ &\ge w - 2^{-2r}(w-1) \sum_{\substack{a \in \mathfrak{B}_r \\ \|a\| < w}} \left|\widehat{B}(a)\right|^2 .\end{aligned}$$

Using the bound of Theorem 29.4 and the bound (10.2) we see that

$$\begin{aligned}2^{-2r} \sum_{\substack{a \in \mathfrak{B}_r \\ \|a\| < w}} \left|\widehat{B}(a)\right|^2 &= O\left(r^{3/2} 2^{-r\delta/2} \sum_{\substack{a \in \mathfrak{B}_r \\ \|a\| < w}} 1\right) \\ &= O\left(r^{3/2} 2^{-r\delta/2} \sum_{i=0}^{w-1} \binom{r}{i}\right) = 2^{r(H(w/r) - \delta/2 + o(1))}.\end{aligned}$$

We now see that one can choose some $w \sim \vartheta r$ with

$$2^{r(H(w/r)-\delta/2+o(1))} = o(1)$$

which yields the desired result. □

As before, Theorems 29.1, 29.2 and 29.5, combined with Lemma 7.5, can be used to estimate the real and real approximate degrees of the corresponding Boolean functions.

We define modulo p functions as functions which take their values in the range $0 \le f(x) \le p-1$ for any $x = 0, \dots, 2^r - 1$.

We now obtain some trade-off relations between the depths of arithmetic and Boolean circuits computing modulo p functions.

We can define the Boolean depth $D_B(f)$ of a modulo p function f as the smallest d for which there exist $r+1$ circuits $C_j \in \mathrm{BC}(d)$, $j = 1, \dots, r+1$, such that the binary vector

$$(C_1(u_1, \dots, u_r), \dots, C_{r+1}(u_1, \dots, u_r)), \qquad 0 \le x \le 2^r - 1,$$

gives the bit representation of $f(x)$, where $x = u_1 \dots u_r$ is the bit representation of x, $r = \lfloor \log p \rfloor$.

Accordingly, we can define the arithmetic depth $D_A(f)$ of a modulo p function f as the smallest d for which there exists a circuit $C \in \mathrm{AC}_p(d)$ such that

$$f(x) = C(x), \qquad 0 \le x \le 2^r - 1.$$

Theorem 29.6. *Suppose that a function $h : \mathbb{F}_p \to \mathbb{F}_p$ is neither a constant nor a linear function modulo p. Then the bound*

$$\max\{D_A(h), D_B(h)\} \geq 0.125 \log r + o(\log r)$$

holds.

Proof. Assume that $D_A(h) \leq 0.4 \log r$. Then h is given by a rational function of $f(X)/g(X) \in \mathbb{F}_p(X)$ with

$$n = \max\{\deg f, \deg g\} \leq 2^{D_A(f)} \leq r^{2/5}.$$

If this is a "proper" rational function (that is, not a polynomial) then we apply Theorem 29.3 and derive that the sensitivity of any Boolean function B coinciding with the second leftmost bit of $h(x)$ is $\Omega(r)$. Hence, from Lemma 7.1, we derive

$$D_B(h) \geq \text{CREW PRAM}\,(B) \geq 0.5 \log r + o(\log r).$$

If this function is a polynomial of degree $n \geq 3$ we apply Theorem 29.1 and derive that the sensitivity of any Boolean function B coinciding with the second leftmost bit of $f(x)$ is $\Omega(r^{1/2})$. Hence, from Lemma 7.1, we conclude

$$D_B(h) \geq \text{CREW PRAM}\,(B) \geq 0.25 \log r + o(\log r).$$

If this function is a polynomial of degree $n = 2$, then we note that $h_0(X) = h(h(X))$ is a polynomial over $\mathbb{F}_p$ of degree $\deg h = 4$. Therefore, we have the inequality

$$D_B(h) \geq 0.5 D_B(h_0) \geq 0.125 \log r + o(\log r),$$

which finishes the proof. □

In the partial case of the modular inversion $f(X) \equiv 1/X \pmod p$ a result similar to Theorem 29.6 has been obtained in [205].

For division-free arithmetic a stronger bound can be obtained.

We can define the division-free arithmetic depth $D_{DFA}(f)$ of a modulo p function f as the smallest d for which there exists a circuit $C \in \mathrm{DFAC}_p(d)$ such that

$$f(x) = C(x), \qquad 0 \leq x \leq 2^r - 1.$$

Theorem 29.7. *Suppose that a function $f : \mathbb{F}_p \to \mathbb{F}_p$ is neither a constant nor a linear function modulo p. Then for any $\varepsilon > 0$ and sufficiently large p at least one of the bounds*

$$D_{DFA}(f) \geq (0.5 - \varepsilon) r \qquad \textit{and} \qquad D_B(f) \geq (0.125 - \varepsilon) \log r$$

holds.

Proof. Assume that $D_{DFA}(f) \leq (0.5 - \varepsilon)r/2$. Let $n \geq 2$ be the degree of f as a polynomial over $\mathbb{F}_p$. Obviously

$$n \leq 2^{D_{DFA}(f)} \leq 2^{(1-2\varepsilon)r/2}$$

provided that p is large enough.

If $n \geq 3$, then, from Theorem 29.1, we derive that the sensitivity of any Boolean function B coinciding with the second leftmost bit of $f(x)$ is $\Omega(r^{1/2})$. Hence, from Lemma 7.1, we derive

$$D_B(f) \geq \text{CREW PRAM}\,(B) \geq 0.25 \log r + o(\log r).$$

If $n = 2$, then we note that $f_0(X) = f(f(X))$ is of degree $\deg f_0 = 4$. Therefore, we have the inequality

$$D_{DFA}(f) \geq 0.5 D_{DFA}(f_0) \geq 0.125 \log r + o(\log r),$$

which finishes the proof. □

Finally, we use Theorem 29.4 in order to obtain a trade-off result between the depth of arithmetic circuits and the size/depth of unbounded fan-in Boolean circuits of the class $\text{UBC}(d, S)$, defined in Chapter 10.

Theorem 29.8. *Suppose that a function $h : \mathbb{F}_p \to \mathbb{F}_p$ is neither a constant nor a linear function modulo p. Then either*

$$D_A(h) \geq 0.1r$$

or for any circuit $C \in \text{UBC}(d, S)$ computing this function

$$d \log \log S \geq (1 + o(1)) \log r,$$

where $r = \lfloor \log p \rfloor$.

Proof. Assume that $D_A(h) < 0.1r$. Then h is given by a rational function of $f(X)/g(X) \in \mathbb{F}_p(X)$ with

$$n = \max\{\deg f, \deg g\} \leq 2^{D_A(f)} \leq p^{1/10}.$$

Let $B(U_1, \dots, U_r)$ be a Boolean function such that $B(u_1, \dots, u_r)$ is equal to the rightmost bit of $h(x)$, where $x = u_1 \dots u_r$ is the bit representation of x.

Assume that B is computed by a circuit $C \in \text{UBC}(d, S)$. As in the proof of Theorem 10.5, we make use of Lemma 7.2. It is easy to verify that the equation

$$H(\vartheta) = 1/5, \qquad 0 < \vartheta < 1/2,$$

has the unique root $\vartheta = 0.031\ldots$, where, as before, $H(\gamma)$ is the binary entropy function (1.4).

We put $w = \lceil 0.03r \rceil$. Then, from Theorem 29.4, we see that

$$\begin{aligned} 2^{-2r} \sum_{\substack{a \in \mathfrak{B}_r \\ \|a\| \le w}} \left|\widehat{B}(a)\right|^2 &\le 25w \binom{r}{w} 2^{-2r} n^{1/2} p^{7/4} \log^{3/2} p \\ &\le 25w \binom{r}{w} 2^{-2r} p^{9/5} \log^{3/2} p \\ &= O\left(2^{-r/5 + H(w/r)r + o(r)}\right) = o(1). \end{aligned}$$

Finally, from the Parseval identity (7.1), we derive

$$2^{-2r} \sum_{\substack{a \in \mathfrak{B}_r \\ \|a\| > w}} \left|\widehat{B}(a)\right|^2 = 1 + o(1)$$

and the desired estimate follows. □

Chapter 30

Polynomial Approximation, Permanents and Noisy Exponentiation in Finite Fields

Here we present the results of [520] on the problem of finding an m-sparse polynomial

$$f(X) = \sum_{j=1}^{m} \alpha_j X^{e_j} \in \mathbb{F}_p[X]$$

from approximates to $\lfloor f(t) \rfloor_l$ at polynomially many points $t \in \mathbb{F}_p$ selected uniformly at random. The results are non-trivial only if the prime p is sufficiently large.

In [221] a similar but different situation has been considered when $f(X)$ is a sparse polynomial over the field $\mathbb{F}_{q^r}$ and the exact values of the trace $\mathrm{Tr}_{\mathbb{F}_{q^r}/\mathbb{F}_q}(f(t))$ are given at polynomially many random $t \in \mathbb{F}_{q^r}$. The algorithm of [221], which is also based on Lemma 2.6, works for any underlying field $\mathbb{F}_q$, for example, it applies to the case $q = 2$ as well.

In turn these results can be applied to studying the security of pseudorandom numbers of [391, 430] formed from the least significant bits of residues of polynomials modulo prime numbers.

The results of [221, 520] may be helpful for constructing new hard core functions, in particular, to generalizations of the results of [408, 409], see also [212, 213, 220, 298]. They can also be used to study *polynomial hash functions*, see [260, 412].

The above problems are also related to

- "noisy" polynomial interpolation and list decoding algorithms for polynomial and algebraic-geometry codes [48, 243, 244, 270, 446, 488, 504, 543, 544, 555];

- learning and reconstructing polynomials with "noisy" queries [12, 13, 175, 217, 300, 294, 295, 296];
- sparse polynomial interpolation [114, 240, 320, 558, 571];

and to some other important questions.

We start with the following two results of [520].

Theorem 30.1. *Let p be a sufficiently large n-bit prime and let $m \ge 1$ be an integer. We define*

$$\mu = (mn)^{1/2}, \quad k = \lceil \mu + \log n \rceil, \quad d = 4\lceil \mu \rceil, \quad \eta = 0.5\mu + 3.$$

There exists a deterministic polynomial time algorithm $\mathcal{A}$ such that for any polynomial

$$f(X) = \sum_{j=1}^{m} \alpha_j X^{e_j} \in \mathbb{F}_p[X]$$

of degree $\deg f \le 2^{-2\eta} p^{1/2}$ with known exponents $1 \le e_1 < \ldots < e_m \le p-2$, given $2d$ integers

$$t_i \qquad \text{and} \qquad s_i = \mathrm{MSB}_{k,p}\left(f(t_i)\right), \qquad i = 1, \ldots, d,$$

its output satisfies

$$\Pr_{t_1,\ldots,t_d \in \mathbb{F}_p}\left[\mathcal{A}(t_1, \ldots, t_d; s_1, \ldots, s_d) = (\alpha_1, \ldots, \alpha_m)\right] = 1 + O\left(p^{-1}\right)$$

if $t_1, \ldots, t_d$ are chosen uniformly and independently at random from $\mathbb{F}_p$.

Proof. We follow the same arguments as in the proof of Lemma 6.3 which we briefly outline here for the sake of completeness. We refer to the first d vectors in the matrix (6.5) as p-vectors and we refer to the other m vectors as power-vectors. Let us consider the vector $\mathbf{s} = (s_1, \ldots, s_d, s_{d+1}, \ldots, s_{d+m})$ where

$$s_{d+j} = 0, \qquad j = 1, \ldots, m.$$

Taking a linear combination of the jth power-vectors of the matrix (6.5) with the coefficients α_j, $j = 1, \ldots, m$, and subtracting a certain linear combination of the p-vectors, we obtain a lattice point

$$\mathbf{u}_f = (u_1, \ldots, u_d, \alpha_1/2^{k+1}, \ldots, \alpha_m/2^{k+1}) \in \mathcal{L}_{k,\mathbf{e},p}(t_1, \ldots, t_d)$$

such that

$$|u_i - s_i| < p2^{-k-1}, \qquad i = 1, \ldots, d+m,$$

where $u_{d+j} = \alpha_j/2^{k+1}$, $j = 1, \ldots, m$. Therefore,

$$\sum_{i=1}^{d+m} (u_i - s_i)^2 \le (d+m)2^{-2k-2}p^2.$$

We can assume that $m \le n$ because in the opposite case $k > n$ and the result is trivial. Therefore $d+m = O(k)$. Now we can use Lemma 6.1 to find in polynomial time a lattice vector $\mathbf{v} = (v_1, \ldots, v_d, v_{d+1}, \ldots, v_{d+m}) \in \mathcal{L}_{k,\mathbf{e},p}(t_1, \ldots, t_d)$ such that

$$\begin{aligned}\sum_{i=1}^{d} (v_i - s_i)^2 &= 2^{o(d+m)} \min\left\{\sum_{i=1}^{d+m} (z_i - s_i)^2, \quad \mathbf{z} = (z_1, \ldots, z_{d+m})\right\} \\ &= 2^{o(k)} \min\left\{\sum_{i=1}^{d+m} (z_i - s_i)^2, \quad \mathbf{z} = (z_1, \ldots, z_{d+m})\right\} \\ &\le 2^{-k+o(k)}(d+m)p^2 \le 2^{-2k+o(k)}p^2 \le 2^{-2\eta-1}p^2,\end{aligned}$$

provided that p is large enough. We also have

$$\sum_{i=1}^{d} (u_i - s_i)^2 \le d2^{-2k-2}p \le 2^{-2\eta-2}p.$$

Therefore,

$$\sum_{i=1}^{d} (u_i - v_i)^2 \le 2^{-2\eta}p^2.$$

Applying Lemma 6.7, we see that $\mathbf{v} = \mathbf{u}_f$ with probability $1 + O\left(p^{-1}\right)$, and therefore the coefficients of f can be recovered in polynomial time. □

Similarly Lemma 6.8 implies the following statement:

Theorem 30.2. *Let p be a sufficiently large n-bit prime and let $m \ge 1$ be an integer. Then for any $\varepsilon > 0$ and $1 \ge \rho > 0$ the following statement holds with*

$$k = \lceil (1 - \rho/m + \varepsilon)n \rceil + 6, \qquad d = \lceil (m+1)/\varepsilon \rceil.$$

There exists a deterministic polynomial time algorithm $\mathcal{A}$ such that for any polynomial

$$f(X) = \sum_{j=1}^{m} \alpha_j X^{e_j} \in \mathbb{F}_p[X]$$

with known exponents $1 \le e_1 < \ldots < e_m \le p-2$ such that

$$\max_{0 \le i < j \le m} \gcd(e_j - e_i, p-1) \le p^{1-\rho},$$

where $e_0 = 0$, *given* $2d$ *integers*

$$t_i \qquad \text{and} \qquad s_i = \mathrm{MSB}_{k,p}\left(f(t_i)\right), \qquad i = 1, \ldots, d,$$

its output satisfies

$$\Pr_{t_1,\ldots,t_d \in \mathbb{F}_p} \left[\mathcal{A}\left(t_1, \ldots, t_d; s_1, \ldots, s_d\right) = (\alpha_1, \ldots, \alpha_m)\right] = 1 + O\left(p^{-1}\right),$$

if $t_1, \ldots, t_d$ *are chosen uniformly and independently at random from* $\mathbb{F}_p$.

We remark that Theorem 30.2 implies that if $\rho = 1$, then the polynomial f can be recovered from $\lceil 2(m+1)/\varepsilon \rceil$ approximations with absolute error $O(p^{1/m-\varepsilon})$.

Now we describe the result from [221] concerning the case when $f(X)$ is a sparse polynomial over the field $\mathbb{F}_{q^r}$ and the exact values of the trace $\mathrm{Tr}_{\mathbb{F}_{q^r}/\mathbb{F}_q}(f(t))$ are given at polynomially many random $t \in \mathbb{F}_{q^r}$.

For an integer $m \geq 1$ and an exponent vector $\mathbf{e} = (e_1, \ldots, e_m) \in \mathbb{Z}^m$ with $0 \leq e_1 < \ldots < e_m \leq q^r - 2$, we denote by $\mathcal{P}_m(\mathbf{e})$ the following class of m-sparse polynomials

$$f(X) = \sum_{j=1}^{m} \alpha_j X^{e_j} \in \mathbb{F}_{q^r}[X].$$

We always assume that the exponent vector $\mathbf{e}$ is such that the following nm elements

$$e_j q^i, \qquad j = 1, \ldots, m, \; i = 0, \ldots, r-1, \tag{30.1}$$

are pairwise distinct modulo $q^m - 1$.

For a polynomial $f(X) \in \mathbb{F}_{q^r}[X]$, let $\mathcal{TR}_f$ denote an oracle which for every call selects (uniformly at random) $t \in \mathbb{F}_{q^r}^*$ and outputs t and $\mathrm{Tr}_{\mathbb{F}_{q^r}/\mathbb{F}_q}(f(t))$.

We show how given an oracle $\mathcal{TR}_f$ one can reconstruct the polynomial $f \in \mathcal{P}_m(\mathbf{e})$ in probabilistic polynomial time. We remark that the condition (30.1) is necessary, as it ensures the oracle calls will determine uniquely a polynomial, whereas if, for instance, $e_1 q^j \equiv e_2 q^k \pmod{q^r - 1}$ with $j > k$, then the two distinct polynomials

$$f_1(X) = \sum_{j=1}^{m} A_j X^{e_j} \quad \text{and} \quad f_2(X) = \left(A_1 + A_2^{q^{r+k-j}}\right) X^{e_1} + \sum_{j=3}^{m} A_j X^{e_j}$$

satisfy $\mathrm{Tr}_{\mathbb{F}_{q^r}/\mathbb{F}_q}(f_1(t)) = \mathrm{Tr}_{\mathbb{F}_{q^r}/\mathbb{F}_q}(f_2(t))$ for all $t \in \mathbb{F}_{q^r}^*$ and therefore $\mathcal{TR}_{f_1} = \mathcal{TR}_{f_2}$.

Theorem 30.3. *There exists an algorithm which, for any* $f \in \mathcal{P}_m(\mathbf{e})$, *with the exponent vector* $\mathbf{e} = (e_1, \ldots, e_m)$ *satisfying (30.1), makes the expected number of* $O(m^2 r^2)$ *calls of the oracle* $\mathcal{TR}_f$ *and finds* f *in probabilistic polynomial time.*

Proof. Let $N = mr$ and let ν_k, $k = 1, \dots, N$ be the non-negative residues modulo $q^r - 1$ of the integers (30.1) (ordered in an arbitrary order). For $s = 1, \dots, N$ we consider the polynomial matrices

$$M_s(X_1, \dots, X_s) = (X_r^{\nu_k})_{r,k=1}^s .$$

We remark that if $\mathrm{Tr}_{\mathbb{F}_{q^r}/\mathbb{F}_q}(f(t_1)), \dots, \mathrm{Tr}_{\mathbb{F}_{q^r}/\mathbb{F}_q}(f(t_N))$ are known and also

$$\det M_N(t_1, \dots, t_N) \neq 0 \tag{30.2}$$

then we obtain a non-singular system of N linear equations for the coefficients of f (and their conjugates) which can be solved in polynomial time.

Now we show how to design a sequence of calls of the oracle $\mathcal{TR}_f$ so that the resulting sequence $t_1, \dots, t_N$ satisfies the condition (30.2).

At first step we make an oracle call which returns us some $t \in \mathbb{F}_{q^r}^*$ and $\mathrm{Tr}_{\mathbb{F}_{q^r}/\mathbb{F}_q}(f(t))$. We put $t_1 = t$. In particular

$$M_1(t_1) = t_1^{\nu_1} \neq 0.$$

Let $s \geq 2$. Assume $t_1, \dots, t_{s-1}$, selected from the oracle calls at the first $s-1$ steps, satisfy

$$\det M_{s-1}(t_1, \dots, t_{s-1}) \neq 0.$$

Since the integers (30.1) are pairwise distinct, the polynomial

$$\det M_s(X_1, \dots, X_s) \in \mathbb{F}_q[X_1, \dots, X_s]$$

is nonzero, and thus from Lemma 2.5 we see that there are at least $(q^r - 1)/s$ values of $t \in \mathbb{F}_{q^r}^*$ with

$$\det M_s(t_1, \dots, t_{s-1}, t) \neq 0.$$

Thus such a value of t will be returned by the oracle $\mathcal{TR}_f$, together with the value of $\mathrm{Tr}_{\mathbb{F}_{q^r}/\mathbb{F}_q}(f(t))$ after the expected number of s calls. Therefore, after the expected number of

$$\sum_{s=1}^{N} s = \frac{mr(mr+1)}{2}$$

oracle calls we obtain $t_1, \dots, t_N$ satisfying the condition (30.2) for which the values of $\mathrm{Tr}_{\mathbb{F}_{q^r}/\mathbb{F}_q}(f(t_1)), \dots, \mathrm{Tr}_{\mathbb{F}_{q^r}/\mathbb{F}_q}(f(t_N))$ are known and the result follows. □

Let

$$f(X_1, \dots, X_m) \in \mathbb{F}_p[X_1, \dots, X_m]$$

be a homogeneous polynomial of degree n over $\mathbb{F}_p$.

Let us denote by $\mathcal{POL}_{f,k}$ an oracle which, given any $x_1, \dots, x_m \in \mathbb{F}_p$, outputs $\mathrm{MSB}_{k,p}(f(x_1, \dots, x_m))$, where the function $\mathrm{MSB}_{k,p}$ is defined by (1.1).

Our next statement shows that if f is hard to evaluate then it is hard to construct the above oracle $\mathcal{POL}_{f,k}$.

Theorem 30.4. *Let Q be a sufficiently large integer. The following statement holds with $\vartheta = 1/3$ for all primes $p \in [Q, 2Q]$, and with $\vartheta = 0$ for all primes $p \in [Q, 2Q]$ except at most $Q^{5/6+\varepsilon}$ of them. For any $\varepsilon > 0$ and any homogeneous polynomial*

$$f(X_1, \dots, X_m) \in \mathbb{F}_p[X_1, \dots, X_m]$$

of degree $n \le p^{1-\vartheta-\varepsilon}$ over $\mathbb{F}_p$ there exists a probabilistic polynomial in n and $\log p$ time algorithm which for any $x_1, \dots, x_m \in \mathbb{F}_p$, makes $O\left(\log^{1/2} p\right)$ calls of the oracle $\mathcal{POL}_{f,k}$ with $k = \left\lceil \log^{1/2} p + \log\log p \right\rceil$ and evaluates $f(x_1, \dots, x_m)$ correctly with probability $1 + O\left(p^{-1}\right)$.

Proof. Given $x_1, \dots, x_m \in \mathbb{F}_p$, let us select $\lambda \in \mathbb{F}_p^*$ uniformly at random and use the oracle $\mathcal{POL}_{f,k}$ with the input $\lambda x_1, \dots, \lambda x_m$ to evaluate

$$\mathrm{MSB}_{k,p}\left(f(\lambda x_1, \dots, \lambda x_m)\right) = \mathrm{MSB}_{k,p}\left(\lambda^n f(x_1, \dots, x_m)\right).$$

Using Lemmas 3.17 and 4.1 together with Lemma 6.3, we obtain the desired result. □

An interesting example of a multivariate polynomial which is hard to evaluate is given by the *permanent* $\operatorname{per} X$ of a matrix X over $\mathbb{F}_p$. We recall that if $X = (x_{ij})_{i,j=1}^n$ then

$$\operatorname{per} X = \sum_{\sigma \in \mathcal{S}_n} \prod_{i=1}^{n} x_{i,\sigma(i)}$$

where σ runs over all elements of the symmetric group $\mathcal{S}_n$. We remark that for arbitrary matrices many "approximability" and "non-approximability" results have been obtained, see [90, 177, 286, 346] and references therein. However for special classes of matrices very little seems to be known, see [37, 124, 125]. Our approach certainly applies to general matrices as well but in this case a much stronger result is given by Theorem 1.7 of [177]. However the method of proof of Theorem 1.7 does not apply to structured matrices. Indeed, the transformation described in the proof of Theorem 5.2 of [177], which underlies the method of [177], does not preserve such structure properties as being symmetric or circulant.

The result of Theorem 30.4 applies to such classes of matrices as symmetric matrices. It also applies to other "structured" matrices, such as circulant, Hankel and Toeplitz matrices, see [126]. For all these classes of matrices over $\mathbb{F}_p$ it has been shown in [126] that if computing their permanent is hard then approximating the permanent is hard as well.

We say that a class $\mathcal{M}_n$ of $n \times n$ matrices with entries from $\mathbb{F}_p$ is *homogeneous* if for any $X = (x_{ij})_{i,j=1}^n \in \mathcal{M}_n$ and any $\lambda \in \mathbb{F}_p^*$ we also have $X_\lambda = (\lambda x_{ij})_{i,j=1}^n \in \mathcal{M}_n$.

Let $\mathcal{PER}_{\mathcal{M}_n,k}$ denote an oracle which, given any matrix $X \in \mathcal{M}_n$, outputs $\mathrm{MSB}_{k,p}(\operatorname{per} X)$, where the function $\mathrm{MSB}_{k,p}$ is defined by (1.1).

Using the same arguments as in the proof of Theorem 30.4 and the identity $\operatorname{per} X_\lambda = \lambda^n \operatorname{per} X$, one derives.

Theorem 30.5. *Let Q be a sufficiently large integer. The following statement holds with $\vartheta = 1/3$ for all primes $p \in [Q, 2Q]$, and with $\vartheta = 0$ for all primes $p \in [Q, 2Q]$ except at most $Q^{5/6+\varepsilon}$ of them. For any $\varepsilon > 0$ and any homogeneous class of matrices $\mathcal{M}_n$ over $\mathbb{F}_p$ of dimension $n \le p^{1-\vartheta-\varepsilon}$ there exists a probabilistic polynomial in n and $\log p$ time algorithm which for any $X \in \mathcal{M}_n$, makes $O\left(\log^{1/2} p\right)$ calls of the oracle $\mathcal{PER}_{\mathcal{M}_n,k}$ with $k = \left\lceil \log^{1/2} p + \log\log p \right\rceil$ and evaluates* $\operatorname{per} X$ *correctly with probability $1 + O\left(2^{-\log^{1/2} p}\right)$.*

We remark that although traditionally it is considered that the input size for an $n \times n$ matrix X over $\mathbb{F}_p$ is about $n^2 \log p$, some matrices admit a much shorter description. For example, an s-sparse circulant matrix, with only s non-zero entries per row, can be described by only $O(s \log np)$ bits. For such matrices it is enough to specify s pairs (m_ν, x_ν), $\nu = 1, \ldots, s$, where m_ν, $1 \le m_\nu \le n$, is the position of the νth non-zero entry $x_\nu \in \mathbb{F}_p$ in the first row. In this case, provided the oracle $\mathcal{PER}_{\mathcal{M}_n,k}$ accepts such a description, the algorithm of Theorem 30.5 becomes polynomial in $s \log np$. In fact in this form, one can consider an analogue of Theorem 30.5 for the determinant $\det X$ as well. Indeed, although the determinant is an "easy" function for dense matrices, it is not clear whether for s-sparse circulants it can be computed in time polynomial in $s \log np$. Moreover, an analogue of Theorem 30.5 and its modification for matrices with "short description" holds for a much wider class of matrix functions which are know as *immanants* and whose complexity has been studied as well, see [86, 87]. Those are expressions of the form

$$\operatorname{imm}_\chi X = \sum_{\sigma \in \mathcal{S}_n} \chi(\sigma) \prod_{i=1}^{n} x_{i,\sigma(i)}$$

where $\chi : \mathcal{S}_n \to \mathbb{C}$ is an irreducible character of the symmetric group $\mathcal{S}_n$. The trivial character $\chi(\sigma) = 1$ corresponds to the permanent, the alternating character $\chi(\sigma) = \operatorname{sign} \sigma$ corresponds to the determinant. This approach can also be applied to showing the hardness of modular approximation of several other polynomial functions arising and algebra and graph theory, such as the *cycle format polynomials* and the *factor polynomials*, see Section 3.3 of [87].

As before, we remark that using the result of [6] one can get slightly stronger versions of Theorems 30.1, 30.4 and 30.5.

Finally we present a result from [275] which shows that Lemma 6.10 implies a "correction" algorithm for "noisy" exponentiation black-boxes. This result is similar to those known for algebraic functions over finite fields [300].

Let $g \in \mathbb{F}_p$ be a fixed element of multiplicative order τ.

Let $\mathcal{EXP}_\Delta$ denote an oracle which, given any $v \in [0, \tau - 1]$, outputs V such that $\left| \lfloor g^v \rfloor_p - V \right| \le \Delta$.

Theorem 30.6. *Let $\varepsilon > 0$ and $1 \ge \rho > 1/2$ be fixed real numbers and let g be of multiplicative order $\tau \ge 0.5p^\rho$, where p is a sufficiently large prime. We define*

$$\Delta = p^{(2\rho-1)/5-\varepsilon}.$$

Given an oracle $\mathcal{EXP}_\Delta$, there exists a probabilistic polynomial time algorithm $\mathcal{A}$ such that for any $s \in [0, \tau - 1]$ it makes the expected number $O(\varepsilon^{-1})$ of calls of the oracle and computes $\lfloor g^s \rfloor_p$ with probability at least $1 + O(p^{-1})$.

Proof. Let $\mathcal{G} \subseteq \mathbb{F}_p^*$ be the multiplicative group generated by g. First of all we remark that when $v \in [0, p-2]$ is selected uniformly at random, the value of $\lfloor g^v \rfloor_p$ is uniformly distributed in $\mathcal{G}$ and the cardinality $|G| = \tau \ge 0.5p^\rho$. Now making calls of the oracle $\mathcal{EXP}_\Delta$ with $v = s$ and with $v = u_i$ and $v = s + u_i$, $i = 1, \ldots, d$, for $d = \lceil 4\varepsilon^{-1}/5 \rceil$ values of $u_i \in [0, p-2]$ chosen uniformly at random, we apply the algorithm of Lemma 6.10 to derive the result. □

We remark, that the algorithm of Theorem 30.6 does not rely on the knowledge of τ.

Chapter 31

Special Polynomials and Boolean Functions

In this chapter we show how to apply the techniques of this book to various questions about permutation polynomials, powers x^d, Zech's logarithm, primitive root testing and symmetric Boolean functions.

Let $f(X) \in \mathbb{F}_q[X]$ be a non-linear permutation polynomial (that is, the mapping $x \to f(x)$ is bijective on $\mathbb{F}_q$). One can also consider the inverse mapping $h(f(x)) = x$, which obviously is also bijective and, like any mapping over $\mathbb{F}_q$, is given by some polynomial $h(X) \in \mathbb{F}_q[X]$. It is mentioned in [398], Problem 10, that apparently not both of these polynomials can be of small degree. Here we show that this is really the case.

Theorem 31.1. *Let* $f(X), h(X) \in \mathbb{F}_q[X]$ *be two permutation polynomials of degree* $\deg f = d$ *and* $\deg h = D$ *which generate relatively inverse mappings. If* $\max\{d, D\} > 1$, *then* $dD \geq q$.

Proof. We see that $h(f(x)) = x$ for $x \in \mathbb{F}_q[X]$. Therefore the polynomial $h(f(X)) - X$ is of degree $dD > 1$ and has at least q zeros. □

Similar arguments have been used in [431] to study the lattice structure of non-linear pseudorandom numbers given by (26.7), see Theorem 26.4.

We now consider polynomial approximations of powers x^d in finite fields. The question is related to the complexity of *modular powering*, or of *finite field powering*, that is, computing x^d in a residue ring or in a finite field.

From the equation $x^{-1} = x^{q-2}$ which holds for any $x \in \mathbb{F}_q^*$ and the congruence $x^{-1} \equiv x^{\varphi(m)-1} \pmod m$ which holds for any $x \in \mathbb{Z}_m^*$, we see that modular inversion is a special partial case of modular powering.

Efficient parallel algorithms for modular and finite field powering and inversion in the Boolean model of computation can be found in [2, 3, 180, 200, 201, 202, 203, 351, 394]; see also [380].

The method of proof of Theorem 21.2 immediately yields the following result.

Theorem 31.2. *Let d be an integer with $0 \le d \le q-1$ and let $f(X) \in \mathbb{F}_q[X]$ be a polynomial of degree n such that $f(X) \neq a_n X^n$. Assume that $f(x) = x^d$, $x \in S$, for a set $S \subseteq \mathbb{F}_q$. Then*

$$n \ge \frac{(|S|-1)^2}{2(q-1)}.$$

Proof. Let $T \ge |S|-1$ denote the number of solutions of the equation $f(x) = x^d$, $x \in \mathbb{F}_q^*$. It is enough to show that $T \le (2n(q-1))^{1/2}$.

Obviously, T^2 is the number of solutions of the system of equations

$$f(x) = x^d, \quad f(y) = y^d, \qquad x, y \in \mathbb{F}_q^*,$$

which after the substitution $y = xz$ reduces to the equivalent system

$$f(x) = x^d, \quad f(xz) = (xz)^d, \qquad x, z \in \mathbb{F}_q^*.$$

From here we derive $f(xz) = z^d f(x)$. For those values of z for which the polynomial $F_z(X) = f(Xz) - z^d f(X)$ is not identical to zero there are at most n corresponding values of x; for other values of z there are at most $q-1$ corresponding values of x. Thus, $T^2 \le (q-1)n + (q-1)Q$, where Q is the number of $z \in \mathbb{F}_q^*$ such that $F_z(X)$ is identical to zero. Taking into account that $f(X)$ contains at least two non-zero monomials, say $a_n X^n$ and $a_m X^m$, $a_n a_m \neq 0$, $0 \le m < n$, we find that this is possible only if $z^m = z^d$ and $z^n = z^d$. In particular, $z^m = z^n$; therefore $Q \le n$ and the result follows. □

The polynomial $f(X) = X$ shows that the condition $f(X) \neq a_n X^n$ is necessary. Indeed, for an odd q, we have $x = x^{(q+1)/2}$ for the set of $(q-1)/2$ quadratic residues x of $\mathbb{F}_q$ (and of course also for $x = 0$).

Using the estimate of Theorem 31.2, one can extend some lower bounds from [200] on the depth of arithmetic circuits for powering in a finite field over sets of size of order $q^{1/2}$, while the results of [200] are applicable to sets of size at least d. For the most interesting case of $d = (q-1)/2$ we see that any arithmetic circuit over $\mathbb{F}_q$ correctly testing whether an element $x \in S$ is a quadratic residue should be of depth $\Omega(\log q)$ provided that $|S| \ge q^{1/2+\varepsilon}$ for some fixed $\varepsilon > 0$ and S contains both quadratic residues and non-residues. From Theorem 2.5 of [200] the same result follows only for sets of cardinality $|S| > (1/2+\varepsilon)q$.

Let g be a primitive root of $\mathbb{F}_q$. For $x \in \{0, \ldots, q-2\}$, *Zech's logarithm* $Z(x)$ is defined by the relation

$$g^{Z(x)} = g^x + 1, \qquad 0 \le Z(x) \le q-2,$$

if $g^x \neq -1$ and $Z(x) = 0$ if $g^x = -1$. (When q is odd we will have $g^x = -1$ if $x = (q-1)/2$; when q is even, $g^x = -1$ if $x = 0$.)

We start with an analogue of Theorems 8.8, 8.9 and 9.1.

Theorem 31.3. *Let $0 \le N < N + H \le q - 2$ and let $(u(x))$ be an integer linear recurrence sequence of order n such that*

$$Z(x) \equiv u(x) \pmod{q-1}, \qquad N+1 \le x \le N+H.$$

Then

$$n \ge \frac{H}{3\log q} - 1.$$

Proof. We follow the same lines as in the proof of Theorem 9.1. Let us define the linear recurrence sequences $(w_i(x))$, $i = 1, \ldots, n$, as in Lemma 2.3.

We consider n-dimensional vectors

$$W(x) = (w_1(x), \ldots, w_n(x)), \qquad x = 1, 2, \ldots.$$

Let $L = \lfloor (H-1)/3 \rfloor$. Consider 2^L linear combinations

$$\sum_{l=1}^{L} \gamma_l W(l), \qquad (\gamma_1, \ldots, \gamma_L) \in \mathfrak{B}_L.$$

Assume that

$$2^L > (q-1)^n. \tag{31.1}$$

Then there are two linear combinations which coincide modulo $q-1$. Therefore we have the identity

$$\sum_{l=1}^{L} \alpha_l W(l) \equiv \sum_{l=1}^{L} \beta_l W(l) \pmod{q-1},$$

with two distinct binary vectors $(\alpha_1, \ldots, \alpha_L) \in \mathfrak{B}_L$ and $(\beta_1, \ldots, \beta_L) \in \mathfrak{B}_L$. Thus from Lemma 2.3 we derive that

$$\sum_{l=1}^{L} \alpha_l u(x+l) \equiv \sum_{l=1}^{L} \beta_l u(x+l) \pmod{q-1}, \qquad x = 1, 2, \ldots.$$

Therefore for $x \in [N+1, N+H-L]$ we have

$$\sum_{l=1}^{L} \alpha_l Z(x+l) \equiv \sum_{l=1}^{L} \beta_l Z(x+l) \pmod{q-1}$$

which is equivalent to the following equation over $\mathbb{F}_q$

$$\prod_{l=1}^{L} \left(g^{x+l}+1\right)^{\alpha_l} = \prod_{l=1}^{L} \left(g^{x+l}+1\right)^{\beta_l}$$

for all except at most L values of $x \in [N+1, N+H-L]$ for which $g^{x+l} = -1$, $l = 1, \ldots, L$.

However, the polynomials on left- and right-hand sides of this congruence are distinct and are of degree at most L, and thus they cannot coincide at $H-2L > L$ points. The obtained contradiction shows that the assumption (31.1) is false and the result follows. □

In particular, Theorem 31.3, implies the lower bound

$$L(H) = \Omega(H \log^{-1} q)$$

on the linear complexity profile of Zech's logarithm, modulo $q-1$.

Our next result is an analogue of Theorem 9.2.

Theorem 31.4. *Let $-1 \le N < N+H \le q-2$ and let $d > 1$ be a divisor of $q-1$. Let $(u(x))$ be an integer linear recurrence sequence of order n such that*

$$Z(x) \equiv u(x) \pmod d, \qquad x \in S,$$

for a set $S \subseteq \{N+1, \ldots, N+H\}$ of cardinality $|S| = H - s$. Then

$$n \ge \frac{H}{2s+4+q^{1/2}\log q} - 1.$$

Proof. We see that for at least $H-n-(n+1)(H-|S|)-(n+1) \ge H-(n+1)(s+2)$ values of $x = N+1, \ldots, N+H$,

$$Z(x+i) \equiv u(x+i) \pmod d, \qquad i = 0, \ldots, n,$$

and

$$g^{x+i} \ne -1, \qquad i = 0, \ldots, n.$$

Put $c_n = -1$ and assume that the sequence $(u(x))$ satisfies (1.2). Then we see that

$$\sum_{i=0}^{n} c_i Z(x+i) \equiv 0 \pmod d \tag{31.2}$$

for at least $H-(n+1)(s+2)$ values of $x = N+1, \ldots, N+H$. The congruence (9.2) is equivalent to the statement that the product

$$(g^x+1)^{c_0}(g^{x+1}+1)^{c_1}\ldots(g^{x+n}+1)^{c_n}$$

is a dth power residue modulo p. Thus, for a non-trivial character χ of $\mathbb{F}_p^*$ of order d, we have

$$\left| \sum_{x=N+1}^{N+H} \chi\left((g^x+1)^{c_0}(g^{x+1}+1)^{c_1}\dots(g^{x+n}+1)^{c_n}\right)\right| \geq H - 2(n+1)(s+2).$$

On the other hand, because $c_n = -1$, Lemma 3.3 can be applied. Therefore,

$$H - 2(n+1)(s+2) \leq (n+1)q^{1/2}\log q$$

and the result follows. □

Now we obtain a lower bound on the sensitivity of Boolean functions deciding if a given number is a primitive root modulo p.

Theorem 31.5. *Let a Boolean function $B(U_1, \dots, U_r)$ be such that for any x, $1 \leq x \leq 2^r - 1$,*

$$B(u_1, \dots, u_r) = \begin{cases} 0, & \text{if } x \text{ is not a primitive root modulo } p, \\ 1, & \text{if } x \text{ is a primitive root modulo } p, \end{cases}$$

where $x = u_1 \dots u_r$ is the bit representation of x and $r = \lfloor \log p \rfloor$. Then

$$\sigma(B) \geq 0.25r + o(r).$$

Proof. We put

$$k = \left\lfloor 0.25r - r\log^{-1/2} r \right\rfloor.$$

It is easy to see that the theorem will be proved if we show that there exist y, $1 \leq y \leq 2^{r-k}$, such that $2^k y$ is a primitive root modulo p but $2^k y + 2^i$, $i = 0, \dots, k-1$, are quadratic residues. Indeed in this case $\sigma(B) \geq k$.

Quite similarly to Exercise 12.c of Chapter 6 of [553] we obtain that the number N of such y can be expressed as

$$N = \sum_{d|p-1} \frac{\mu(d)}{2^k d} \sum_{y=1}^{2^{r-k}} \prod_{i=0}^{k-1} \left(1 + \chi(2^k y + 2^i)\right) \sum_{\lambda=0}^{d-1} \mathbf{e}_d\left(\lambda \operatorname{ind}(2^k y)\right),$$

where $\mu(m)$ is the Möbius function and $\chi(z)$ is the quadratic character modulo p.

For each divisor d of $p-1$ the inner sum contains $d2^k$ character sums of the form

$$\sum_{y=1}^{2^{r-k}} \prod_{i=0}^{m} \chi(2^k y + 2^{j_i})\, \mathbf{e}_d\left(\lambda \operatorname{ind}(2^k y)\right),$$

where $0 \le m \le k-1$, $0 \le j_0 < \ldots < j_m \le k-1$. The sum corresponding to $m=0$ equals 2^{r-k}. To other terms Lemma 3.4 can be applied; thus each of them does not exceed $kp^{1/2}\log p$. Therefore

$$N = \sum_{d|p-1} \frac{\mu(d)}{2^k d}\left(2^{r-k} + O(d2^k kp^{1/2}\log p)\right).$$

We make use of the known identities

$$\sum_{d|m} |\mu(d)| = 2^{\nu(m)}, \qquad \sum_{d|m} \frac{\mu(d)}{d} = \frac{\varphi(m)}{m}$$

(see Sections 2.b and 4.b of Chapter 2 of [553]). which yield the bound

$$N = \frac{2^r\varphi(p-1)}{2^{2k}p} + O(2^{\nu(p-1)}kp^{1/2}\log p) = \frac{2^r\varphi(p-1)}{2^{2k}p} + O(2^{\nu(p-1)}p^{1/2}\log^2 p).$$

From the estimates (5.1) and (5.2) we derive

$$\begin{aligned} N &\ge 2^{r-2k+O(\log\log r)} - 2^{r/2+O(r\log^{-1} r)} \\ &= 2^{r/2+2r\log^{-1/2} r+O(\log\log r)} - 2^{r/2+O(r\log^{-1} r)} > 0, \end{aligned}$$

provided that p is large enough. □

As in Chapter 10, one may combine Theorem 31.5 with Lemma 7.1 and obtain the lower bound $0.5\log\log p + O(1)$ on the CREW PRAM complexity of primitive root testing modulo p.

There are also some applications of the methods of this book to explicit construction of symmetric Boolean functions which cannot be approximated by polynomials of small degree; see [30], for example, which demonstrates how to use such functions for obtaining non-trivial lower bounds in the computational complexity theory.

First of all we fix a normal basis

$$\omega, \omega^2, \ldots, \omega^{2^{r-1}}$$

of $\mathbb{F}_{2^r}$ over $\mathbb{F}_2$ (see [343, 508]) and identify the Boolean cube $\mathfrak{B}_r$ and $\mathbb{F}_{2^r}$ as follows:

$$(x_1, \ldots, x_r) \in \mathfrak{B}_r \longleftrightarrow x = x_1\omega + \ldots + x_r\omega^{2^{r-1}} \in \mathbb{F}_{2^r}.$$

Taking into account the identity

$$\begin{aligned} x^2 &= \left(x_1\omega + \ldots + x_r\omega^{2^{r-1}}\right)^2 = x_1\omega^2 + \ldots + x_r\omega^{2^r} \\ &= x_r\omega + x_1\omega^2 + \ldots + x_{r-1}\omega^{2^{r-1}}, \end{aligned}$$

we see that any non-constant symmetric Boolean function $B(x_1, \dots, x_r)$ defined in terms of some property of x satisfies the functional equation $B(x) = B(x^2)$ over $\mathbb{F}_{2^r}$. Thus if a polynomial $f(X) \in \mathbb{F}_{2^r}[X]$ coincides with $B(x)$ for at least $2^r - s$ values of $x \in \mathbb{F}_{2^r}$ and $B(x)$ is not a constant for these values, then $f(x) = f(x^2)$ for at least $2^r - 2s$ points and f is not constant. Thus $2 \deg f \geq 2^r - 2s$ and $\deg f \geq 2^{r-1} - s$.

Part VII

Concluding Remarks and Open Questions

In this book we have considered the discrete logarithm only over prime fields. This is because there is an obvious bijective mapping between the residue ring $\mathbb{Z}/(p-1)$ and the multiplicative group $\mathbb{F}_p^*$, sending x to $x+1$ for $x \in \{0,1,\dots,p-2\} = \mathbb{Z}/(p-1)$. For $\mathbb{Z}/(q-1)$ and $\mathbb{F}_q^*$, where $q = p^r$ is a prime power, a similar map also exists [399] (via p-adic expansions and representation of elements of $\mathbb{F}_q$ with respect to some fixed basis). An alternative is to consider representations of the discrete logarithm via multivariate functions (on coordinates of its argument). In both cases many details become quite messy. Nevertheless, apparently many of the results of this book can be extended to arbitrary finite fields $\mathbb{F}_q$. Several such extensions can be found in [322, 323, 324, 381, 445, 560, 561].

One can see that Theorems 8.1, 8.2, 12.1 and 12.2 can easily be extended to rational functions (with somewhat stronger bounds than those following from Theorems 8.5, 8.6, 12.4 and 12.5).

Note that using the same "symmetrization" trick as is used in the proofs of Theorems 10.6, 11.9, 13.8, 23.12 and 29.3, one can replace $p^{1/2}\log p$ by $p^{1/2}$ in Theorems 8.4, 9.2, 10.1 and 31.4 with slightly worse constants (the constants which we obtain in this book are not the best possible anyway). We use this trick for Theorems 11.9, and 29.3 because it actually simplifies their proofs, and in particular, leads to the estimate of $\sigma(B)$ in Theorem 29.3 which is of the correct order (obviously $\sigma(B) \le r$). It is also the only possible way to prove Theorems 10.6 and 13.8. In a very general form this trick is described in [107].

Also, instead of Lemma 3.5 one could use a more general estimate from [459] of character sums with algebraic functions (including rational functions of course). That would allow one to generalise Theorem 8.4 to the case of approximation by algebraic and rational functions.

As we have mentioned, one can obtain the upper bound $\deg f \le |S| - 1$ on the smallest possible degrees of polynomials involved in Theorems 8.1, 8.2, 12.1 and 12.2. Moreover, Theorems 8.3 and 12.3 show that this bound is precise for almost all sets.

Question VII.1. *Find "interesting" examples of sets S for which the aforementioned trivial upper bound can be improved.*

Unfortunately, a modulo $p-1$ analogue of the explicit representation (12.1) is not known. Moreover, it is obvious that for p large enough such a representation does not hold with a polynomial with integer coefficients (it would imply that, say, all even numbers have the same value of quadratic character modulo p). However this is obviously possible with a polynomial with rational coefficients. Theorem 9.1 provides the lower bound $\Omega(p\log^{-1}p)$ on the degree of such a polynomial while it is reasonable to expect a lower bound of order p.

Question VII.2. *Find an explicit expression for a polynomial $f(X) \in \mathbb{Q}[X]$ of the smallest possible degree such that*

$$\operatorname{ind} x \equiv f(x) \pmod{p-1}, \qquad x = 1,\dots,p-1.$$

It is also clear that the constants in Theorems 8.8, 8.9, 9.1, 11.4 and 31.3 can be improved. In fact the bounds of Theorems 8.9, 9.1 and 31.3 are close to the best possible (up to $\log p$). On the other hand, the bounds of Theorem 8.8 and 11.4 do not seem to be tight.

In this book we have been dealing with representations of the discrete logarithm modulo p and modulo a divisor d of $p-1$. Representations of the discrete logarithm modulo p in the residue ring modulo another integer $M \neq p$ are of interest as well.

Question VII.3. *Obtain analogues of the results above for congruences of the form*

$$\operatorname{ind} x \equiv f(x) \pmod M, \qquad x \in S,$$

with an arbitrary integer M, *some "simple" functions* $f(X)$ *and various sets* $S \subseteq \{1, \dots, p-1\}$, *where* $\operatorname{ind} x$ *denotes the discrete logarithm modulo* p.

If $M \geq p-1$, then the value of $\operatorname{ind} x$ can be recovered unambiguously. If $M < p-1$, then complete recovery is not possible but this congruence can still provide some non-trivial information about $\operatorname{ind} x$. For example, if $M = 2^k$, then such a representation gives us k rightmost bits of $\operatorname{ind} x$. The case $k = 1$ is covered by Theorem 9.2 and its improvement for $k > 1$ would be very interesting.

The case when M is prime can possibly be handled along the same lines as we use in this book but if M is composite some additional considerations are required. Indeed, although analogues of our auxiliary results (bounds for the number of zeros of polynomials [303, 307], bounds of character sums [117, 118, 119, 120, 121, 122, 283, 284, 285, 533, 539] and so on) are known modulo an arbitrary composite M as well, they are substantially weaker and, therefore, lead to much weaker statements. Nevertheless, this question can apparently be approached, at least for some interesting values of M such as $M = 2^k$. We recall that Theorems 8.8 and 8.9 give lower bounds for the linear complexity of the discrete logarithm modulo an arbitrary integer $M \geq 1$.

Along with Boolean functions covered by Theorem 10.1 one can also consider the Boolean function

$$B(u_1, \dots, u_r) = \begin{cases} 0, & \text{if } \operatorname{ind} x \leq (p-3)/2, \\ 1, & \text{if } \operatorname{ind} x \geq (p-1)/2, \end{cases} \qquad \text{(VII.1)}$$

where $x = u_1 \dots u_r$ is the bit representation of x, $1 \leq x \leq 2^r - 1$, $r = \lfloor \log p \rfloor$. Obviously, this function has to do with the first bits of $\operatorname{ind} x$. The method of proof of Theorem 10.1 can be applied to this function but produces a rather weak result (the lower bound $\Omega(\log p)$ for the number of terms and no non-trivial bound on the degree).

Question VII.4. *Obtain a non-trivial lower bound on the degree of Boolean functions satisfying (VII.1).*

Theorem 10.1 gives a lower bound on the sparsity in the basis $\{+, \times\}$ over $\mathbb{F}_2$. Obtaining a lower bound on the sparsity in the basis $\{\vee, \wedge, \neg\}$, that is, on the number of monomials in the minimal disjunctive normal form for B, is an important question of independent interest.

Question VII.5. *Obtain a non-trivial lower bound on the number of monomials* DNF (B) *in the minimal disjunctive normal form for the Boolean function* B *of Theorem 10.1.*

The bound (10.6) provides such a result for at least one of B and $\neg B$ but does not say anything about them individually.

A randomized version of Theorem 10.1 would be of interest as well.

Question VII.6. *Obtain a non-trivial lower bound on the sparsity and degree of a Boolean function* $B(U_1, \dots, U_r, V_1, \dots, V_s)$ *such that for any* x, $1 \le x \le 2^r - 1$,

$$B(u_1, \dots, u_r, v_1, \dots, v_s) \equiv \operatorname{ind} x \pmod 2,$$

where $x = u_1 \dots u_r$ *is the bit representation of* x, $r = \lfloor \log p \rfloor$, *for at least* $\alpha 2^s$ *binary vectors* $(v_1, \dots, v_s)$ *with some constant* $\alpha > 0.5$.

The most important case is apparently $s = r^{O(1)}$.

One of the possible ways to refine the bound (10.5) is to consider slightly more complicated than $\mathcal{P}^r_{k,s}$ classes $\mathcal{Q}^r_{k,s}$, see [492, 493].

Question VII.7. *Obtain an analogue of Theorem 10.6 for complexity classes* $\mathcal{Q}^r_{k,s}$.

Theorem 10.2 provides the lower bound $\Omega(\log \log p)$ on the depth of a straight-line deterministic Boolean circuit solving the discrete logarithm problem modulo p.

Question VII.8. *Obtain analogues of Theorem 10.2 for branching and randomized Boolean circuits.*

In this book we have considered Boolean circuits with bounded and unbounded fan-in but only bounded fan-in arithmetic circuits. The same method enables us to consider arithmetic circuits with unbounded fan-in with respect to addition and to get trade-off results between the size and the depth of such circuits. Such circuits are of interest and have been considered in the literature [198]. On the other hand, arithmetic circuits with totally unbounded fan-in are, apparently, more difficult to study.

Question VII.9. *Obtain an analogue of Theorem 29.8 for both arithmetic and Boolean circuits with unbounded fan-in.*

It is interesting to compare Theorem 10.5 with one of the results of [537]. In that paper, for any non-decreasing function $\psi(z)$ satisfying the inequality $\psi(z) \leq (z/\log z)^{1/2}$ for $z \geq 2$, *probabilistic* Boolean circuits with bounded fan-in of depth

$$d = O\left(\psi^2(\log p)\log\log^2 p + \log^{1/2} p \log\log^{3.5} p + \frac{\log p}{\psi(\log p)}\right),$$

and of size

$$S = \exp\left(O\left(\frac{\log p}{\psi(\log p)}\right)\right).$$

and with unbounded fan-in of depth

$$d = O(\psi^2(\log p)\log\log^2 p + \log\log^3 p),$$

and of size

$$S = \exp\left(O\left(\frac{\log p}{\psi(\log p)}\right)\right).$$

are constructed for computing the discrete logarithm modulo p. Selecting $\psi(z) = z^{1/2}(\log z)^{-3/2}$, we obtain a probabilistic bounded fan-in circuit of depth

$$d = O\left((\log p \log\log p)^{2/3}\right)$$

and of size

$$S = \exp\left(O\left((\log p \log\log p)^{2/3}\right)\right).$$

The results of Chapter 10 provide lower bounds on the complexity of testing if a given number is a quadratic residue modulo p. One may also ask about the complexity of square root extracting, provided that it is known that x is a quadratic residue modulo p; see [4, 508]. For example, for a quadratic residue a, let us denote by $\mathrm{sqrt}_p(x)$ the unique integer z defined by the conditions

$$z^2 \equiv x \pmod p, \qquad 1 \leq z \leq (p-1)/2.$$

Question VII.10. *Obtain lower bounds on the parallel arithmetic and Boolean complexities of computing* $\mathrm{sqrt}_p(x)$.

The method we used in the proof of Theorem 10.1 leads to the study of the system of quadratic congruences

$$4z^2 \equiv 2^k x \pmod p, \quad (2z_i+1)^2 \equiv 2^k x + 2^i \pmod p, \qquad i = 1, \dots, k,$$

in variables

$$1 \leq z, z_1, \dots z_k \leq (p-3)/4, \qquad 1 \leq x \leq 2^{r-k},$$

where $r = \lfloor \log p \rfloor$ and $k < r$ is some integer parameter.

The result of Theorem 11.5 is rather weak. It can probably be sharpened in several particular cases.

Question VII.11. *Improve Theorem 11.5 for real polynomials $f(X) \in \mathbb{R}[X]$ satisfying the inequality*

$$|\operatorname{ind} x - f(x)| < 1/2, \qquad x \in S,$$

for various sets $S \subseteq \{1, \dots, p-1\}$.

One can also consider functions of a more general form than polynomials, approximating $\operatorname{ind} x$ on bits of x.

It has been shown in [10] that the highest order Fourier coefficient

$$\Psi(B) = \widehat{B}(1, \dots, 1) = \sum_{(u_1, \dots, u_r) \in \mathfrak{B}_r} (-1)^{B(u_1, \dots, u_r) + u_1 + \dots + u_r}$$

of a Boolean function B plays a very important role in lower bounds on the complexity of B in various models of computation. In particular, if $\Psi(B) \neq 0$, then $\deg B = r$.

Question VII.12. *Prove that $\Psi(B) \neq 0$ for almost all (or infinitely many) values of p for the Boolean function of Theorem 10.1 and thus improve the bound (10.3).*

We remark that the desired property is equivalent to the following "obviously true" statement. The numbers $R_0(p)$ and $R_1(p)$ of quadratic residues in the interval $[0, 2^r - 1]$, $r = \lfloor \log p \rfloor$, which have even and odd sums of binary digits, respectively, satisfy $R_0(p) \neq R_1(p)$ and $R_0(p) \neq R_1(p) + 1$ (and thus "declaring" $x = 0$ as a quadratic residue or non-residue cannot balance these quantities). Certainly this may happen for some primes p but we believe that such primes are very rare.

Other Boolean functions related to number-theoretic problems, for example, those considered in Theorems 13.1, 13.2, 13.4, 28.1, 28.4, and others, are of interest as well, of course.

Question VII.13. *Obtain lower bounds on the degree and the number of monomials of a rational function*

$$f(X_1, \dots, X_r) \in \mathbb{R}(X_1, \dots, X_r)$$

such that $|f(u_1, \dots, u_r) - 1| \leq 1/3$ if x is a quadratic residue modulo p and $|f(u_1, \dots, u_r)| \leq 1/3$ otherwise, where $x = u_1 \dots u_r$ is the bit representation of x, $1 \leq x \leq 2^r - 1$, $r = \lfloor \log p \rfloor$.

One can also ask about iterations of threshold representations. Several classes are known in the literature [80, 211, 265, 481] but our method does not seem to work for any of them. Even the second iteration would be of interest.

Question VII.14. *Do there exist real numbers* $\alpha_1, \dots, \alpha_s$ *and polynomials*

$$f_i(X_1, \dots, X_r) \in \mathbb{R}(X_1, \dots, X_r), \qquad i = 1, \dots, s,$$

with

$$\max_{i=1,\dots,s} \deg f_i = o(r), \qquad s = 2^{o(r)},$$

and such that

$$\sum_{i=1}^{s} \alpha_i B_i(u_1, \dots, u_r) \geq 0$$

if x *is a quadratic residue modulo* p *and*

$$\sum_{i=1}^{s} \alpha_i B_i(u_1, \dots, u_r) < 0$$

otherwise, where

$$B_i(u_1, \dots, u_r) = \begin{cases} 0, & \text{if } f_i(u_1, \dots, u_r) \geq 0, \\ 1, & \text{if } f_i(u_1, \dots, u_r) < 0, \end{cases}$$

and $x = u_1 \dots u_r$ *is the bit representation of* x, $1 \leq x \leq 2^r - 1$, $r = \lfloor \log p \rfloor$ *?*

We expect that such a representation is not possible. We note that a similar question for the third iteration is apparently very hard because of the very high computational power of the corresponding class of functions, even if one imposes some additional conditions on the size and the number of coefficients involved in such functions [265].

We have already mentioned that polynomials of low additive complexity form another family of "easily computable" functions over any ring. Unfortunately, neither the method of proof [473, 474], nor the method of [479] more precisely, the method of proof of the underlying estimates [293, 480] on the number of real zeros of systems of sparse polynomial equations over the reals, cannot be extended to finite fields (although p-adic generalisations are known [348]). These bounds have already produced a large number of results on computational complexity of various classes of functions [237].

One of the drawbacks of Theorem 11.1 and Theorem 11.2 is that it is useful only for sets S of large cardinality, $|S| > 0.75p$.

Question VII.15. *Obtain lower bounds on the additive complexity of polynomials* $f(X) \in \mathbb{R}[X]$ *and* $f(X) \in \mathbb{Z}[X]$ *for which* $\operatorname{ind} x = f(x)$, $x \in S$, *for sets* $S \in \{0, \dots, p-1\}$ *of size* $|S| \geq p^{\alpha}$ *with some constant* $\alpha < 1$.

We also remark that it would be interesting to consider bivariate real polynomials $F(X, Y) \in \mathbb{R}[X, Y]$ for which

$$F(x, \operatorname{ind} x) = 0, \qquad x \in S,$$

for various sets $S \in \{0, \dots, p-1\}$.

Certainly, obtaining a finite field analogue of those estimates would be very important.

Question VII.16. *Obtain a non-trivial upper bound for the number of zeros over* $\mathbb{F}_q$ *of a polynomial* $f(X) \in \mathbb{F}_q[X]$ *of additive complexity at most* k *in terms of* k *and some other intrinsic characteristics of* f.

For instance, this question is related to studying arithmetic circuits with unbounded fan-in, in particular to Question VII.9.

As the very first step to such a general bound one can try to estimate the number of zeros of polynomials of the shape

$$f(X) = \sum_{i=1}^{t} a_i (X + b_i)^{r_i} \in \mathbb{F}_q[X].$$

The example of the polynomial

$$f(X) = X^{(q-1)/2} + 1$$

of additive complexity $C_{\pm}(f) = 1$ and having $(q-1)/2$ zeros shows that some additional conditions should be imposed.

Such a result would be a very natural generalisation of the bound on the number of zeros of a t-sparse polynomial

$$f(X) = \sum_{i=1}^{t} a_i X^{r_i} \in \mathbb{F}_q[X]$$

provided by Lemma 2.6.

One can also try to obtain a non-trivial upper bound on the number of solutions of a system of sparse polynomial equations. Appropriate generalisations of Lemmas 2.5 and 2.6 and the bound (2.1) would be very useful.

Lemma 5.2 provides a necessary tool to approach the following problem.

Question VII.17. *Extend the results of [216, 255] to the discrete logarithm modulo an arbitrary integer.*

Some analogues of the bounds of Theorems 12.2 and 12.5 for bivariate polynomials f satisfying $g^{xy} = f(g^x, g^y)$ and trivariate polynomials F satisfying $F(g^x, g^y, g^{xy}) = 0$ for $(x, y) \in W$, where $W \subseteq [N+1, N+H]^2$ have been given in [167]. However, the method which has been used to derive an analogue of Theorem 12.5 is somewhat less involved than that of Theorem 12.5 itself and thus leads to a weaker result. For the case of the equation $g^{xy} = f(g^x, g^y)$ several results are also given in [560].

Question VII.18. *Extend the method of proof of Theorem 12.5 to equations of the form* $F(g^x, g^y, g^{xy}) = 0$ *for* $(x, y) \in W$, *where* $W \subseteq [0, q-2]^2$.

Obtaining analogues of other results of Chapter 12 is of interest as well.

Question VII.19. *It would be very important to obtain analogues of Theorems 12.8, 14.1, 14.3, 15.1, 18.2, 19.1 and 19.3 for the case when the corresponding oracle returns correct values only for a certain portion of inputs and returns wrong (but consistent) results for other inputs (instead of the error message or a random element, thus wrong outputs cannot be immediately identified).*

Theorems 12.8, 14.1, 14.3, 15.1, 18.2, 19.1 and 19.3 are based on various modifications of the hidden number problem to which we have applied the technique originated from [64] (see also the surveys [524, 525]). The paper [65] introduces another approach, giving much stronger results but in a different model of computation.

Question VII.20. *Study how the technique of [65] can be applied to the modifications of the hidden number problem considered in Chapter 6.*

In particular, as the first step, one can try to obtain an analogue of Lemma 6.3 for *non-uniform* algorithms to which the technique of [65] is designed to apply.

Following [450] we define the *differential uniformity* Δ_f of a mapping $f : \mathbb{F}_q \to \mathbb{F}_q$ as

$$\Delta_f = \max_{a,b \in \mathbb{F}_q} N_f(a, b)$$

where $N_f(a, b)$ is the number of solutions of the equation $f(u + a) = f(u) + b$, $u \in \mathbb{F}_q$.

Question VII.21. *Obtain a non-trivial upper bound on differential uniformity of the Diffie–Hellman mapping* $f(u) = u^{\operatorname{ind} u}$, $u \in \mathbb{F}_q^*$.

Our bound of the combinatorial discrepancy of B, obtained in the proof of Theorem 13.4, combined with Lemma 2.2 of [15] can be used to derive a linear lower bound on ε-*distributional communication complexity*, which is defined in a similar way, however the communicating parties are allowed to make mistakes on at most $\varepsilon 2^{2r}$ inputs $x, y \in \mathfrak{B}_r$.

The same arguments which have been used in the proof of Lemma 3.22 can also be used to estimate more general sums

$$S_a(\mathcal{X}_1, \ldots, \mathcal{X}_k) = \sum_{x_1 \in \mathcal{X}_1} \cdots \sum_{x_k \in \mathcal{X}_k} \mathbf{e}_p\left(a\vartheta^{x_1 \ldots x_k}\right)$$

with $k \geq 2$ sets $\mathcal{X}_1, \ldots, \mathcal{X}_k \subseteq \mathbb{Z}_t$ and thus study the *multi-party communication complexity* of the Boolean function B given by (13.3).

One can also study the probabilistic communication complexity, which is defined analogously with respect to probabilistic communication protocols whose probability of error does not exceed $1/3$.

One of the possible approaches to studying probabilistic communication complexity, which follows from Theorem 2.1 of [469], could be via bounds on Fourier coefficients of Theorem 13.5. It has been shown in [299] that a combination of some results of [184] and [503] with the bound of the combinatorial discrepancy, obtained in the proof of Theorem 13.4, gives a linear lower bound on the probabilistic communication complexity of the Diffie–Hellman secret key.

It has also been discovered in [310] that the probabilistic communication complexity and several other complexity characteristics of B, depend on the spectral norm

$$\lambda(2M(B) - I_{2^r}) = \max\{\| (M(B) - I_{2^r}) z\|_2 \mid z \in \mathbb{R}^{2^r}, \|z\|_2 = 1\},$$

of the matrix $2M(B) - I_{2^r}$, where the matrix $M(B)$ is given by (7.3), $\|z\|_2$ denotes the Euclidean norm of $z \in \mathbb{R}^{2^r}$ and I_{2^r} is the $2^r \times 2^r$ unit matrix.

In fact, the approach of [299], combined with some other complexity theory results, can also be used to obtain an exponential lower bound of the form

$$\lambda(2M(B) - I_{2^r}) \geq 2^{23/24r+o(r)}$$

for the Boolean function B given by (13.3) and some other similar functions, see [299] for a precise statement.

The function of several variables $g^{x_1 \ldots x_k}$ also deserves further study. Probably for any fixed k its properties are more or less the same as in the case of $k = 2$. However when k grows it may exhibit some new features.

Question VII.22. *Obtain analogues of the explicit formula (13.5) for the distribution of Diffie–Hellman secret key $g^{x_1 \ldots x_k}$ which is used in multiparty communication.*

In particular, it would be interesting to find for which values of k (if such exists) the distribution becomes significantly different from uniform.

It might be worth noting that Lemma 4.5 implies that $r(k)$, defined by (16.1), takes exponentially many distinct values. Thus the Digital Signature Algorithm indeed generates exponentially many distinct signatures. Certainly this fact has never been doubted in practice but Lemma 4.5 provides its rigorous confirmation. On the other hand, the lower bound q^{δ} implied by Lemma 4.5 on the number of distinct values of $r(k)$ falls far below the expected value of order about q.

Obtaining such a lower bound is a very challenging question which probably requires some advanced number-theoretic tools. Although studying the function

$N(\rho)$ for "individual" values of ρ seems very difficult, one can possibly obtain better bounds on the average value

$$N = \sum_{\rho=0}^{q-1} N(\rho)^2.$$

We note that because

$$\sum_{\rho=0}^{q-1} N(\rho) = q$$

the Cauchy inequality implies that there are at least q^2/N distinct values of $r(k)$.

Question VII.23. *Prove that $N = O(q^{1+\varepsilon})$ for any $\varepsilon > 0$.*

The following question has already been mentioned in Section 11.71 of [389]. The security of some signature schemes, of the same type as the ElGamal signature or Digital Signature Algorithm, depends on the hardness of finding a solution of the congruence $x^x \equiv a \pmod p$, $1 \le x \le p-1$. As the first step in this direction, which is of independent interest, one can study the distribution of values $x^x \pmod p$, $1 \le x \le p-1$.

Question VII.24. *Obtain non-trivial upper bounds for the number of solutions of the congruence $x^x \equiv a \pmod p$, $1 \le x \le p-1$ and for the exponential sums*

$$\sum_{x=1}^{p-1} \mathbf{e}_p\left(cx^x\right), \qquad 1 \le c \le p-1.$$

In Chapter 13 we have described the results of [64, 65, 222] which demonstrate that finding some small portion of bits of the Diffie–Hellman secret key (over a prime field $\mathbb{F}_p$) is essentially equivalent to the whole problem of breaking the Diffie–Hellman key exchange protocol. Similar results are also known for other cryptographic constructions, see [105, 106, 145, 149, 159, 182, 210, 216, 220, 254, 255, 298, 342, 411, 455, 457, 496, 518] as well as Chapters 13, 18 and 15.

For the RSA cryptosystem several similar results (but of slightly different spirit) are known as well. For example, it is shown in [55] that if we are given a certain fraction of the most significant bits of the RSA private key d, then the whole key can be recovered in polynomial time. We recall that for a given *RSA public key* e with $\gcd(e, \varphi(m)) = 1$, the RSA public key $d_m(e)$ is defined by the congruence

$$ed_m(e) \equiv 1 \pmod{\varphi(m)}, \qquad 1 \le d_m(e) \le \varphi(m) - 1. \qquad \text{(VII.2)}$$

It is also known that small RSA private keys are vulnerable. More precisely, for $d_m(e) \le 0.3m^{1/4}$ one can apply a very simple but ingenious deterministic polynomial time attack from [559]. For $d_m(e) \le m^{\alpha}$, with any $\alpha < 1 - 2^{-1/2} =$

0.292... one can apply a more sophisticated (but heuristic) attack from [54], see also [49, 162, 263, 276, 377, 552]. These results hold only if the RSA public key e is selected in its natural range $1 \le e \le \varphi(m) - 1$, with $\gcd(e, \varphi(m)) = 1$, (sometimes e should even be smaller, but sometimes the upper bound can be slightly extended). Thus, selecting e artificially large, for example, in the range, say $[m^2, m^2 + \varphi(m) - 1]$, provides a protection against the aforementioned attacks. Nevertheless, they indicate that small RSA private keys and a partial key exposure could be very dangerous. A partial exposure of factors of m is dangerous as well, see [130, 131] for the traditional case where $m = pl$ is a product of two distinct primes and [56] for the case $m = p^r l$ with some integer $r \ge 2$. The aforementioned results are based on the lattice basis reduction technique. The main difference between these results and the bit security results known for the Diffie–Hellman key exchange protocol is that we have some information only for one value of d (rather than an oracle which can be queried with various inputs). The result of [56] has been generalised in [274].

The method of [130], which is essentially a method of finding small roots of univariate and bivariate congruences, has been refined and advanced in [273, 290], see also [131]. This method is rigorous for univariate polynomials and given a polynomial $f \in \mathbb{Z}[X]$ of degree $n \ge 1$ and an integer modulus M, in polynomial time it finds all roots of the congruence $f(x) \equiv 0 \pmod{M}$ in the interval $0 \le x \le M^{1/n}$. It is easy to see that large intervals, say $0 \le x \le M^{1/n+\varepsilon}$, may contain exponentially many roots and thus no polynomial time algorithm for finding all of them is possible. For example, the polynomial $f(X) = X^n$ has at least M^ε roots $x = 2^k j$, $j = 0, \dots, \lfloor M^\varepsilon \rfloor$, modulo $M = 2^{kn}$ in the interval $0 \le x \le M^{1/n+\varepsilon}$. Unfortunately in the case of bivariate polynomials the method still remains heuristic. However a remarkable progress has recently been achieved in [377], where in a special situation, having important applications to the RSA cryptosystem, such a rigorous result is obtained. It is interesting to note that the method of [130] gives an alternative to the approach of [307] for obtaining a tight upper bound on the number of roots of polynomial congruences.

A heuristic lattice reduction based attack on a certain server-aided RSA signature scheme has been proposed in [386]. This attack has been simplified, extended and made rigorous in [416]. It is interesting that the underlying problem is the only known example of a cryptographic problem which is *provably secure* in the generic model [388] but is *provably weak* in the traditional model of computation [416].

Lattice reduction has also played a crucial role in the attack of [393] on the so-called *padded RSA signature scheme*, see also [75, 208, 209, 331]. However, it is still unclear what is the natural limit of this approach. Several attacks obtained within this approach remain purely heuristic and it would be of ultimate interest to find their rigorous confirmations.

Certainly, it would be very important to understand if there are analogues of the aforementioned results for the discrete logarithm and for the Diffie–Hellman key exchange protocol.

As the first step in this direction it would be very important to study the following two questions from [160].

Question VII.25. *Are there an absolute constant* $c > 0$ *and a function* $f(k)$ *such that for any* k *one can decide whether* $\operatorname{ind} x \le \log^k p$, $x \in \mathbb{F}_p^*$, *in time bounded by* $f(k)\log^c p$.

Question VII.26. *Are there an absolute constant* $C > 0$ *and a function* $h(k)$ *such that for any* k *one can decide whether the bit representation of* $\operatorname{ind} x$, $x \in \mathbb{F}_p^*$, *contains more than* k *zeros, in time bounded by* $h(k)\log^C p$.

Following the usual RSA scenario, we assume that *Bob* selects the modulus m, the RSA private key d and computes the RSA public key e from the congruence (VII.2), which he communicates to *Alice*, and thus to possible attackers as well. Because small private keys are potentially dangerous, *Alice* would like to know that d is large and "safe".

Question VII.27. *Can Bob convince Alice that d is large enough without revealing any information about d and factors of m?*

In other words, we ask about an efficient *zero-knowledge* proof, see [214, 353, 389, 541], of the statement that d is large enough, say $d \ge m^{1-\varepsilon}$. In particular, it is quite possible that the technique of Chapter 29 can be applied to obtain lower bounds on the length of such proofs.

Coming back to the discussion of the Diffie–Hellman key exchange protocol, we remark that it would be interesting to study what happens on the diagonal subsum of the sum (3.9).

Question VII.28. *Obtain a non-trivial upper bound on the sum*

$$S(a,c) = \sum_{x=1}^{p-1} \mathbf{e}_p(ag^x + cg^{x^2})$$

for $\gcd(a,c,p) = 1$, *where* g *is a primitive root modulo* p.

Unfortunately the approach of [94, 95], presented in Lemma 3.21, is not applicable to such one-dimensional sums. In the case $a = 0$ non-trivial bounds for $S(0,c)$ are given in [187, 344], see Lemma 3.24, but the methods of these papers do not seem to apply to the general case. We remark that bounds of the sums with $a = 0$ have found some unexpected applications to cryptography, see [523] (whose main result is presented in Theorem 14.3). Generalizations to arbitrary composite moduli analogues of these results are given in [189]. In fact, the case $a = 0$ is a special partial case of the result of [187] which in particular implies the bounds

$$\max_{1\le c\le p-1} \left| \sum_{x=1}^{t} \mathbf{e}_p\left(c\vartheta^{x^n}\right) \right| \ll \begin{cases} t^{3/4}p^{1/8+\varepsilon}, & \text{if } n = 2 \text{ or if } t \text{ is cube-free,} \\ t^{1-\alpha(n)}p^{\alpha(n)/2+\varepsilon}, & \text{for any } n \ge 3, \end{cases}$$

where t is the multiplicative order of g modulo p and $\alpha(n)$ is given by

$$\alpha(n) = \frac{\lceil n/2 \rceil - 1}{n \lceil n/2 \rceil + 1}.$$

Sums with general polynomials in the exponent are of ultimate interest as well.

Question VII.29. *Obtain a non-trivial upper bound on the sum*

$$S_f(c) = \sum_{x=1}^{t-1} \mathbf{e}_p\left(cg^{f(x)}\right)$$

for $\gcd(c,p) = 1$, *where* f *is a polynomial with integer coefficients of degree* $\deg f = d$ *and* g *is of multiplicative order* t *modulo* p.

Using the arguments of [195] one can probably estimate such sums "on average" over all $g \in \mathbb{F}_p^*$ of multiplicative order t modulo p.

It would also be interesting to study the distribution of the triples $\left(g^x, g^y, g^{F(x,y)}\right)$ or the pairs $\left(g^x, g^{f(x)}\right)$ with integer polynomials F and f. Such triples and pairs have been introduced in [297]. Pairs $\left(g^x, g^{x^{-1}}\right)$, $x \in \mathbb{Z}_t^*$, where as before t is the multiplicative order of g modulo p are of interest as well, see [356, 487].

In particular, it would be very interesting to estimate the following exponential function analogue of *Kloosterman sums*.

Question VII.30. *Obtain a non-trivial upper bound on the sum*

$$K_g(a,b) = \sum_{x \in \mathbb{Z}_t^*} \mathbf{e}_p\left(ag^x + bg^{x^{-1}}\right)$$

for $\gcd(a,b,p) = 1$, *where* g *is of multiplicative order* t *modulo* p.

Even if cryptographic applications are not immediately clear, to have non-trivial upper bounds on sums as elegant as $K_g(a,b)$ is of ultimate number theoretic interest. In [527], using some results of [94], the bound on the average value

$$\max_{\gcd(a,b,p)=1} \frac{1}{\varphi(t)} \sum_{\mathrm{ord}_p g = t} |K_g(a,b)| \ll p^{1/8+\varepsilon} t^{5/6}$$

has been given, which is non-trivial for $t \ge p^{3/4+\delta}$ with an arbitrary fixed $\delta > 0$.

It is shown in [542] that the function ψ_p given by (13.1) has a unique polynomial representation via a polynomial of degree $p-1$ over $\mathbb{F}_p$. Interestingly enough its coefficients can be expressed via *Bernoulli numbers*. Partial representations would be of interest as well.

Question VII.31. *Obtain lower bounds on the degree and sparsity of polynomials f such that $\psi_p(x) \equiv f(x) \pmod p$ for elements $x \in S$ of various sets $S \subseteq \{0, 1, \dots, p-1\}$.*

Several other similar functions defined in [542] are of interest as well.

Question VII.32. *Extend Theorem 29.1 to the case of quadratic modulo p polynomials $f(X) \in \mathbb{Z}[X]$.*

In particular, this would lead to a more direct treatment of the case of quadratic polynomials in Theorems 29.6 and 29.7 and to their possible refinements. We note that for the partial case $f(X) = X^2$ one can use Theorem 29.2 but the proof of that theorem cannot be extended to quadratic polynomials having a "large" leading coefficient.

The constants 0.6 and 1/16 in Theorem 29.3 are not the best possible ones but we still do not know how to obtain a non-trivial result for rational functions of degree essentially greater than $r^{1/2}$.

Question VII.33. *Extend Theorem 29.3 to the case of rational functions of larger degree; say of order $2^{\alpha r}$ with some $\alpha > 0$.*

Such a result would immediately lead to an improvement of Theorem 29.6. In particular, we believe that a result similar to that of Theorem 29.7 holds for general arithmetic circuits modulo p, not only division-free ones. The main obstacle is obtaining a better version of Lemma 2.11.

Question VII.34. *Extend Theorems 29.1 and 29.3 to the case of functions modulo an arbitrary integer M.*

There are two classes of moduli M for which this question is especially interesting. The first one is the class of moduli of the form $M = 2^k$ which corresponds to the "computer" arithmetic. The second one is the class of square-free moduli M having only small prime divisors (of order $\log^{O(1)} M$, say). The latter class is interesting because smooth square-free moduli admit a very efficient parallel algorithm relying on the Chinese Remainder Theorem. Thus for such moduli there are good chances of matching upper and lower bounds of complexity of various functions. For example, in [205] this was done for the modular inversion. In particular, in that paper an infinite sequence of moduli M is constructed for which the lower bound $\Omega(\log\log M)$ on the CREW PRAM complexity of modular inversion obtained there matches the upper bound $\Omega(\log\log M)$ previously known for such special moduli. Moduli of the form $M = 2^m$ are of special interest as well. For example, in [205] a lower bound $\Omega(\log\log M)$ on the CREW PRAM complexity for the following problem posed by Allan Borodin (see Open Question 7.2 of [200]) is obtained. The problem is: given r-bit positive integers m, x, d, compute the mth bit of x^d. It is easy to see that the problem is related to the distribution of powers x^d modulo 2^m, and thus is related to modular powering.

Extending Theorem 23.11 to arbitrary values of e is a very natural and interesting question.

Question VII.35. *For the power generator (23.1), study the distribution of s-tuples*

$$(u_n, \dots, u_{n+s-1}), \qquad n = 1, \dots, N,$$

in the s-dimensional unit cube, where $1 \le N \le t$ and t is the period of this sequence.

In the case of the power generator modulo a power of a fixed prime number, such a result about s-dimensional distribution has been obtained in [188]. The method is based on bounds of exponential sums with sparse polynomials from [507] and does not seem to work for square-free moduli.

One may also hope to obtain stronger bounds on the discrepancy "on average" over all initial values of the power generator. A similar question can also be studied for the pseudorandom number generators of Chapters 26 and 27.

Of course Theorem 23.13 allows us to derive lower bounds on the linear complexity of the power generator modulo any square-free integer m without too many prime divisors.

Question VII.36. *Obtain a non-trivial lower bound on the linear complexity of the power generator (23.1) modulo an arbitrary integer m.*

In particular, it would be of interest to obtain a non-trivial lower bound for large powers of small prime numbers. Studying the distribution and the period of the power generator (23.1) modulo an arbitrary integer m is of interest as well. In particular, it would be important to obtain an analogue of Theorem 23.2 for the values of $\lambda(\lambda(m))$ at arbitrary integers m.

Question VII.37. *Is it true that for almost all integers $m \ge 1$ the value of $\lambda(\lambda(m))$ is of order close to m.*

Obtaining analogues of the results of Chapter 23 for the *exponential generator*

$$v_n \equiv g^{v_{n-1}} \pmod p, \quad 1 \le v_n \le p-1, \qquad n = 1, 2, \dots,$$

where g is a primitive root modulo a prime p, is a very interesting but probably hard challenge. This sequence is also believed to have good pseudorandom properties [138, 319] although almost no non-trivial theoretical results about the distribution of this generator are known. Fixed points of this generator have been studied in [92, 116, 565], this question is also know as the *Brizolis problem*, see Section F9 of [248]. It is quite possible that the same technique can be applied to studying cycles of length two (but probably not to longer cycles). see [271]. It seems quite plausible that the bounds of character sums along an elliptic curve of [301] can be applied to an analogues of the Brizolis problem on elliptic curves.

It has also been shown in [194] that exponential sums with exponential functions may help to extract some other non-trivial information about the exponential generator. Indeed, let us consider the sequence of binary strings $\left(\sigma_n^{(k)}\right)$ generated by the k rightmost bits of v_n, $n = 1, 2, \ldots$. Let the sequence (v_n) be purely periodic with period t and let τ be the period of the sequence $\left(\sigma_n^{(k)}\right)$. Because there are at most $O(p2^{-k})$ integers v with given k rightmost bits, then obviously $\tau \geq 2^k t/p$. It has been shown in [194] that a variant of Lemmas 3.13 and 3.15 implies that $\tau \gg 2^{2k}t/p$. This, rather weak, improvement of the trivial bound may give some hope for the possibility of obtaining more interesting results about the exponential generator.

Obtaining multidimensional analogues of Theorems 24.1 and 24.2 would be of interest as well.

Question VII.38. *For the Naor–Reingold pseudorandom function, given by the equation (24.1), study the distribution of s-tuples*

$$(f_{\mathbf{a}}(x), \ldots, f_{\mathbf{a}}(x+s-1)), \qquad n = 1, \ldots, N,$$

where $1 \leq N \leq 2^r$.

Unfortunately Theorem 24.3 does not cover the most interesting case $n \sim \log l \sim \log p$.

Question VII.39. *Extend Theorem 24.3 to a wider range of parameters, including the case $n \sim \log l \sim \log p$.*

As we mentioned, Theorem 24.3 is given in [517] which sharpens the bound of [235]. On the other hand, the lower bound of [22] of the non-linear complexity uses the method of [235].

Question VII.40. *Improve the result of [22] by using the method of proof of Theorem 24.3.*

Studying number theoretic properties of the pseudorandom number generators from [149, 159, 406] is of ultimate interest as well.

It would be interesting to increase the value of k in Theorem 25.1. and to make it logarithmically close to L, which is probably a natural limit for results of such kind.

Question VII.41. *Prove that for some constant $c > 0$ and $k = \lfloor L - c \log L \rfloor$ any (or almost any) string of k consecutive digits appears in the g-ary expansion of $1/M$ for exponentially many different values of $M < g^L$.*

We have already mentioned the papers [11, 59, 145, 155, 266, 369, 371, 458, 475] studying the distribution and some other pseudorandom properties of the quadratic character χ modulo p. Yet another question has recently been considered in the series of papers [141, 142, 143, 282]. The question is: given an oracle which for each $x \in \mathbb{F}_p$ computes $\chi(x+s)$, find s. A similar question has also been considered in [59]. In the above papers [141, 142, 143, 282] an efficient quantum algorithm has been designed for this problem. A more general question of recovering an unknown monic polynomial $f(X) \in \mathbb{F}_p[X]$ of degree $\deg f = d$, given an oracle which for each $x \in \mathbb{F}_p$ computes $\chi(f(x))$ has been studied in [486]. Using the Weil bound naively one can easily design a deterministic algorithm which solves this problem in $O\left(p^{d+1/2+\varepsilon}\right)$. It has been shown in [486] that in fact the Weil bound allows to design a deterministic algorithm of complexity $O\left(p^{d+\varepsilon}\right)$.

Recently an interesting link has been found [144] between the discrete logarithm problem and approximation of the argument $\gamma_q(\chi)$ of Gauss sums

$$\sum_{x \in \mathbb{F}_q} \chi(x)\, \mathbf{e}_p\left(\mathrm{Tr}\,(x)\right) = q^{1/2} \exp\left(i\gamma_q(\chi)\right)$$

corresponding to nontrivial multiplicative characters χ of $\mathbb{F}_q$, where $q = p^r$ and $\mathrm{Tr}\,(x)$ is a trace of $x \in \mathbb{F}_q$ in $\mathbb{F}_p$. A quantum polynomial time approximation algorithm has been designed. It has also been shown that any approximation algorithm leads to a polynomial time algorithm (on a classical computer) for the discrete logarithm problem in $\mathbb{F}_q$ (by using arguments similar to those used in the bit security proofs of RSA cryptosystem).

We remark that the bound of Theorem 26.3 for the discrepancy of the polynomial generator is much weaker than the bound of Theorem 26.1 for the discrepancy of the inversive generator, despite that they have been obtained by the same method. Of course this happens because iterations of the function $(aX+b)/(cX+b)$ remain of the same form while iterations of a polynomial of degree $d \geq 2$ have exponentially growing degree. A natural way to overcome this difficulty (for some special polynomials) would be to avoid using the Weil bound. This approach has been outlined in [128] and has led to a very interesting notion of $\mathcal{R}$-orthomorphism of $\mathbb{F}_q$. A polynomial $f(X) \in \mathbb{F}_q[X]$ is called an *orthomorphism* of $\mathbb{F}_q$ if both $f(X)$ and $f(X) - X$ are permutation polynomials over $\mathbb{F}_q$. This notion is well known in combinatorics and underlies some constructions of *Latin squares*, see Chapter 22 of [347]. It has also been used in cryptography [501]. Finally, for a set $\mathcal{R}$ of positive integers, we say that a polynomial $f(X) \in \mathbb{F}_q[X]$ is an *$\mathcal{R}$-orthomorphism* of $\mathbb{F}_q$ if for all $r \in \mathcal{R}$ the rth iteration of f is an orthomorphism of $\mathbb{F}_q$. It is shown in [128] that $\mathcal{R}$-orthomorphism with a large set $\mathcal{R}$ produce good polynomial generators. Unfortunately very few constructions of $\mathcal{R}$-orthomorphism are known, and in all of them the cardinality of $\mathcal{R}$ is of order $\log q$, see [113, 128].

Question VII.42. *Find constructions of $\mathcal{R}$-orthomorphism of $\mathbb{F}_q$ with a set $\mathcal{R}$ of cardinality $|\mathcal{R}| \geq q^\delta$ for some fixed $\delta > 0$.*

The results of Chapter 26 about the lattice structure of the polynomial generator apply only to sequences of period $t = p$ but it is not known how to guarantee this condition. On the other hand, the results of [247], although applying to the general case, are rather weak.

Question VII.43. *Improve the results of [247] on the linear complexity profile of the polynomial generator (26.7).*

We also believe that the quadratic exponential generator (26.10), deserves more a detailed study.

Question VII.44. *Extend Theorem 26.5 to the discrepancy of the quadratic exponential generator (26.10), in parts of the full period.*

Theorem 27.2 gives a rather weak statement about the iterative subset sum generator. One can try to use the exponential sum technique to study some other properties of this generator but so far no results in this direction have been obtained.

Question VII.45. *Study the period, linear complexity and distribution of the iterative subset sum generator (27.2) "on average" over all vectors* $\mathbf{z} \in \mathbb{Z}_t^r$ *and seeds* $w_0 \in [0, t-1]$.

Although the following question seems quite easy it has not been worked out yet.

Question VII.46. *Obtain an analogue of the asymptotic formula of Theorem 28.3 for the average sensitivity of the Boolean function (28.2).*

Obtaining analogues of the results of Chapter 28 for the Boolean function (28.3) is a very interesting question as well.

One can also ask whether the techniques of [49, 54, 55, 162, 276, 552, 559] can be applied to recovering the RSA private exponent $d_m(e)$, which is defined from the congruence (VII.2), if it is known that the bit representation of $d_m(e)$ contains less than k non-zeros. That is, one can ask analogues of Question VII.26 for $d_m(e)$. Another interesting question is obtaining analogues of results [64, 65, 222, 223] or similar results known for cryptographic constructions, see [105, 145, 149, 159, 182, 210, 216, 220, 254, 255, 298, 342, 411, 455, 457, 496, 518], for the RSA private exponent $d_m(e)$.

Question VII.47. *Let* $m = pl$ *be an RSA modulus. Given an oracle* $\mathcal{INV}_k$ *which for any* e *with* $\gcd(e, \varphi(m)) = 1$ *returns* $k \le \lfloor \log m \rfloor$ *most significant bits of* $d_m(e)$, *find an efficient algorithm to compute the entire value of* $d_m(e)$.

It would very interesting to find such an algorithm which works with $k = o(\log m)$.

It also would be interesting to show that the triples of fractional parts

$$\left(\left\{\frac{e}{\varphi(m)}\right\}, \left\{\frac{x}{m}\right\}, \left\{\frac{x^{d_m(e)}}{m}\right\}\right)$$

are uniformly distributed in the 3-dimensional unit cube. Thus one can study the following RSA-related analogues of the sums (3.9).

Question VII.48. *Obtain a non-trivial upper bound on the following double exponential sums*

$$T(a,b,c) = \sum_{\substack{e=1 \\ \gcd(e,\varphi(m))=1}}^{\varphi(m)} \sum_{\substack{x=1 \\ \gcd(x,m)=1}}^{m} \mathbf{e}\left(a\frac{e}{\varphi(m)} + b\frac{x}{m} + c\frac{x^{d_m(e)}}{m}\right).$$

In [461] an analogue of the Diffie–Hellman indistinguishability assumption has been introduced for the pairs

$$\left(\lfloor x^e \rfloor_m, \lfloor (x+1)^e \rfloor_m\right), \qquad x \in \mathbb{Z}_m^*.$$

It is conjectured that such pairs cannot be distinguished from the pairs (y, z) of random elements $y, z \in \mathbb{Z}_m^*$ unless the RSA cryptosystem modulo m can be broken.

Thus it is natural to ask about an analogue of the results of Theorems 13.6, 13.7, 13.8, and thus of Lemma 3.21. For "individual" values of e this seems to be extremely hard but maybe one can approach it "on average" over all admissible values of e.

Question VII.49. *Obtain a non-trivial upper bound on the double exponential sums*

$$U(a,b) = \sum_{\substack{e=1 \\ \gcd(e,\varphi(m))=1}}^{\varphi(m)} \left| \sum_{\substack{x=1 \\ \gcd(x,m)=1}}^{m} \mathbf{e}_m\left(ax^e + b(x+1)^e\right)\right|.$$

The following question is motivated by the well-known fact that there are some "favourite" values of e, such as $e = 3$ for example.

Question VII.50. *Study the distribution of the $d_m(e)$ given by (VII.2), when e is fixed but m varies.*

We have already mentioned several papers where, under certain plausible assumptions, a proof is given that breaking the Diffie–Hellman key exchange protocol is polynomial time equivalent to solving the discrete logarithm problem. In this context, the result of [66] looks quite unusual. It is shown in that paper that, for some algorithms of a certain kind, breaking the RSA cryptosystem is easier than integer factorisation, (unless factorisation can be done in polynomial time). Several other recent results about the security of the RSA cryptosystem and its various modifications are given in [254, 388, 410], see also [220].

In [58], one can find an attacks on the ElGamal and on RSA encryption of "short" messages (which corresponds to the naive application of these cryptosystems to transfer a key for a private key cryptosystem (such a key is usually much smaller than the size of the RSA modulus). This approach is based on results about the distribution of smooth integers and a certain heuristic assumption about the distribution of residues of exponential functions. It seems plausible that using bounds of character sums in the style of Section 22 and other similar results and techniques, one can obtain rigorous versions of some results of [58].

It is well known that smooth integers appear in design and analysis of many cryptographic algorithms, including the famous *number field sieve*, see [1, 4, 16, 74, 82, 127, 136, 203, 326, 328, 352, 379, 389, 395, 451, 462, 463, 465, 494, 508, 541]. For applications of smooth integers to some other cryptographic problems, see [24, 58, 331, 466]. A computational approach to finding smooth integers in small intervals has been proposed in [53]. This approach is based on lattice reduction and some ideas of [215, 243] on "noisy" Chinese Remaindering, see also [48, 529, 530]. For an exhaustive survey of known results about the distribution of smooth integers see [262]. Another very complete overview of this area, which also exhibits many links with other areas of mathematics, and also includes some more recent results, has been given in [228].

A very interesting survey of the known "non-factorisation based" attacks on the RSA cryptosystem has been presented in [52]. In particular, it is known [49, 52, 54, 55, 162, 276, 540, 552, 559] that if some information about the bits of the decryption exponent d is available, then the corresponding encryption is vulnerable to various attacks.

Let us denote by $\mathcal{N}_{r,k}(m)$ the set of $e \in [1, 2^r - 1]$ with exactly k non-zero binary digits and such that $\gcd(e, \varphi(m)) = 1$. Using (5.8) and (5.9), it has been shown in [26] that for almost all m the cardinality $N_{r,k}(m)$ of $\mathcal{N}_{n,k}(M)$ is close to its expected value

$$N_{r,k}(m) \sim 2\binom{r-1}{k-1}\frac{\varphi(\varphi(m))}{\varphi(m)}$$

provided that $k \geq r^{2/3+\varepsilon}$, when m runs through the set of RSA moduli $m = pl$, where p and l are two prime numbers. A very similar result also holds when m runs through the set of all integer moduli.

Question VII.51. *Prove that the set of values of $d_m(e)$, which is defined from the congruence (VII.2), for $e \in \mathcal{N}_{r,k}(m)$ is asymptotically uniformly distributed modulo $\varphi(m)$.*

We hope that the method of [196], combined with the Hua Loo Keng method of estimating of exponential sums, maybe in the form of [117, 118, 119, 120, 121, 122,

283, 284, 285, 539], can be applied to this problem. Nevertheless, so far there are several obstacles that we have not been able to overcome.

The paper [268] exploits the fact that in $\mathbb{F}_{2^m}$ squaring is essentially for free. Thus computing g^k costs only $\sigma - 1$ multiplications in $\mathbb{F}_{2^m}$, where σ is the sum of binary digits of $k \geq 1$. The method of [268] of generating random powers of $g \in \mathbb{F}_{2^m}$ is based on this idea, suggests to generate random powers as $g^{k_1 \ldots k_s}$ with randomly chosen integers $1 \leq k_1, \ldots, k_s \leq 2^m - 2$, having σ non-zero binary digits each, thus this computation costs $s(\sigma - 1)$ multiplications in $\mathbb{F}_{2^m}$. It would be interesting to have some security assurance that computing the discrete logarithm of such powers is not easier that for random powers of g, in particular an analogue of Theorems 22.1, 22.3 and 22.4 would be very useful.

Question VII.52. *Study the distribution of the products*

$$\prod_{i=1}^{s} \sum_{\nu=1}^{\sigma} 2^{r_{i,\nu}} \pmod{2^m - 1},$$

where $0 \leq r_{i,\nu} \leq m - 1$ *are chosen uniformly and independently at random,* $i = 1, \ldots, s$, $\nu = 1, \ldots, \sigma$.

The case $s\sigma = o(m)$ is especially interesting and important.

Similar questions are also of interest for polynomial rings $\mathbb{F}_q[X]/F(X)$ where $F(X) \in \mathbb{F}_q[X]$. For example, one can study the inverses of polynomials $f \in \mathbb{F}_q[X]/F(X)$ which are sparse or have small coefficients (or both). Such polynomials appear in the settings of the NTRU cryptosystem, see [267, 269], and are also of independent intrinsic interest. Some results about the uniformity of distribution of the inverses of polynomials with some restrictions on the coefficients or of small degree are given in [23, 25, 28] but the sets of NTRU polynomials are too "thin" to be covered by the technique of [23, 25, 28].

For $a, b \in \mathbb{F}_p$ we define

$$\mu(a, b) = \min_{u \in \mathbb{F}_p^*} \max \left\{ \lfloor au^4 \rfloor_p, \lfloor bu^6 \rfloor_p \right\}.$$

Using the Weil bound one can easily show that $\mu(a, b) = O(p^{3/4})$ for any $a, b \in \mathbb{F}_p$ which are not both zero, see [115, 185]. The question has the following motivation. Given an elliptic curve $y^2 = x^3 + ax + b$ over $\mathbb{F}_p$ one can try to choose from the family of isomorphic curves $y^2 = x^3 + au^4x + bu^6$, $u \in \mathbb{F}_p^*$, the one with the smallest possible coefficients (which facilitates the group operation on this curve). It has been shown in [27] that for almost all $a, b \in \mathbb{F}_p$ a stronger bound $\mu(a, b) = O(p^{2/3})$ holds. Thus almost all elliptic curves have a rather "compact" representation. Probably both bounds can be refined but simple counting arguments show that the bound $\mu(a, b) = o(p^{1/2})$ cannot be valid for almost all $a, b \in \mathbb{F}_p$.

Question VII.53. *Design an efficient algorithm of finding a multiplier* $u \in \mathbb{F}_p^*$ *for which* $\max\left\{\lfloor au^4 \rfloor_p, \lfloor bu^6 \rfloor_p\right\}$ *is close to* $\mu(a, b)$.

We have already remarked that inversion in $\mathbb{F}_{2^r}$ which is one of the building blocks of *Rijndael*, see [139]. The papers [135, 178, 400] indicate that this simple algebraic structure could be a potential weakness of this cipher. It would interesting to clarify whether the methods of this book, and in particular of Chapter 29, can provide some insight on this issue. For example, one can certainly study some mixing properties and various complexity characteristics of the Boolean functions

$$x^{-1} = (B_1(x), \dots, B_r(x)),$$

where $x = (x_1, \dots, x_r) \in \mathbb{F}_{2^r}^*$, associated with inversion in $\mathbb{F}_{2^r}$. Indeed if

$$x = x_1\omega_1 + \dots + x_r\omega_r$$

where $\omega_1, \dots, \omega_r$ is some basis of $\mathbb{F}_{2^r}$ over $\mathbb{F}_2$. Then we

$$B_i(x) = \mathrm{Tr}_{\mathbb{F}_{2^r}/\mathbb{F}_2}(\vartheta_i x^{-1}), \qquad i = 1, \dots, r$$

where $\vartheta_1, \dots, \vartheta_r$ is the dual basis to $\omega_1, \dots, \omega_r$; see [343]. Thus we can use the Weil bound and other exponential sum technique to study these Boolean functions, see also [450] for some other properties of inversion which are of cryptographic interest. In particular, it has been noted in [450] that the Weil bound implies

$$|\widehat{B}_i(a)| \le 2^{r/2+1}, \qquad a \in \mathfrak{B}_r$$

for the Fourier coefficients of each function $B_i(x)$, $i = 1, \dots, r$. As before, we can use Theorem 5.1 of [79] or Theorem 3 of [80] to conclude that each of the functions $B_i(x)$, $i = 1, \dots, r$, has at least $2^{r/2-1}$ monomials in their polynomial representations over $\mathbb{F}_2$.

It has been shown in Chapter 30 that, from approximate values of the residues of sparse polynomials with known exponents at polynomially many points $t \in \mathbb{F}_p$ selected uniformly at random, one can reconstruct the whole polynomial. Further modifications and extensions of these results would be of interest as well. It is easy to see that the same method applies to the case where some of the exponents $e_1, \dots, e_m$ are negative (and thus f is a rational function). Both our basic tools, Lemma 3.1 and Lemma 2.6 work for such rational functions as well. One can also obtain analogues of our results for multivariate polynomials.

Question VII.54. *Obtain analogues of Theorems 30.1, 30.2 and 30.3 in the case where the values of the exponents* $e_1, \dots, e_m$ *are unknown.*

It would also be important to extend the result of Theorem 30.1 to sparse polynomials f of arbitrary degree $\deg f \le p - 1$. The main ingredient that is missing is a result about the uniformity of distribution of the values of polynomials of large degree. Certainly Theorem 30.2 applies to such polynomials but it yields a rather weak statement.

Question VII.55. *Obtain an analogue of Theorem 30.3 for the case where the norm* $\mathrm{Nm}_{\mathbb{F}_{q^r}/\mathbb{F}_q}(f(t))$ *(rather than the trace) is given for polynomially many randomly selected values of* $t \in \mathbb{F}_{q^r}$.

Because the norm is multiplicative, one can only try to recover f up to a factor $a \in \mathbb{F}_{q^r}$ with $\mathrm{Nm}_{\mathbb{F}_{q^r}/\mathbb{F}_q}(a) = 1$.

Obtaining lower bounds on the complexity of linear algebra problems is a very interesting area as well.

For example, let $r = n^2$. Then one can define the Boolean function

$$B(u_1, \dots, u_r) = \begin{cases} 0, & \text{if } \mathcal{A}(u_1, \dots, u_r) \text{ is singular over } \mathbb{F}_2, \\ 1, & \text{if } \mathcal{A}(u_1, \dots, u_r) \text{ is non-singular over } \mathbb{F}_2, \end{cases}$$

where $\mathcal{A}(u_1, \dots, u_r)$ is the $n \times n$ matrix

$$\mathcal{A}(u_1, \dots, u_r) = \left(u_{i+n(j-1)}\right)_{i,j=1}^{n},$$

defined over $\mathbb{F}_2$.

Studying various polynomial representations and approximations of the determinant is a very important question with many links to some very deep problems of complexity theory [239, 241, 449, 534].

It is shown in [239] that for any finite field $\mathbb{F}_q$ there is a constant $\delta > 0$ such that, in any representation of the form

$$\det\left(X_{ij}\right)_{i,j=1}^{n} = \sum_{l=1}^{L} \prod_{m=1}^{M} L_{i,m}(X_{11}, \dots, X_{nn})$$

with LM linear forms $L_{i,m}(X_{11}, \dots, X_{nn}) \in \mathbb{F}_q[X_{11}, \dots, X_{nn}]$, the number of terms $L \gg \exp(\delta n)$. Further developments in this direction can be found in [241].

Lower bounds on the circuit complexity of matrix multiplication can be found in [470, 471], see also [7, 43, 264].

Discrete logarithms in residue rings of composite numbers have been considered in the literature as well [4]. Apparently the method developed in this book combined with results of [117, 118, 119, 120, 121, 122, 283, 284, 285, 303, 307, 533, 539] can lead to some non-trivial results for that case as well.

Finally, we mention that many of the results of this book can be generalised to the discrete logarithm, the Diffie–Hellman key exchange protocol and other related constructions over elliptic curves, including the pseudorandom number generators proposed in [218, 251]. Some results of this kind have recently been obtained in [32, 63, 77, 169, 170, 199, 218, 219, 261, 288, 321, 324, 325, 418, 512, 528], while [301] provides a necessary tool in the form of a bound of exponential sums similar to those of Lemmas 3.13 and 3.15. However these questions have not been studied systematically as yet.

Bibliography

[1] L. M. Adleman, 'The function field sieve', *Lect. Notes in Comp. Sci.*, Springer-Verlag, Berlin, **877** (1994), 108–121.

[2] L. M. Adleman, M.-D. Huang and K. Kompella, 'Efficient checkers for number-theoretic computations', *Inform. and Comp.*, **121** (1995), 93–102.

[3] L. M. Adleman and K. Kompella, 'Using smoothness to achieve parallelism', *Proc. 20th ACM Symp. on Theory of Comp.*, ACM, 1988, 528–538.

[4] L. M. Adleman and K. S. McCurley, 'Open problems in number theoretic complexity, 2', *Lect. Notes in Comp. Sci.*, Springer-Verlag, Berlin, **877** (1994), 291–322.

[5] M. Agrawal, N. Kayal and N. Saxena, 'PRIMES is in P', *Preprint*, 2002, 1–9.

[6] M. Ajtai, R. Kumar and D. Sivakumar, 'A sieve algorithm for the shortest lattice vector problem', *Proc. 33rd ACM Symp. on Theory of Comput.*, ACM, 2001, 601–610.

[7] E. Allender, R. Beals and M. Ogihara, 'The complexity of matrix rank and feasible systems of linear equations', *Comp. Compl.*, **8** (1999), 99–126.

[8] E. Allender, H. Buhrman, M. Koucký, D. Ronneburger and D. van Melkebeek, 'Power from random strings', *Proc. 43st IEEE Symp. on Found. of Comp. Sci.*, IEEE, 2002, (to appear).

[9] E. Allender, M. Saks and I. E. Shparlinski, 'A lower bound for primality', *J. of Comp. and Syst. Sci.*, **62** (2001), 356–366.

[10] E. Allender, A. Bernasconi, C. Damm, J. von zur Gathen, M. Saks and I. E. Shparlinski, 'On the complexity of some arithmetic problems over $\mathbf{F}_2[T]$', *Comp. Compl.*, (to appear). 23

[11] M. Anshel and D. Goldfeld, 'Zeta functions, one-way functions, and pseudorandom number generators', *Duke Math. J.*, **88** (1997), 371–390.

[12] S. Ar, R. Lipton, R. Rubinfeld and M. Sudan, 'Reconstructing algebraic functions from erroneous data', *SIAM J. Comput.*, **28** (1999), 487–510.

[13] S. Arora and M. Sudan, 'Improved low-degree testing and its applications', *Electronic Colloq. on Comp. Compl.*, Univ. of Trier, **TR1997-003** (1997), 1–35.

[14] L. Babai, T. P. Hayes and P. G. Kimmel, 'The cost of missing bit: Communication complexity with help', *Combinatorica*, **21** (2001), 455–488.

[15] L. Babai, N. Nisan and M. Szegedy, 'Multiparty protocols, pseudorandom generators for logspace and time–space trade-offs', *J. Comp. and Syst. Sci.*, **45** (1992), 204–232.

[16] E. Bach and J. Shallit, *Algorithmic number theory*, MIT Press, 1996.

[17] E. Bach and J. Sorenson, 'Sieve algorithms for perfect power testing', *Algorithmica*, **9** (1993), 313–328.

[18] R. C. Baker and G. Harman, 'Shifted primes without large prime factors', *Acta Arith.*, **83** (1998), 331-361.

[19] A. Balog, 'The prime k-tuplets conjecture on average', *Analytic Number Theory*, Progress in Mathematics **85**, Birkhäuser, Boston, 1990, 47–75.

[20] W. Banks, A. Conflitti, J. B. Friedlander and I. E. Shparlinski, 'Exponential sums with Mersenne numbers', *Preprint MPI-2002-96*, Max Planck Inst. of Math., Bonn, 2002, 1–23.

[21] W. Banks, A. Conflitti and I. E. Shparlinski, 'Character sums over integers with restricted g-ary digits', *Illinois J. Math.*, (to appear).

[22] W. Banks, F. Griffin, D. Lieman and I. E. Shparlinski, 'Non-linear complexity of the Naor–Reingold pseudo-random function', *Lect. Notes in Comp. Sci.*, Springer-Verlag, Berlin, **1787** (2000), 53–59.

[23] W. Banks, A. Harcharras and I. E. Shparlinski, 'Short Kloosterman sums for polynomials over finite fields', *Canad J. Math.*, (to appear).

[24] W. Banks, A. Harcharras and I. E. Shparlinski, 'Smooth values of shifted primes in arithmetic progressions', *Preprint*, 2002, 1–21.

[25] W. Banks and I. E. Shparlinski, 'Distribution of inverses in polynomial rings', *Indag. Math.*, **12** (2001), 303–315.

[26] W. Banks and I. E. Shparlinski, 'On the number of sparse RSA exponents', *J. Number Theory*, **95** (2002), 340–350.

[27] W. Banks and I. E. Shparlinski, 'Average normalizations of elliptic curves', *Bull. Austral. Math. Soc.*, (to appear).

[28] W. Banks and I. E. Shparlinski, 'A variant of NTRU with non-invertible polynomials', *Proc. Indocrypt'02*, Hyderabad, India, 2002, Springer-Verlag, Berlin, (to appear).

[29] W. Banks and I. E. Shparlinski, 'Smooth values of the Euler function', *Preprint*, 2002, 1–11.

[30] D. A. M. Barrington and H. Straubing, 'Lower bounds for modular counting by circuits with modular gates', *Lect. Notes in Comp. Sci.*, Springer-Verlag, Berlin, **911** (1995), 60–71.

[31] R. Beals, H. Buhrman, R. Cleve, M. Mosca and R. de Wolf, 'Quantum lower bounds by polynomials', *J. ACM*, **48** (2001), 778–797.

[32] P. Beelen and J. Doumen, 'Pseudorandom sequences from elliptic curves', *Finite Fields with Applications to Coding Theory, Cryptography and Related Areas*, Springer-Verlag, Berlin, 2002, 37–52.

[33] A. Bernasconi, 'On the complexity of balanced Boolean functions', *Lect. Notes in Comp. Sci.*, Springer-Verlag, Berlin, **1203** (1997), 253–263.

[34] A. Bernasconi, 'Combinatorial properties of classes of functions hard to compute in constant depth', *Lect. Notes in Comp. Sci.*, Springer-Verlag, Berlin, **1449** (1998), 339–348.

[35] A. Bernasconi and B. Codenotti, 'Measures of Boolean function complexity based on harmonic analysis', *Lect. Notes in Comp. Sci.*, Springer-Verlag, Berlin, **778** (1994), 63–72.

[36] A. Bernasconi, B. Codenotti and J. Simon, 'On the Fourier analysis of Boolean functions', *Preprint*, 1996, 1–24.

[37] A. Bernasconi, B. Codenotti, V. Crespi and G. Resta, 'How fast can one compute the permanent of circulant matrices?', *Linear Algebra and its Appl.*, **292** (1999), 15–37.

[38] A. Bernasconi, C. Damm and I. E. Shparlinski, 'On the average sensitivity of testing square-free numbers', *Lect. Notes in Comp. Sci.*, Springer-Verlag, Berlin, **1627** (1999), 291–299.

[39] A. Bernasconi, C. Damm and I. E. Shparlinski, 'The average sensitivity of square-freeness', *Comp. Compl.*, **9** (2000), 39–51.

[40] A. Bernasconi, C. Damm and I. E. Shparlinski, 'Circuit and decision tree complexity of some number theoretic problems', *Inform. and Comp.*, **168** (2001), 113–124.

[41] A. Bernasconi and I. E. Shparlinski, 'Circuit complexity of testing squarefree numbers', *Lect. Notes in Comp. Sci.*. Springer-Verlag, Berlin, **1563** (1999), 47–56.

[42] D. J. Bernstein, 'Detecting perfect powers in essentially linear time', *Math. Comp.*, **67** (1998), 1253–1283.

[43] A. Beygelzimer and M. Ogihara, 'On the enumerability of the determinant and the rank', *Electronic Colloq. on Comp. Compl.*, Univ. of Trier, **TR2002-016** (2002), 1–14.

[44] E. Biham, D. Boneh and O. Reingold, 'Breaking generalized Diffie–Hellman modulo a composite is no easier than factoring', *Inform. Proc. Letters*, **70** (1999), 83–87.

[45] G. Blakley and I. Borosh, 'Rivest–Shamir–Adleman public key cryptosystems do not always conceal messages', *Computers and Math. with Applications*, **5** (1979), 169–178.

[46] G. Blake and T. Garefalakis, 'On the security of the digital signature algorithm', *Designs, Codes and Cryptography*, **26** (2002), 87–96.

[47] D. Bleichenbacher, W. Bosma and A. K. Lenstra, 'Some remarks on Lucas-based cryptosystems', *Lect. Notes in Comp. Sci.*, Springer-Verlag, Berlin, **963** (1995), 386–396.

[48] D. Bleichenbacher and P. Q. Nguyen, 'Noisy polynomial interpolation and noisy Chinese remaindering', *Lect. Notes in Comp. Sci., Springer-Verlag*, Berlin, **1807** (2000), 53–69.

[49] J. Blömer and A. May, 'Low exponent RSA revisited', *Lect. Notes in Comp. Sci.*, Springer-Verlag, Berlin, **2146** (2001), 4–19.

[50] L. Blum, M. Blum and M. Shub, 'A simple unpredictable pseudo-random number generator', *SIAM J. Comp.*, **15** (1986), 364–383.

[51] D. Boneh, 'The decision Diffie–Hellman Problem', *Lect. Notes in Comp. Sci.*, Springer-Verlag, Berlin, **1423** (1998), 48–63.

[52] D. Boneh, 'Twenty years of attacks on the RSA cryptosystem', *Notices Amer. Math. Soc.*, **46** (1999), 203–213.

[53] D. Boneh, 'Finding smooth integers in short intervals using CRT decoding', *J. Comp. and Syst. Sci.*, **64** (2002), 768–784.

[54] D. Boneh and G. Durfee, 'Cryptanalysis of RSA with private key d less than $N^{0.292}$', *IEEE Trans. Inform. Theory*, **46** (2000), 1339–1349.

[55] D. Boneh, G. Durfee and Y. Frankel, 'An attack on RSA given a small fraction of the private key bits', *Lect. Notes in Comp. Sci.*, Springer-Verlag, Berlin, **1514** (1998), 25–34.

[56] D. Boneh, G. Durfee and N. A. Howgrave-Graham, 'Factoring $N = p^r q$ for large r', *Lect. Notes in Comp. Sci.*, Springer-Verlag, Berlin, **1666** (1999), 326–337.

[57] D. Boneh, S. Halevi and N. A. Howgrave-Graham, 'The modular inversion hidden number problem', *Lect. Notes in Comp. Sci.*, Springer-Verlag, Berlin, **2248** (2001), 36–51.

[58] D. Boneh, A. Joux and P. Q. Nguyen, 'Why textbook ElGamal and RSA encryption are insecure', *Lect. Notes in Comp. Sci.*, Springer-Verlag, Berlin, **1976** (2000), 30–43.

[59] D. Boneh and R. Lipton, 'Algorithms for black-box fields and their applications to cryptography', *Lect. Notes in Comp. Sci.*, Springer-Verlag, Berlin, **1109** (1996), 283–297.

[60] D. Boneh and R. Lipton, 'Effect of operators on straight line complexity', *Proc. 5th Israel Symp. on the Theory of Comp. Systems*, IEEE, 1997, 1–5.

[61] D. Boneh and R. Lipton, 'Lower bounds for algebraic algorithms based on the hardness of factoring', *Preprint*, 1997, 1–17.

[62] D. Boneh and R. Naor, 'Timed commitments', *Lect. Notes in Comp. Sci.*, Springer-Verlag, Berlin, **1880** (2000), 236–254.

[63] D. Boneh and I. E. Shparlinski, 'On the unpredictability of bits of the elliptic curve Diffie–Hellman scheme', *Lect. Notes in Comp. Sci.*, Springer-Verlag, Berlin, **2139** (2001), 201–212.

[64] D. Boneh and R. Venkatesan, 'Hardness of computing the most significant bits of secret keys in Diffie–Hellman and related schemes', *Lect. Notes in Comp. Sci.*, Springer-Verlag, Berlin, **1109** (1996), 129–142.

[65] D. Boneh and R. Venkatesan, 'Rounding in lattices and its cryptographic applications', *Proc. 8th Annual ACM-SIAM Symp. on Discr. Algorithms*, ACM, 1997, 675–681.

[66] D. Boneh and R. Venkatesan, 'Breaking RSA may not be equivalent to factoring', *Lect. Notes in Comp. Sci.*, Springer-Verlag, Berlin, **1403** (1998), 59–71.

[67] R. B. Boppana, 'The average sensitivity of bounded-depth circuits', *Inform. Proc. Letters*, **63** (1997), 257–261.

[68] R. B. Boppana and M. Sipser, 'The complexity of finite functions', *Handbook of Theoretical Comp. Sci., Vol. A*, Elsevier, Amsterdam, 1990, 757–804.

[69] W. Bosma, J. Hutton and Eric Verheul, 'Looking beyond XTR', *Proc. AsiaCrypt'02*, Springer-Verlag, Berlin, (to appear).

[70] V. Boyko, M. Peinado and R. Venkatesan, 'Speeding up discrete log and factoring based schemes via precomputations', *Lect. Notes in Comp. Sci.*, Springer-Verlag, Berlin, **1403** (1998), 221–234.

[71] J. Boyar, G. S. Frandsen and C. Sturtivant, 'An arithmetic model of computation equivalent to threshold circuits', *Theor. Comp. Sci.*, **93** (1992), 303–319.

[72] Y. Brandman, A. Orlitsky, and J. Hennessy, 'A spectral lower bound technique for the size of decision trees and two-level AND/OR circuits', *IEEE Trans. Computers*, **39** (1990), 282–287.

[73] J. J. Brennan and B. Geist, 'Analysis of iterated modular exponentiation: The orbit of x^α mod N', *Designs, Codes and Cryptography*, **13** (1998), 229–245.

[74] R. P. Brent, 'Recent progress and prospects for integer factorisation algorithms', *Lect. Notes in Comp. Sci.*, Springer-Verlag, Berlin, **1858** (2000), 3–23.

[75] E. Brier, C. Clavier, J.-S. Coron and D. Naccache, 'Cryptanalysis of RSA signatures with fixed-pattern padding', *Lect. Notes in Comp. Sci.*, Springer-Verlag, Berlin, **2139** (2001), 433–439.

[76] A. E. Brouwer, R. Pellikaan and E. R. Verheul, 'Doing more with fewer bits', *Lect. Notes in Comp. Sci.*, Springer-Verlag, Berlin, **1716** (1999), 321–332.

[77] D. R. L. Brown, 'Generic groups, collision resistance, and ECDSA', *Cryptology ePrint Archive, Report 2002/026*, 2002, 1–21.

[78] D. R. L. Brown and A. J. Menezes, 'A small subgroup attack on a key agreement protocol of Arazi', *Research Report CORR 2001-50*, Faculty of Math., Univ. Waterloo, Waterloo, 2001, 1–5.

[79] J. Bruck, 'Harmonic analysis of polynomial threshold functions', *SIAM J. Discr. Math.*, **3** (1990), 168–177.

[80] J. Bruck and R. Smolensky, 'Polynomial threshold functions, $\mathcal{AC}^0$ functions, and spectral norms', *SIAM J. Comp.*, **21** (1992), 33–42.

[81] N. H. Bshouty, Y. Mansour, B. Schieber and P. Tiwari, 'Fast exponentiation using the truncation operations', *Comp. Compl.*, **2** (1992), 244–255.

[82] J. Buchmann and D. Weber, 'Discrete logarithms: Recent progress', *Proc. International Conference on Coding Theory, Cryptography and Related Areas, Guanajuato 1998*, Springer-Verlag, Berlin, 2000, 42–56.

[83] H. Buhrman and R. de Wolf, 'Communication complexity lower bounds by polynomials', *Proc. 16th IEEE Conf. on Comp. Compl.*, IEEE, 2001, 120–130.

[84] H. Buhrman and R. de Wolf, 'Complexity measures and decision tree complexity: A survey', *Theor. Comp. Sci.*, **288** (2002), 21–43.

[85] D. A. Burgess, 'The distribution of quadratic residues and non-residues', *Mathematika*, **4** (1957), 106–112.

[86] P. Bürgisser, 'The computational complexity of immanants', *SIAM J. Comp.*, **30** (2000), 1023–1040.

[87] P. Bürgisser, *Completeness and reduction in algebraic complexity theory*, Springer-Verlag, Berlin, 2000.

[88] P. Bürgisser, M. Clausen and M. A. Shokrollahi, *Algebraic complexity theory*, Springer-Verlag, Berlin, 1996.

[89] J.-Y. Cai, F. Green and T. Thierauf, 'On the correlation of symmetric functions', *Math. Syst. Theory*, **29** (1996), 245–258.

[90] J.-Y. Cai, A. Pavan and D. Sivakumar, 'On the hardness of permanent', *Lect. Notes in Comp. Sci.*, Springer-Verlag, Berlin, **1563** (1999), 90–99.

[91] J. L. Camenisch and M. A. Stadler, 'Efficient group signature schemes for large groups', *Lect. Notes in Comp. Sci.*, Springer-Verlag, Berlin, **1294** (1997), 410–424.

[92] M. Campbell and C. Pomerance, 'Explicit estimates on some problems concerning primitive roots', *Preprint*, 2002.

[93] R. Canetti, 'Towards realizing random oracles: Hash functions that hide all partial information', *Lect. Notes in Comp. Sci.*, Springer-Verlag, Berlin, **1294** (1997), 455–469.

[94] R. Canetti, J. B. Friedlander, S. V. Konyagin, M. Larsen, D. Lieman and I. E. Shparlinski, 'On the statistical properties of Diffie–Hellman distributions', *Israel J. Math.*, **120** (2000), 23–46.

[95] R. Canetti, J. B. Friedlander and I. E. Shparlinski, 'On certain exponential sums and the distribution of Diffie–Hellman triples', *J. London Math. Soc.*, **59** (1999), 799–812.

[96] E. R. Canfield, P. Erdős and C. Pomerance, 'On a problem of Oppenheim concerning "Factorisatio Numerorum"', *J. Number Theory*, **17** (1983), 1–28.

[97] A. Canteaut, C. Carlet, P. Charpin and C. Fontaine, 'Propagation characteristics and correlation-immunity of highly nonlinear Boolean functions', *Lect. Notes in Comp. Sci.*, Springer-Verlag, Berlin, **1807** (1997), 507–522.

[98] A. Canteaut and M. Videau, 'Degree of composition of highly nonlinear functions and applications to higher order differential cryptanalysis', *Lect. Notes in Comp. Sci.*, Springer-Verlag, Berlin, **2332** (2002), 518–533.

[99] C. Carlet, 'On cryptographic propagation criteria for Boolean functions', *Inform. and Comp.*, **151** (1999), 32–56.

[100] C. Carlet, 'On the coset weight divisibility and nonlinearity of resilient correlation-immune functions', *Proc. Intern. Conf. on Sequences and their Applications, Bergen 2001*, Springer-Verlag, London, 2002, 131–144.

[101] C. Carlet, 'A larger class of cryptographic Boolean functions via a study of the Maiorana–McFarland construction', *Lect. Notes in Comp. Sci.*, Springer-Verlag, Berlin, **2442** (2002), 549–564.

[102] C. Carlet, 'On cryptographic complexity of Boolean functions', *Finite Fields with Applications to Coding Theory, Cryptography and Related Areas*, Springer-Verlag, Berlin, 2002, 53–69.

[103] C. Carlet and P. Sarkar, 'Spectral domain analysis of correlation immune and resilient Boolean functions', *Finite Fields and Their Appl.*, **8** (2002), 120–130.

[104] C. Carlet and Y. Tarannikov , 'Covering sequences of Boolean functions and their cryptographic significance', *Designs, Codes and Cryptography*, **25** (2002), 263–279.

[105] D. Catalano, R. Gennaro and N. A. Howgrave-Graham, 'Paillier's trapdoor function hides up to $O(n)$ bits', *J. Cryptology*, (to appear).

[106] D. Catalano, P. Q. Nguyen and J. Stern, 'The hardness of Hensel Lifting: The case of RSA and discrete logarithm', *Proc. Asiacrypt'02*, Springer-Verlag, Berlin, (to appear).

[107] J. H. H. Chalk, 'Polynomial congruences over incomplete residue systems modulo k', *Proc. Kon. Ned. Acad. Wetensch.*, **A92** (1989), 49–62.

[108] P. Charpin and E. Pasalic, 'On propagation characteristics of resilient functions', *INRIA Research Report RR-4537*, INRIA, 2002, 1–27.

[109] M. Chateauneuf and A. C. H. Ling and D. R. Stinson, 'Slope packings and coverings, and generic algorithms for the discrete logarithm problem', *Research Report CORR 2001-60*, Faculty of Math., Univ. Waterloo, Waterloo, 2001, 1–12.

[110] L. Chen and F. W. Fu, 'On the construction of new resilient functions from old ones', *IEEE Trans. Inform. Theory*, **45** (1999), 2077–2082.

[111] J. H. Cheon, 'Nonlinear vector resilient functions', *Lect. Notes in Comp. Sci.*, Springer-Verlag, Berlin, **2139** (2001), 458–469.

[112] M. A. Cherepnev, 'On the connection between the discrete logarithms and the Diffie–Hellman problem', *Diskretnaja Matem.*, **6** (1996), 341–349 (in Russian).

[113] W.-S. Chou and H. Niederreiter, 'Monomials and binomials over finite fields as $\mathcal{R}$-orthomorphism', *Preprint*, 2001, 1–10.

[114] M. Clausen, A. Dress, J. Grabmeier and M. Karpinski, 'On zero testing and interpolation of k-sparse multivariate polynomials over finite fields', *Theor. Comp. Sci.*, **84** (1991), 151–164.

[115] M. Ciet, J.-J. Quisquater and F. Sica, 'Elliptic curve normalization', *Crypto Group Technical Report Series CG-2001/2*, Univ. Catholique de Louvain, Belgium, 2001, 1–13.

[116] C. Cobeli and A. Zaharescu, 'An exponential congruence with solutions in primitive roots', *Rev. Roumaine Math. Pures Appl.*, **44** (1999), 15–22.

[117] T. Cochrane, 'Exponential sums modulo prime powers', *Acta Arith.*, **101** (2002), 131–149.

[118] T. Cochrane, C. Li and Z. Y. Zheng, 'Upper bounds on character sums with rational function entries', *Acta Math. Sinica.* , (to appear).

[119] T. Cochrane and Z. Y. Zheng, 'Pure and mixed exponential sums', *Acta Arith.*, **91** (1999), 249–278.

[120] T. Cochrane and Z. Y. Zheng, 'Exponential sums with rational function entries', *Acta Arith.*, **95** (2000), 67–95.

[121] T. Cochrane and Z. Y. Zheng, 'On upper bounds of Chalk and Hua for exponential sums', *Proc. Amer. Math. Soc.*, **129** (2001), 2505–2516.

[122] T. Cochrane and Z. Y. Zheng, 'A survey on pure and mixed exponential sums modulo prime powers', *Proc. Illinois Millennial Conf. on Number Theory*, (to appear).

[123] B. Codenotti, V. Crespi, and G. Resta, 'On the permanent of certain $(0, 1)$ Toeplitz matrices', *Linear Algebra and its Appl.*, **267** (1997), 65–100.

[124] B. Codenotti and G. Resta, 'On the permanent of certain circulant matrices', *Algebraic combinatorics and computer science*, Springer–Italia, Milan, 2001. 513–532.

[125] B. Codenotti and G. Resta, 'Computation of sparse circulant permanents via determinants', *Linear Algebra and its Appl.*, 2002, (to appear).

[126] B. Codenotti and I. E. Shparlinski, 'Non-approximability of the permanent of structured matrices over finite fields', *Preprint*, 2002, 1–11.

[127] H. Cohen, *A course in computational algebraic number theory*, Springer-Verlag, Berlin, 1997.

[128] S. Cohen, H. Niederreiter, I. E. Shparlinski and M. Zieve, 'Incomplete character sums and a special class of permutations', *J. Théorie des Nombres Bordeaux*, **13** (2001), 53–63.

[129] A. Conflitti and I. E. Shparlinski, 'On the multidimensional distribution of the subset sum generator of pseudorandom numbers', *Math. Comp.*, (to appear).

[130] D. Coppersmith, 'Small solutions to polynomial equations, and low exponent RSA vulnerabilities', *J. Cryptology*, **10** (1997), 233–260.

[131] D. Coppersmith, 'Small solutions of small degree polynomials', *Lect. Notes in Comp. Sci.*, Springer-Verlag, Berlin, **2146** (2001), 20–31.

[132] D. Coppersmith, H. Krawczyk and Y. Mansour, 'The shrinking generator', *Lect. Notes in Comp. Sci.*, Springer-Verlag, Berlin, **773** (1994), 22–39.

[133] D. Coppersmith and I. E. Shparlinski, 'On polynomial approximation of the discrete logarithm and the Diffie–Hellman mapping', *J. Cryptology*, **13** (2000), 339–360.

[134] J.-S. Coron, D. M'Raïhi, and C. Tymen, 'Fast generation of pairs $(k, [k]P)$ for Koblitz elliptic curves', *Lect. Notes in Comp. Sci.*, Springer-Verlag, Berlin, **2259** (2001), 151–164.

[135] N. Courtois and J, Pieprzyk, 'Cryptanalysis of block ciphers with overdefined systems of equations', *Cryptology ePrint Archive, Report 2002/044*, 2002, 1–35.

[136] R. Crandall and C. Pomerance, *Prime numbers: A Computational perspective*, Springer-Verlag, Berlin, 2001.

[137] T. W. Cusic, 'Properties of the x^2 mod N pseudorandom number generator', *IEEE Trans. Inform. Theory*, **41** (1995), 1155–1159.

[138] T. W. Cusic, C. Ding and A. Renvall, *Stream ciphers and number theory*, Elsevier, Amsterdam, 1998.

[139] J. Daemen and V. Rijmen, *The design of Rijndael*, Springer-Verlag, 2002.

[140] Z. Dai, J. Yang, G. Gong and P. Wang, 'On the linear complexity of generalized Legendre sequence', *Proc. Intern. Conf. on Sequences and their Applications, Bergen 2001*, Springer-Verlag, London, 2002, 145–153.

[141] W. van Dam, 'Quantum algorithms for weighing matrices and quadratic residues', *Algorithmica*, (to appear).

[142] W. van Dam, and S. Hallgren, 'Efficient quantum algorithms for shifted quadratic character problem', *Preprint*, 2001, 1–15.

[143] W. van Dam, S. Hallgren and L. Ip, 'Quantum algorithms for some hidden shift problems', *Proc. 14th ACM-SIAM Symp. on Discr. Algorithms*, ACM, 2003, (to appear).

[144] W. van Dam and G. Seroussi, 'Efficient quantum algorithms for estimating Gauss sums', *HP Labs Technical Report HPL-2002-208* , HP Labs, 2002, 1–11.

[145] I. B. Damgård, 'On the randomness of Legendre and Jacobi sequences', *Lect. Notes in Comp. Sci.*, Springer-Verlag, Berlin, **403** (1990), 163–172.

[146] I. B. Damgård and M. Koprowski, 'Generic lower bounds for root extraction and signature schemes in general groups', *Lect. Notes in Comp. Sci.*, Springer-Verlag, Berlin, **2332** (2002), 256–271.

[147] C. Damm, 'A note on spectral lower bound arguments for decision trees', *Bull. of the EATCS*, **65** (1998), 136–138.

[148] C. Damm, M. Krause, C. Meinel and S. Waack, 'On relations between counting communication complexity classes', *J. Comp. and Syst. Sci.*, (to appear).

[149] N. Dedić, L. Reyzin and S. Vadhan, 'A simple and sfficient pseudorandom generator based on hardness of factoring', *Proc. 3rd Conf. on Security in Communication Networks*, Springer-Verlag, Berlin, (to appear).

[150] P. Deligne, *Cohomologie 'etale (SGA* $4\frac{1}{2}$*)*, Lect. Notes in Math., Springer-Verlag, Berlin, **569** (1977).

[151] M. Dietzfelbinger, M. Kutyłowski and R. Reischuk, 'Exact lower time bounds for computing Boolean functions on CREW PRAMs', *J. Comp. and Syst. Sci.*, **48** (1994), 231–254.

[152] M. Dietzfelbinger, M. Kutyłowski and R. Reischuk, 'Feasible time-optimal algorithms for Boolean functions on exclusive-write parallel random access machine', *SIAM J. Comp.*, **25** (1996), 1196–1230.

[153] W. Diffie and M. Hellman, 'New directions in cryptography', *IEEE Trans. Inform. Theory*, **22** (1976), 644–654.

[154] C. Ding, 'Linear complexity of generalized cyclotomic binary sequences of order 2', *Finite Fields and Their Appl.*, **3** (1997), 159–174.

[155] C. Ding, 'Pattern distributions of Legendre sequences', *IEEE Trans. Inform. Theory*, **44** (1998), 1693–1699.

[156] C. Ding and T. Helleseth, 'On cyclotomic generator of order r', *Inform. Proc. Letters*, **66** (1998), 21–25.

[157] C. Ding, T. Helleseth and K. Y. Lam, 'Duadic sequences of prime length', *Discr. Math.*, **218** (2000), 33–49.

[158] C. Ding, T. Helleseth and W. Shan, 'On the linear complexity of Legendre sequences', *IEEE Trans. Inform. Theory*, **44** (1998), 1276–1278.

[159] Y. Dodis, 'Efficient construction of (distributed) verifiable random functions', *Proc. 6th Intern. Workshop on Public Key Cryptography, PKC'03*, Miami, USA, 2003, Springer-Verlag, Berlin, (to appear).

[160] R. Downey and M. R. Fellows, *Parameterized complexity*, Springer-Verlag, Berlin, 1998.

[161] M. Drmota and R. Tichy, *Sequences, discrepancies and applications*, Springer-Verlag, Berlin, 1997.

[162] G. Durfee and P. Q. Nguyen, 'Cryptanalysis of the RSA Schemes with short secret exponent from Asiacrypt'99', *Lect. Notes in Comp. Sci.*, Springer-Verlag, Berlin, **1976** (2000), 14–29.

[163] J. Eichenauer-Herrmann and F. Emmerich, 'Compound inversive congruential pseudorandom numbers: An average-case analysis', *Math. Comp.*, **65** (1996), 215–225.

[164] J. Eichenauer-Herrmann, F. Emmerich and G. Larcher, 'Average discrepancy, hyperplanes, and compound pseudorandom numbers', *Finite Fields Appl.*, **3** (1997), 203–218.

[165] J. Eichenauer-Herrmann, E. Herrmann and S. Wegenkittl, 'A survey of quadratic and inversive congruential pseudorandom numbers', *Lect. Notes in Statistics*, Springer-Verlag, Berlin, **127** (1998), 66–97.

[166] E. El Mahassni, P. Q. Nguyen and I. E. Shparlinski, 'The insecurity of Nyberg–Rueppel and other DSA-like signature schemes with partially known nonces', *Lect. Notes in Comp. Sci.*, Springer-Verlag, Berlin, **2146** (2001), 97–109.

[167] E. El Mahassni and I. E. Shparlinski, 'Polynomial representations of the Diffie–Hellman mapping', *Bull. Aust. Math. Soc.*, **63** (2001), 467–473.

[168] E. El Mahassni and I. E. Shparlinski, 'On some uniformity of distribution properties of ESIGN', *Proc. Intern. Workshop on Coding and Cryptography, Paris 2001*, INRIA, 2001, 189–196 (see also *Electronic Notes in Discr. Math.*, Vol.6, Elsevier, 2001).

[169] E. El Mahassni and I. E. Shparlinski, 'On the uniformity of distribution of congruential generators over elliptic curves', *Proc. Intern. Conf. on Sequences and their Applications, Bergen 2001*, Springer-Verlag, London, 2002, 257–264.

[170] E. El Mahassni and I. E. Shparlinski, 'On the uniformity of distribution of the elliptic curve ElGamal signature', *Finite Fields and Their Appl.*, **8** (2002), 589–596.

[171] A. Enge and P. Gaudry, 'A general framework for subexponential discrete logarithm algorithms', *Acta Arith.*, **102** (2002), 83–103.

[172] P. Erdős and R. Murty, 'On the order of $a \pmod p$', *Proc. 5th Canadian Number Theory Association Conf.*, Amer. Math. Soc., Providence, RI, 1999, 87–97.

[173] P. Erdős and C. Pomerance, 'On the normal number of prime factors of $\varphi(n)$', *Rocky Mountain J. Math.*, **15** (1985), 343–352.

[174] P. Erdős, C. Pomerance and E. Schmutz, 'Carmichael's lambda function', *Acta Arith.*, **58** (1991), 363–385.

[175] F. Ergün, S. R. Kumar and R. Rubinfeld, 'Approximate checking of polynomials and functional equations', *SIAM J. Comput.*, **31** (2002), 550–576.

[176] M. Fedorova and Y. Tarannikov, 'On the constructing of highly nonlinear resilient Boolean functions by means of special matrices', *Lect. Notes in Comp. Sci.*, Springer-Verlag, Berlin, **2247** (2001), 254–266.

[177] U. Feige and C. Lund, 'On the hardness of computing the permanent of random matrices', *Comp. Compl.*, **6** (1996/1997), 101–132.

[178] N. Ferguson, R. Schroeppel and D. Whiting, 'A simple algebraic representation of Rijndael', *Lect. Notes in Comp. Sci.*, Springer-Verlag, Berlin, **2259** (2001), 103–111.

[179] F. E. Fich, 'The complexity of computation on the parallel random access machine', *Synthesis of parallel algorithms* . Morgan Kaufmann Publ., San Mateo, CA, 1993, 843–899.

[180] E. Fich and M. Tompa, 'The parallel complexity of exponentiating polynomials over finite fields', *J. ACM*, **35** (1988). 651–667.

[181] E. Filiol and C. Fontaine, 'Highly nonlinear balanced Boolean functions with a good correlation-immunity', *Lect. Notes in Comp. Sci.*, Springer-Verlag, Berlin, **1403** (1998), 475–488.

[182] R. Fischlin and C. P. Schnorr, 'Stronger security proofs for RSA and Rabin bits', *J. Cryptology*, **13** (2000), 221–244.

[183] M. Flahive and H. Niederreiter, 'On inversive congruential generators for pseudorandom numbers', *Finite Fields, Coding Theory, and Advances in Communications and Computing*. Marcel Dekker, New York, 1993, 75–80.

[184] J. Forster, 'A linear lower bound on the unbounded error probablistic communication complexity', *Proc. 16th IEEE Conf. on Comp. Compl.*, IEEE, 2001, 100–106.

[185] E. Fouvry and M. R. Murty, 'On the distribution of supersingular primes', *Canad. J. Math.*, **48** (1996), 81–104.

[186] G. S. Frandsen, M. Valence and D. A. M. Barrington, 'Some results on uniform arithmetic circuit complexity', *Math. System Theory*, **27** (1994), 105–124.

[187] J. B. Friedlander, J. Hansen and I. E. Shparlinski, 'On character sums with exponential functions', *Mathematatika*, **47** (2000), 75–85.

[188] J. B. Friedlander, J. Hansen and I. E. Shparlinski, 'On the distribution of the power generator modulo a prime power', *Proc. DIMACS Workshop on Unusual Applications of Number Theory, 2000*, Amer. Math. Soc. (to appear).

[189] J. B. Friedlander, S. V. Konyagin and I. E. Shparlinski, 'Some doubly exponential sums over $\mathbb{Z}_m$', *Acta Arith.*, (to appear).

[190] J. B. Friedlander, M. Larsen, D. Lieman and I. E. Shparlinski, 'On correlation of binary M-sequences', *Designs, Codes and Cryptography*, **16** (1999), 249–256.

[191] J. B. Friedlander, D. Lieman and I. E. Shparlinski, 'On the distribution of the RSA generator', *Proc. Intern. Conf. on Sequences and their Applications, Singapore 1998*, Springer-Verlag, London, 1999, 205–212.

[192] J. B. Friedlander, C. Pomerance and I. E. Shparlinski, 'Period of the power generator and small values of Carmichael's function', *Math. Comp.*, **70** (2001), 1591–1605 (see also **71** (2002), 1803-1806).

[193] J. B. Friedlander, C. Pomerance and I. E. Shparlinski, 'Small values of the Carmichael function and cryptographic applications', *Proc. Workshop on Cryptography and Computational Number Theory, Singapore 1999*, Birkhäuser, 2001, 25–32.

[194] J. B. Friedlander and I. E. Shparlinski, 'On the distribution of the power generator', *Math. Comp.*, **70** (2001), 1575–1589.

[195] J. B. Friedlander and I. E. Shparlinski, 'Double exponential sums over thin sets', *Proc. Amer. Math. Soc.*, **129** (2001), 1617–1621.

[196] J. B. Friedlander and I. E. Shparlinski, 'On the distribution of Diffie–Hellman triples with sparse exponents', *SIAM J. Discr. Math.*, **14** (2001), 162–169.

[197] A. M. Frieze, J. Håstad, R. Kannan, J. C. Lagarias and A. Shamir, 'Reconstructing truncated integer variables satisfying linear congruence', *SIAM J. Comp.*, **17** (1988), 262–280.

[198] A. Gál and A. Wigderson, 'Boolean complexity classes vs. their arithmetic analogues', *Random Structures and Algorithms*, **9** (1996), 99–111.

[199] S. D. Galbraith, H. J. Hopkins and I. E. Shparlinski, 'Secure bilinear Diffie-Hellman bits', *Cryptology ePrint Archive, Report 2002/155*, 2002, 1–8.

[200] J. von zur Gathen, 'Computing powers in parallel', *SIAM J. Comp.*, **16** (1987), 930–945.

[201] J. von zur Gathen, 'Algebraic complexity theory', *Annual Review of Comp. Sci.*, **3** (1988), 317–347.

[202] J. von zur Gathen, 'Efficient and optimal exponentiation in finite fields', *Comp. Compl.*, **1** (1991), 360–394.

[203] J. von zur Gathen and J. Gerhard, *Modern computer algebra*, Cambridge University Press, Cambridge, 1999.

[204] J. von zur Gathen and G. Seroussi, 'Boolean circuits versus arithmetic circuits', *Inform. and Comp.*, **91** (1991), 142–154.

[205] J. von zur Gathen and I. E. Shparlinski, 'The CREW PRAM complexity of modular inversion', *SIAM J. Comp.*, **29** (1999), 1839–1857.

[206] J. von zur Gathen, I. E. Shparlinski and A. Sinclair, 'Finding points on curves over finite fields', *SIAM J. Comp.*, (to appear).

[207] C. Gentry, 'Key recovery and message attack on NTRU-composite', *Lect. Notes in Comp. Sci.*, Springer-Verlag, Berlin, **2045** (2001), 182–194.

[208] M. Girault and J.-F. Misarsky, 'Selective forgery of RSA signatures using redundancy', *Lect. Notes in Comp. Sci.*, Springer-Verlag, Berlin, **1233** (1997), 495–507.

[209] M. Girault and J.-F. Misarsky, 'Cryptoanalysis of countermeasures proposed for repairing ISO 9796', *Lect. Notes in Comp. Sci.*, Springer-Verlag, Berlin, **1807** (2000), 81–90.

[210] R. Gennaro, 'An improved pseudo-random generator based on discrete log', *Lect. Notes in Comp. Sci.*, Springer-Verlag, Berlin, **1880** (2000), 469–481.

[211] M. Goldmann, 'Communication complexity and lower bounds for simulating threshold circuits', *Theoretical advances in neural computing and learning*, Kluwer Acad. Publ., Dordrecht, 1994, 85–125.

[212] M. Goldmann and M. Näslund, 'The complexity of computing hard core predicates', *Lect. Notes in Comp. Sci.*, Springer-Verlag, Berlin, **1294** (1997), 1–15.

[213] M. Goldman, M. Näslund and A. Russell, 'Complexity bounds on general hard-core predicates', *J. Cryptology*, **14** (2001), 177–195.

[214] O. Goldreich, *Modern cryptography, probabilistic proofs and pseudorandomness*, Springer-Verlag, Berlin, 1999.

[215] O. Goldreich, D. Ron, and M. Sudan, 'Chinese remaindering with errors', *IEEE Transactions on Information Theory*, **46** (2000), 1330–1338.

[216] O. Goldreich and V. Rosen, 'On the security of modular exponentiation with application to the construction of pseudorandom generators ', *Electronic Colloq. on Comp. Compl.*, Univ. of Trier, **TR2002-049** (2002), 1–28.

[217] O. Goldreich, R. Rubinfeld and M. Sudan, 'Learning polynomials with queries: The highly noisy case', *SIAM J. Discr. Math.*, **13** (2000), 535–570.

[218] G. Gong, T. A. Berson and D. R. Stinson, 'Elliptic curve pseudorandom sequence generators', *Lect. Notes in Comp. Sci.*, Springer-Verlag, Berlin, **1758** (2000), 34–49.

[219] G. Gong and C. C. Y. Lam, 'Linear recursive sequences over elliptic curves', *Proc. Intern. Conf. on Sequences and their Applications, Bergen 2001*, Springer-Verlag, London, 2002, 182–196.

[220] M. I. González Vasco and M. Näslund, 'A survey of hard core functions', *Proc. Workshop on Cryptography and Computational Number Theory, Singapore 1999*, Birkhäuser, 2001, 227–256.

[221] M. I. González Vasco, M. Näslund and I. E. Shparlinski, 'The hidden number problem in extension fields and its applications', *Lect. Notes in Comp. Sci.*, Springer-Verlag, Berlin, **2286** (2002), 105–117.

[222] M. I. González Vasco and I. E. Shparlinski, 'On the security of Diffie–Hellman bits', *Proc. Workshop on Cryptography and Computational Number Theory, Singapore 1999*, Birkhäuser, 2001, 257–268.

[223] M. I. González Vasco and I. E. Shparlinski, 'Security of the most significant bits of the Shamir message passing scheme', *Math. Comp.*, **71** (2002), 333–342.

[224] M. Goresky, A. Klapper and R. Murty, 'On the distinctness of decimations of l-sequences', *Proc. Intern. Conf. on Sequences and their Applications, Bergen 2001*, Springer-Verlag, London, 2002, 197–208.

[225] M. Goresky, A. Klapper, R. Murty and I. E. Shparlinski, 'On decimations of ℓ-sequences', *Preprint*, 2002, 1–13.

[226] C. Gotsman and N. Linial, 'Spectral properties of threshold functions', *Combinatorica*, **14** (1994), 35–50.

[227] L. Goubin, C. Mauduit and A. Sárközy, 'Construction of large families of pseudo-random binary sequences', *Preprint*, 2001, 1–13.

[228] A. Granville, 'Smooth numbers: Computational number theory and beyond', *Proc. MSRI Conf. Algorithmic Number Theory: Lattices, Number Fields, Curves, and Cryptography, Berkeley 2000*, Cambridge Univ. Press, (to appear).

[229] A. Granville, I. E. Shparlinski and A. Zaharescu, 'On the distribution of rational functions along a curve over $\mathbb{F}_p$ and residue races', *Preprint*, 2002, 1–10.

[230] A. Granville and K. Soundararajan, 'Large character sums', *Jour. Amer. Math. Soc.*, **14** (2001), 365–397.

[231] F. Green, 'Exponential sums and circuits with a single threshold gate and mod-gates', *Theory Comp. Syst.*, **32** (1999), 453–466.

[232] F. Green, 'Complex-number Fourier technique for lower bounds on the mod $-m$ degree', *Comp. Compl.*, **9** (2000), 16–38.

[233] F. Green, 'The correlation between parity and quadratic polynomials mod 3', *Proc. 17th IEEE Conf. on Comp. Compl.*, 2002, IEEE, 2002, (to appear).

[234] F. Griffin, H. Niederreiter and I. E. Shparlinski, 'On the distribution of nonlinear recursive congruential pseudorandom numbers of higher orders', *Lect. Notes in Comp. Sci.*, Springer-Verlag, Berlin, **1719** 1999, 87–93.

[235] F. Griffin and I. E. Shparlinski, 'On the linear complexity of the Naor–Reingold pseudo-random function', *Lect. Notes in Comp. Sci.*,, Springer-Verlag, Berlin, **1726** (1999), 301–308.

[236] F. Griffin and I. E. Shparlinski, 'On the linear complexity profile of the power generator', *IEEE Trans. on Inform. Theory*, **46** (2000), 2159–2162.

[237] D. Grigoriev, 'Lower bounds in the algebraic computational complexity', *Zapiski Nauchn. Semin. Leningr. Otdel. Matem. Inst. Acad. Sci. USSR*, **118** (1982), 25–82 (in Russian).

[238] D. Grigoriev, 'Testing shift-equivalence of polynomials by deterministic, probabilistic and quantum machines', *Theor. Comp. Sci.*, **180** (1997), 217–228.

[239] D. Grigoriev and M. Karpinski, 'An exponential lower bound for depths 3 arithmetic circuits', *Proc. 30th ACM Symp. on Theory of Comp.*, ACM, 1998, 577–582.

[240] D. Grigoriev, M. Karpinski and M. Singer, 'Computational complexity of sparse rational interpolation', *SIAM J. Comput.*, **23** (1994), 1–11.

[241] D. Grigoriev and A. Razborov, 'Exponential lower bounds for depths 3 arithmetic circuits in algebras of functions over finite fields', *Appl. Algebra in Engin., Commun. and Computing*, **10** (2000), 465–487.

[242] R. Gupta and M. R. Murty, 'A remark on Artin's conjecture', *Invent. Math.*, **78** (1984), 127–130.

[243] V. Guruswami, A. Sahai and M. Sudan, '"Soft-decision" decoding of Chinese remainder codes', *Proc. 41st IEEE Symp. on Found. of Comp. Sci.*, IEEE, 2000, 159–168.

[244] V. Guruswami and M. Sudan, 'Improved decoding of Reed-Solomon codes and algebraic-geometric codes', *IEEE Trans. Inform. Theory*, **45** (1999), 1757–1767.

[245] J. Gutierrez and D. Gomes-Perez, 'Iterations of multivariate polynomials and discrepancy of pseudorandom numbers', *Lect. Notes in Comp. Sci.*, Springer-Verlag, Berlin, **2227** (2001), 192–199.

[246] J. Gutierrez, H. Niederreiter and I. E. Shparlinski, 'On the multidimensional distribution of inversive congruential pseudorandom numbers in parts of the period', *Monatsh. Math.*, **129** (2000), 31–36.

[247] J. Gutierrez, I. E. Shparlinski and A. Winterhof, 'On the linear and nonlinear complexity profile of nonlinear pseudorandom number generators', *IEEE Trans. on Information Theory*, (to appear).

[248] R. K. Guy, *Unsolved problems in number theory*, Springer-Verlag, NY, 1994.

[249] M. Gysin and J. Seberry, 'Generalised cycling attacks on RSA and strong RSA primes', *Lect. Notes in Comp. Sci.*, Springer-Verlag, Berlin, **1587** (1999), 149–163.

[250] H. Halberstam and H.–E. Richert, *Sieve methods*, Academic Press, London, 1974.

[251] S. Hallgren, 'Linear congruential generators over elliptic curves', *Preprint CS-94-143*, Dept. of Comp. Sci., Cornegie Mellon Univ., 1994, 1–10.

[252] L. H. Harper, W. N. Hsieh and J. E. Savage, 'A class of Boolean functions with linear combinatorial complexity', *Theor. Comp. Sci.*, **1** (1975), 161–183.

[253] J. Håstad, 'A slight sharpening of LMN', *J. Comp. and Syst. Sci.*, **63** (2001), 498–508.

[254] J. Håstad and M. Näslund, 'The security of individual RSA and discrete log bits', *J. ACM*, (to appear).

[255] J. Håstad, A. W. Schrift and A. Shamir, 'The discrete logarithm modulo a composite hides $O(n)$ bits', *J. Comp. and Syst. Sci.*, **47** (1993), 376–404.

[256] J. Håstad, I. Wegener, N. Wurm and S.-Z. Yi, 'Optimal depth, very small size circuits for symmetric functions in AC^0', *Inform. and Comp.*, **108** (1994), 200–211.

[257] D. R. Heath-Brown, 'The least square-free number in an arithmetic progression', *J. Reine Angew. Math.*, **332** (1982), 204–220.

[258] D. R. Heath-Brown, 'Artin's conjecture for primitive roots', *Quart. J. Math.*, **37** (1986), 27–38.

[259] D. R. Heath-Brown and S. V. Konyagin, 'New bounds for Gauss sums derived from kth powers, and for Heilbronn's exponential sum', *Ouart. J. Math.*, **51** (2000), 221–235.

[260] T. Helleseth and T. Johansson, 'Universal hash functions from exponential sums over finite fields and Galois rings', *Lect. Notes in Comp. Sci.*, Springer-Verlag, Berlin, **921** (1996), 31–44.

[261] F. Hess and I. E. Shparlinski, 'On the linear complexity and multidimensional distribution of congruential generators over elliptic curves', *Preprint*, 2000, 1–13.

[262] A. Hildebrand and G. Tenenbaum, 'Integers without large prime factors', *J. de Théorie des Nombres de Bordeaux*, **5** (1993), 411–484.

[263] M. J. Hinek, M. K. Low and E. Teske, 'On some attacks on multi-prime RSA', *Research Report CORR 2002-11*, Faculty of Math., Univ. Waterloo, Waterloo, 2002, 1–18.

[264] T. M. Hoang and T. Thierauf, 'The complexity of the minimal polynomial', *Lect. Notes in Comp. Sci.*, Springer-Verlag, Berlin, **2387** (2002), 37–46.

[265] T. Hofmeister, 'Depth-efficient threshold circuits for arithmetic functions', *Theoretical advances in neural computing and learning*, Kluwer Acad. Publ., Dordrecht, 1994, 37–84.

[266] J. Hoffstein and D. Lieman, 'The distribution of the quadratic symbol in function fields and a faster mathematical stream cipher', *Proc. Workshop on Cryptography and Computational Number Theory, Singapore 1999*, Birkhäuser, 2001, 59–68.

[267] J. Hoffstein, J. Pipher and J. H. Silverman, 'NTRU: A ring based public key cryptosystem', *Lect. Notes in Comp. Sci.*, Springer-Verlag, Berlin, **1433** (1998), 267–288.

[268] J. Hoffstein and J. H. Silverman, 'Random small Hamming weight products with applications to cryptography', *Preprint*, 2001, 1–16.

[269] J. Hoffstein and J. H. Silverman, 'Optimizations for NTRU', *Proc. the Conf. on Public Key Cryptography and Computational Number Theory, Warsaw, 2000*, Walter de Gruyter, 2001, 77–88.

[270] T. Hoholdt and R. R. Nielsen, 'Decoding Hermitian codes with Sudan's algorithm', *Lect. Notes in Comp. Sci.*, Springer-Verlag, Berlin, **1719** (1999), 260–270.

[271] J. Holden, 'Fixed points and two cycles of the discrete logarithm', *Lect. Notes in Comp. Sci.*, Springer-Verlag, Berlin, **2369** (2002), 405–416.

[272] C. Hooley, 'On Artin's conjecture', *J. Reine Angew. Math.*, **225** (1967), 209–220.

[273] N. A. Howgrave-Graham, 'Finding small roots of univariate modular equations revisited', *Lect. Notes in Comp. Sci.*, Springer-Verlag, Berlin, **1355** (1997), 131–142.

[274] N. A. Howgrave-Graham, 'Approximate integer common divisor', *Lect. Notes in Comp. Sci.*, Springer-Verlag, Berlin, **2146** (2001), 51–66.

[275] N. A. Howgrave-Graham, P. Q. Nguyen and I. E. Shparlinski, 'Hidden number problem with hidden multipliers, timed-release crypto and noisy exponentiation', *Math. Comp.*, (to appear).

[276] N. A. Howgrave-Graham and J.-P. Seifert, 'Extending Wiener's attack in the presence of many decryption exponents', *Lect. Notes in Comp. Sci.*, Springer-Verlag, Berlin, **1740** (1999), 153–166.

[277] N. A. Howgrave-Graham and N. P. Smart, 'Lattice attacks on digital signature schemes', *Designs, Codes and Cryptography*, **23** (2001), 283–290.

[278] T. Hromkovič, *Communication complexity and parallel computing*, Springer-Verlag, Berlin, 1997.

[279] L.-K. Hua, *Abschätzungen von Exponentialsummen und ihre Anwendung in der Zahlentheorie*, Teubner-Verlag, Leipzig, 1959.

[280] R. Impagliazzo and M. Naor, 'Efficient cryptographic schemes provably as secure as subset sum', *J. Cryptology*, **9** (1996), 199–216.

[281] H.-K. Indlekofer and N. M. Timofeev, 'Divisors of shifted primes', *Publ. Math. Debrecen*, **60** (2002), 307–345.

[282] L. Ip, 'Solving shift problems and hidden coset problems using the Fourier transform', *Preprint*, 2002, 1–15.

[283] D. Ismoilov, 'Estimates of complete character sums of polynomials', *Proc. Steklov Math. Inst.*, Moscow, **200** (1992), 171–184 (in Russian).

[284] D. Ismoilov, 'Estimates for complete trigonometric sums', *Proc. Steklov Math. Inst.*, Moscow, **207** (1994), 153–171 (in Russian).

[285] D. Ismoilov, 'On a method of Hua Loo-Keng of estimating complete trigonometric sums', *Adv. in Math. (China)*, **23** (1994), 31–49.

[286] M. Jerrum, A. Sinclair and E. Vigoda, 'A polynomial-time algorithm for the permanent of a matrix with non-negative entries', *Electronic Colloq. on Comp. Compl.*, Univ. of Trier, **TR2000-079** (2000), 1–22.

[287] A. Joux and J. Stern, 'Lattice reduction: A toolbox for the cryptanalyst', *J. Cryptology*, **11** (1998), 161–185.

[288] M. Joye, J.-J. Quisquater and T. Takagi, 'How to choose secret parameters for RSA-type cryptosystems over elliptic curves', *Designs, Codes and Cryptography*, **23** (2001), 297–316.

[289] S. Jukna, A. Razborov, P. Savicky and I. Wegener, 'On **P** versus **NP**$\cap$ *co*-**NP** for decision trees and read-once branching programs', *Comp. Compl.*, **8** (1999), 357–370.

[290] C. S. Jutla, 'On finding small solutions of multivariate polynomial equations', *Lect. Notes in Comp. Sci.*, Springer-Verlag, Berlin, **1403** (1998), 158–170.

[291] R. Kannan, 'Algorithmic geometry of numbers', *Annual Review of Comp. Sci.*, **2** (1987), 231–267.

[292] N. M. Katz, *Gauss sums, Kloosterman sums, and monodromy groups*, Ann. of Math. Studies, **116**, Princeton Univ. Press, 1988.

[293] A. G. Khovanski, *Fewnomials*, Amer. Math. Soc., Providence, RI, 1991.

[294] A. Kiayias and M. Yung, 'Secure games with polynomial expressions', *Lect. Notes in Comp. Sci.*, Springer-Verlag, Berlin, **2076** (2001), 939–950.

[295] A. Kiayias and M. Yung, 'Polynomial reconstruction based cryptography', *Lect. Notes in Comp. Sci.*, Springer-Verlag, Berlin, **2259** (2001), 129–133.

[296] A. Kiayias and M. Yung, 'Cryptographic hardness based on the decoding of Reed-Solomon codes with applications', *Electronic Colloq. on Comp. Compl.*, Univ. of Trier, **TR2002-017** (2002), 1–28.

[297] E. Kiltz, 'A tool box of cryptographic functions related to the Diffie-Hellman function', *Lect. Notes in Comp. Sci.*, Springer-Verlag, Berlin, **2247** (2001), 339–350.

[298] E. Kiltz, 'A primitive for proving the security of every bit and about universal hash functions & hard core bits', *Preprint*, 2001, 1–19.

[299] E. Kiltz, 'On the representation of Boolean predicates of the Diffie–Hellman function', *Preprint*, 2002, 1–17.

[300] M. Kiwi, F. Magniez and M. Santha, 'Exact and approximate testing/correcting of algebraic functions: A survey', *Lect. Notes in Comp. Sci.*, Springer-Verlag, Berlin, **2292** (2002), 30–83.

[301] D. R. Kohel and I. E. Shparlinski, 'Exponential sums and group generators for elliptic curves over finite fields', *Lect. Notes in Comp. Sci.*, Springer-Verlag, Berlin, **1838** (2000), 395–404.

[302] J. F. Koksma, 'Some theorems on diophantine inequalities', *Math. Centrum Scriptum no. 5*, Amsterdam, 1950.

[303] S. V. Konyagin, 'On the number of solutions of an univariate congruence of nth degree', *Matem. Sbornik*, **102** (1979), 171–187 (in Russian).

[304] S. V. Konyagin, 'Bounds of exponential sums over subgroups and Gauss sums', *Preprint*, 2002, 1–25 (in Russian).

[305] S. V. Konyagin, T. Lange, and I. Shparlinski, 'Linear complexity of the discrete logarithm', *Designs, Codes and Cryptography*, (to appear).

[306] S. V. Konyagin and I. Shparlinski, *Character sums with exponential functions and their applications*, Cambridge Univ. Press, Cambridge, 1999.

[307] S. V. Konyagin and T. Steger, 'On the number of solutions of polynomial congruences', *Matem. Zametki*, **55** (1994), no. 1, 73–79 (in Russian).

[308] N. M. Korobov, 'On the distribution of digits in periodic fractions', *Matem. Sbornik*, **89** (1972), 654–670 (in Russian).

[309] N. M. Korobov, *Exponential sums and their applications*, Kluwer Acad. Publ., Dordrecht, 1992.

[310] M. Krause, 'Geometric arguments yield better bounds for the threshold circuits and distributed computing', *Theor. Comp. Sci.*, **156** (1996), 99–117.

[311] M. Krause and S. Lucks, 'On learning versus distinguishing and the minimal hardware complexity of pseudorandom function generators', *Lect. Notes in Comp. Sci.*, Springer-Verlag, Berlin, **2010** (2001), 419–430.

[312] M. Krause and P. Pudlák, 'On computing Boolean functions by sparse real polynomials', *Proc. 36th IEEE Symp. on Found. of Comp. Sci.*, 1995, 682–691.

[313] H. Krawczyk, 'How to predict congruential generators', *J. Algorithms*, **13** (1992), 527–545.

[314] P. Kurlberg, 'On the order of unimodular matrices modulo integers', *Preprint*, 2001, 1–12.

[315] P. Kurlberg and Z. Rudnick, 'On quantum ergodicity for linear maps of the torus', *Comm. Math. Phys.*, **222** (2001), 201–227.

[316] E. Kushilevitz and N. Nisan, *Communication complexity*, Cambridge University Press, Cambridge, 1997.

[317] E. Kushilevitz and A. Rosén, 'A randomness-rounds tradeoff in private computation', *SIAM J. Comp.*, **11** (1998), 61–80.

[318] H. Kuwakado and H. Tanaka, 'On the security of the ElGamal-type signature scheme with small parameters', *IEICE Trans. Fundamentals of Electronics, Commun., and Comp. Sci.*, **E82-A** (1999), 93–97.

[319] J. C. Lagarias, 'Pseudorandom number generators in cryptography and number theory', *Proc. Symp. in Appl. Math.*, Amer. Math. Soc., Providence, RI, **42** (1990), 115–143.

[320] Y. N. Lakshman and B. D. Saunders, 'Sparse polynomial interpolation in non-standard bases', *SIAM J. Comp.*, **24** (1995), 387–397.

[321] C. C. Y. Lam and G. Gong, 'Randomness of elliptic curve sequences', *Research Report CORR 2002-18*, Faculty of Math., Univ. Waterloo, Waterloo, 2002, 1–11.

[322] T. Lange and A. Winterhof, 'Incomplete character sums over finite fields and their applications to the interpolation of the discrete logarithm by Boolean functions', *Acta Arith.*, **101** (2002), 223–229.

[323] T. Lange and A. Winterhof, 'Interpolation of the discrete logarithm in $\mathbb{F}_q$ by Boolean functions and by polynomials in several variables modulo a divisor of $q-1$', *Discr. Appl. Math.*, (to appear).

[324] T. Lange and A. Winterhof, 'Polynomial interpolation of elliptic curve and XTR discrete logarithm', *Lect. Notes in Comp. Sci.*, Springer-Verlag, Berlin, **2387** (2002), 137–143.

[325] P. J. Leadbitter and N. Smart, 'Cryptanalysis of MQV with partially known nonces', *Preprint*, 2002, 1–10.

[326] A. K. Lenstra, 'Primality testing', *Proc. Symp. in Appl. Math.*, Amer. Math. Soc., Providence, RI, **42** (1990), 13–25.

[327] A. K. Lenstra, 'Generating standard DSA signatures without long inversions', *Lect. Notes in Comp. Sci.*, Springer-Verlag, Berlin, **1163** (1996), 57–64.

[328] A. K. Lenstra and H. W. Lenstra, 'Algorithms in number theory', *Handbook of Theoretical Comp. Sci., Vol.A*, Elsevier, Amsterdam, 1990, 673–715.

[329] A. K. Lenstra, H. W. Lenstra and L. Lovász, 'Factoring polynomials with rational coefficients', *Mathematische Annalen*, **261** (1982), 515–534.

[330] A. K. Lenstra and I. E. Shparlinski, 'On the security of a variant of DSA without long inversions', *Lect. Notes in Comp. Sci.*, Springer-Verlag, Berlin, **1992** (2001), 64–72.

[331] A. K. Lenstra and I. E. Shparlinski, 'Selective forgery of RSA signatures with fixed-pattern padding', *Lect. Notes in Comp. Sci.*, Springer-Verlag, Berlin, **2274** (2002), 228–236.

[332] A. K. Lenstra and E. R. Verheul, 'The XTR public key system', *Lect. Notes in Comp. Sci.*, Springer-Verlag, Berlin, **1880** (2000), 1–19.

[333] A. K. Lenstra and E. R. Verheul, 'Key improvements to XTR', *Lect. Notes in Comp. Sci.*, Springer-Verlag, Berlin, **1976** (2000), 220–233.

[334] A. K. Lenstra and E. R. Verheul, 'Fast irreducibility and subgroup membership testing in XTR', *Lect. Notes in Comp. Sci.*, Springer-Verlag, Berlin, **1992** (2001), 73–86.

[335] A. K. Lenstra and E. R. Verheul, 'An overview of the XTR public key system', *Proc. the Conf. on Public Key Cryptography and Computational Number Theory, Warsaw 2000*, Walter de Gruyter, 2001, 151–180.

[336] S. Li, 'On the number of elements with maximal order in the multiplicative group modulo n', *Acta Arith.*, **86** (1998), 113–132.

[337] S. Li, 'On extending Artin's conjecture to composite moduli', *Mathematika*, **46** (1999), 373–390.

[338] S. Li, 'Artin's conjecture on average for composite moduli', *J. Number Theory*, **84** (2000), 93–118.

[339] S. Li and C. Pomerance, 'On generalizing Artin's conjecture on primitive roots to composite moduli', *J. Reine Angew. Math.*, (to appear).

[340] W.-C. W. Li,'Character sums and abelian Ramanujan graphs', *J. Number Theory*, **41** (1992), 199–217.

[341] W.-C. W. Li, *Number theory with applications*, World Scientific, Singapore, 1996.

[342] W.-C. W. Li, M. Näslund and I. E. Shparlinski, 'The hidden number problem with the trace and bit security of XTR and LUC', *Lect. Notes in Comp. Sci.*, Springer-Verlag, Berlin, **2442** (2002), 433–448.

[343] R. Lidl and H. Niederreiter, *Finite fields*, Cambridge University Press, Cambridge, 1997.

[344] D. Lieman and I. E. Shparlinski, 'On a new exponential sum', *Canad. Math. Bull.*, **44** (2001), 87–92.

[345] N. Linial, Y. Mansour and N. Nisan, 'Constant depth circuits, Fourier transform, and learnability', *J. ACM*, **40** (1993), 607-620.

[346] N. Linial, A. Samorodinski and A. Wigderson, 'A deterministic strongly polynomial algorithm for matrix scaling and approximate permanents', *Proc. 30th ACM Symp. on Theory of Comp.*, ACM, 1998, 644–652.

[347] J. H. van Lint and R. M. Wilson, *A Course in Combinatorics*, Cambridge Univ. Press, Cambridge, 1992.

[348] L. Lipshits, 'P-adic zeros of polynomials', *J. Reine Angew. Math.*, **390** (1988), 208–214.

[349] R. Lipton, 'Straight-line complexity and integer factorization', *Lect. Notes in Comp. Sci.*, Springer-Verlag, Berlin, **877** (1994), 71–79.

[350] M. Liskov and R. D. Silverman, 'A statistical limited-knowledge proof for secure RSA keys', *IEEE Working Group P1363: Research Contributions*, 1998, 1–14 (available from `http://grouper.ieee.org/groups/1363/Research/`).

[351] B. E. Litow and G. I. Davida, '$O(\log(n))$ parallel time finite field inversion', *Lect. Notes in Comp. Sci.*, Springer-Verlag, Berlin, **319** (1988), 74–80.

[352] R. Lovorn Bender and C. Pomerance, 'Rigorous discrete logarithm computations in finite fields via smooth polynomials', *Computational Perspectives on Number Theory*, Amer. Math. Soc., Providence, RI, 1998, 221–232.

[353] M. Luby, *Pseudorandomness and cryptographic applications*, Princeton Univ. Press, Princeton, 1996.

[354] F. Luca, 'Some mean values related to average multiplicative orders of elements in finite fields', *Preprint*, 2002, 1–16.

[355] F. Luca and I. E. Shparlinski, 'Average multiplicative orders of elements modulo n', *Acta Arith.*, (to appear).

[356] P. MacKenzie, 'On the security of the SPEKE password-authenticated key exchange protocol', *Cryptology ePrint Archive, Report 2001/57*, 2001, 1–19.

[357] F. J. MacWilliams and N. J. A. Sloane, *The theory of error-correcting codes*, North-Holland, Amsterdam, 1977.

[358] S. Maitra, 'Autocorrelation properties of correlation immune Boolean functions', *Lect. Notes in Comp. Sci.*, Springer-Verlag, Berlin, **2247** (2001), 242–253.

[359] S. Maitra and E. Pasalic, 'Further constructions of resilient Boolean functions with very high nonlinearity ', *IEEE Trans. Inform. Theory*, **48** (2002), 1825–1834.

[360] S. Maitra and P. Sarkar, 'Modifications of Patterson–Wiedemann functions for cryptographic applications', *IEEE Trans. Inform. Theory*, **48** (2002), 278–284.

[361] S. Maitra and P. Sarkar, 'Cryptographically significant Boolean functions with five valued Walsh spectra', *Theor. Comp. Sci.*, **276** (2002), 133–146.

[362] S. Maitra and P. Sarkar, 'Maximum nonlinearity of symmetric Boolean functions on odd number of variables ', *IEEE Trans. Inform. Theory*, **48** (2002), 2626–2630.

[363] S. Maity and T. Johansson, 'Construction of cryptographically important Boolean functions', *Proc. Indocrypt'02*, Hyderabad, India, 2002, Springer-Verlag, Berlin, (to appear).

[364] Y. Mansour, 'Learning Boolean functions via the Fourier transform', *Theoretical advances in neural computing and learning*, Kluwer Acad. Publ., Dordrecht, 1994, 391–424.

[365] Y. Mansour, B. Schieber and P. Tiwari, 'A lower bound for integer greatest common divisor computations', *J. Assoc. Comp. Mach.*, **38** (1991), 453–471.

[366] Y. Mansour, B. Schieber and P. Tiwari, 'Lower bounds for computations with the floor operations', *SIAM J. Comp.*, **20** (1991), 315–327.

[367] W. Mao, 'Timed-release cryptography', *Lect. Notes in Comp. Sci.*, Springer-Verlag, Berlin, **2259** (2001), 342–357.

[368] G. Marsaglia, 'The structure of linear congruential sequences', *Applications of Number Theory to Numerical Analysis*, Academic Press, New York, 1972, 249–285.

[369] C. Mauduit, 'Finite and infinite pseudorandom binary words', *Theor. Comp. Sci.*, **273** (2002), 249–261.

[370] C. Mauduit and A. Sárközy, 'On the arithmetic structure of the integers whose sum of digits is fixed', *Acta Arith.*, **81** (1997), 145–173.

[371] C. Mauduit and A. Sárközy, 'On finite pseudorandom binary sequences 1: Measure of pseudorandomness, the Legendre symbol', *Acta Arith.*, **82** (1997), 365–377.

[372] U. M. Maurer, 'Fast generation of prime numbers and secure public-key cryptographic parameters', *J. Cryptology*, **8** (1995), 123–155.

[373] U. M. Maurer and S. Wolf, 'Lower bounds on generic algorithms in groups', *Lect. Notes in Comp. Sci.*, Springer-Verlag, Berlin, **1403** (1998), 72–84.

[374] U. M. Maurer and S. Wolf, 'On the hardness of the Diffie–Hellman decision problem', *Preprint*, 1998, 1–4.

[375] U. M. Maurer and S. Wolf, 'The relationship between breaking the Diffie–Hellman protocol and computing discrete logarithms', *SIAM J. Comp.*, **28** (1999), 1689–1721.

[376] U. M. Maurer and S. Wolf, 'The Diffie–Hellman protocol', *Designs, Codes and Cryptography*, **19** (2000) 147–171.

[377] A. May, 'Cryptanalysis of unbalanced RSA with small CRT-exponent', *Lect. Notes in Comp. Sci.*, Springer-Verlag, Berlin, **2442** (2002), 242–256.

[378] K. S. McCurley, 'A key distribution system equivalent to factoring *J. Cryptology*, **1** (1988), 95–105.

[379] K. S. McCurley, 'The discrete logarithm problem', *Proc. Symp. in Appl. Math.*, Amer. Math. Soc., Providence, RI, **42** (1990), 49–74.

[380] J. Meidânis, 'Lower bounds for arithmetic problems', *Inform. Proc. Letters*, **38** (1991), 83–87.

[381] W. Meidl and A. Winterhof, 'Lower bounds on the linear complexity of the discrete logarithm in finite fields', *IEEE Trans. Inform. Theory*, **47** (2001), 2807–2811.

[382] W. Meidl and A. Winterhof, 'Linear complexity and polynomial degree of a function over a finite field', *Finite Fields with Applications to Coding Theory, Cryptography and Related Areas*, Springer-Verlag, Berlin, 2002, 229–238.

[383] W. Meidl and A. Winterhof, 'On the linear complexity profile of explicit nonlinear pseudorandom numbers', *Inform. Proc. Letters*, (to appear).

[384] W. Meidl and A. Winterhof, 'A polynomial representation of the Diffie–Hellman mapping', *Appl. Algebra in Engin., Commun. and Computing*, (to appear).

[385] W. Meidl and A. Winterhof, 'On the linear complexity profile of some new explicit nonlinear pseudorandom number generators', *Preprint*, 2002,

[386] J. Merkle, 'Multi-round passive attacks on server-aided RSA protocols', *Proc. 7th ACM Conf. on Computer and Commun. Security*, ACM, 2000, 102–107.

[387] J. Merkle and C. P. Schnorr, 'Perfect, generic pseudo-randomness for cyclic groups', *Preprint*, 1998, 1–12.

[388] J. Merkle and R. Werchner, 'On the security of server-aided RSA protocols', *Lect. Notes in Comp. Sci.*, Springer-Verlag, Berlin, **1431** (1998), 99–116.

[389] A. J. Menezes, P. C. van Oorschot and S. A. Vanstone, *Handbook of applied cryptography*, CRC Press, Boca Raton, FL, 1996.

[390] S. Meyer and J. Sorenson, 'Efficient algorithms for computing the Jacobi symbol', *Lect. Notes in Comp. Sci.*, Springer-Verlag, Berlin, **1122** (1996), 225–240.

[391] S. Micali and C. P. Schnorr, 'Efficient, perfect polynomial random number generators', *J. Cryptology*, **3** (1991), 157–172.

[392] D. Micciancio, 'On the hardness of the shortest vector problem', *PhD Thesis*, MIT, 1998.

[393] J.-F. Misarsky, 'A multiplicative attack using LLL algorithm on RSA signatures with redundancy', *Lect. Notes in Comp. Sci.*, Springer-Verlag, Berlin, **1294** (1997), 221–234.

[394] M. Mňuk, 'A div (n) depth Boolean circuit for smooth modular inverse', *Inform. Proc. Letters*, **38** (1991), 153–156.

[395] P. L. Montgomery, 'A survey of modern integer factorization algorithms', *Centrum vor Wiskunde en Informatica*, Amsterdam, **7** (1994), 337–366.

[396] F. Montoya Vitini, J. Muñoz Masqué and A. Peinado Domínguez, 'Linear complexity of the $x^2 \bmod p$ orbits', *Inform. Proc. Letters*, **72** (1999), 3–7.

[397] C. J. Moreno and O. Moreno, 'Exponential sums and Goppa codes, 1', *Proc. Amer. Math. Soc.*, **111** (1991), 523–531.

[398] G. L. Mullen, 'Permutation polynomials over finite fields', *Finite fields, Coding theory, and Advances in Communications and Computing*, Marcel Dekker, New York, 1993, 131–151.

[399] G. L. Mullen and D. White, 'A polynomial representation for logarithms in $GF(q)$', *Acta Arith.*, **47** (1986), 255–261.

[400] S. Murphy and M. Robshaw, 'Essential algebraic atructure within the AES', *Lect. Notes in Comp. Sci.*, Springer-Verlag, Berlin, **2442** (2002), 1–16.

[401] M. R. Murty, 'On Artin's conjecture', *J. Number Theory*, **16** (1983), 147–168.

[402] M. R. Murty and F. Saidak, 'Non-abelian generalizations of the Erdős–Kac theorem', *Canad. J. Math.*, (to appear).

[403] M. R. Murty and S. Srinivasan, 'Some remarks on Artin's conjecture', *Canad. Math. Bull.*, **30** (1987), 80–84.

[404] M. R. Murty, M. Rosen and J. H. Silverman, 'Variations on a theme of Romanoff', *Intern. J. Math. Soc.*, **7** (1996), 373–391.

[405] M. Naor and O. Reingold, 'Number-theoretic constructions of efficient pseudo-random functions', *Proc 38th IEEE Symp. on Found. of Comp. Sci.*, IEEE, 1997, 458–467.

[406] M. Naor, O. Reingold and A. Rosen, 'Pseudorandom functions and factoring', *SIAM J. Comp.*, **31** (2002), 1383–1404.

[407] A. M. Naranjani, 'On Dirichlet characters of polynomials', *Acta Arith.*, **43** (1984), 245–251.

[408] M. Näslund, 'Universal hash functions & hard core bits', *Lect. Notes in Comp. Sci.*, Springer-Verlag, Berlin, **921** (1995), 356–366.

[409] M. Näslund, 'All bits in $ax+b$ are hard', *Lect. Notes in Comp. Sci.*, Springer-Verlag, Berlin, **1109** (1996), 114–128.

[410] M. Näslund, 'Bit extraction, hard-core bits, and the bit security of RSA', *Ph.D. Thesis*, Dept. Numerical Anal. and Comp. Sci., Royal Inst. of Technology, Stockholm, 1998, 1–175.

[411] M. Näslund, I. E. Shparlinski and W. Whyte, 'On the bit security of NTRU', *Proc. 6th Intern. Workshop on Public Key Cryptography, PKC'03*, Miami, USA, 2003, Springer-Verlag, Berlin, (to appear).

[412] M. Näslund and A. Russell, 'Hard core functions: Survey and new results', *Proc. NordSec'99*, 1999, 305–322.

[413] V. I. Nečaev, 'Complexity of a deterministic algorithm for the discrete logarithm', *Matem. Zametki*, **55** (1994), no. 2, 91–101 (in Russian).

[414] P. Q. Nguyen, 'The dark side of the hidden number problem: Lattice attacks on DSA', *Proc. Workshop on Cryptography and Computational Number Theory, Singapore 1999*, Birkhäuser, 2001, 321–330.

[415] P. Q. Nguyen and D. Pointcheval, 'Analysis and improvements of NTRU encryption paddings', *Lect. Notes in Comp. Sci.*, Springer-Verlag, Berlin, **2442** (2002), 210–225.

[416] P. Q. Nguyen and I. E. Shparlinski, 'On the insecurity of a server-aided RSA protocol', *Lect. Notes in Comp. Sci.*, Springer-Verlag, Berlin, **2248** (2001), 21–35.

[417] P. Q. Nguyen and I. E. Shparlinski, 'The insecurity of the digital signature algorithm with partially known nonces', *J. Cryptology*, **15** (2002), 151–176.

[418] P. Q. Nguyen and I. E. Shparlinski, 'The insecurity of the elliptic curve digital signature algorithm with partially known nonces', *Designs, Codes and Cryptography*, (to appear).

[419] P. Q. Nguyen, I. E. Shparlinski and J. Stern, 'Distribution of modular sums and the security of the server aided exponentiation', *Proc. Workshop on Cryptography and Computational Number Theory, Singapore 1999*, Birkhäuser, 2001, 331–342.

[420] P. Q. Nguyen and J. Stern, 'The hardness of the hidden subset sum problem and its cryptographic implications', *Lect. Notes in Comp. Sci.*, Springer-Verlag, Berlin, **1666** (1999), 31–46.

[421] P. Q. Nguyen and J. Stern, 'Lattice reduction in cryptology: An update', *Lect. Notes in Comp. Sci.*, Springer-Verlag, Berlin, **1838** (2000), 85–112.

[422] P. Q. Nguyen and J. Stern, 'The two faces of lattices in cryptology', *Lect. Notes in Comp. Sci.*, Springer-Verlag, Berlin, **2146** (2001), 146–180.

[423] H. Niederreiter, 'Quasi-Monte Carlo methods and pseudo-random numbers', *Bull. Amer. Math. Soc.*, **84** (1978), 957–1041.

[424] H. Niederreiter, 'The serial test for congruential pseudorandom numbers generated by inversions', *Math. Comp.*, **52** (1989), 135–144.

[425] H. Niederreiter, 'The linear complexity profile and the jump complexity profile of keystream sequences', *Lect. Notes in Comp. Sci.*, Springer-Verlag, Berlin, **473** (1991), 174–188.

[426] H. Niederreiter, *Random number generation and Quasi–Monte Carlo methods*, SIAM Press, 1992.

[427] H. Niederreiter, 'Finite fields, pseudorandom numbers, and quasirandom points', *Finite Fields, Coding Theory, and Advances in Communications and Computing*, Marcel Dekker, New York, 1993, 375–394.

[428] H. Niederreiter, 'New developments in uniform pseudorandom number and vector generation', *Lect. Notes in Statistics*, Springer-Verlag, Berlin, **106** (1995), 87–120.

[429] H. Niederreiter, 'Design and analysis of nonlinear pseudorandom number generators', *Monte Carlo Simulation*, A.A. Balkema Publishers, Rotterdam, 2001, 3–9.

[430] H. Niederreiter and C. P. Schnorr, 'Local randomness in polynomial random number and random function generators', *SIAM J. Comp.*, **13** (1993), 684–694.

[431] H. Niederreiter and I. E. Shparlinski, 'On the distribution and lattice structure of nonlinear congruential pseudorandom numbers', *Finite Fields and Their Appl.*, **5** (1999), 246–253.

[432] H. Niederreiter and I. E. Shparlinski, 'Exponential sums and the distribution of inversive congruential pseudorandom numbers with prime-power modulus', *Acta Arith.*, **92** (2000), 89–98.

[433] H. Niederreiter and I. E. Shparlinski, 'On the distribution of pseudorandom numbers and vectors generated by inversive methods', *Appl. Algebra in Engin., Commun. and Computing*, **10** (2000), 189–202.

[434] H. Niederreiter and I. E. Shparlinski, 'On the distribution of inversive congruential pseudorandom numbers in parts of the period', *Math. Comp.*, **70** (2001), 1569–1574.

[435] H. Niederreiter and I. E. Shparlinski, 'Recent advances in the theory of nonlinear pseudorandom number generators', *Proc. Conf. on Monte Carlo and Quasi-Monte Carlo Methods, 2000*, Springer-Verlag, Berlin., 2002, 86–102.

[436] H. Niederreiter and I. E. Shparlinski, 'On the average distribution of inversive pseudorandom numbers', *Finite Fields and Their Appl.*, **8** (2002), 491–503.

[437] H. Niederreiter and I. E. Shparlinski, 'On the distribution of power residues and primitive elements in some nonlinear recurring sequences', *Bull. Lond. Math. Soc.*, (to appear).

[438] H. Niederreiter and I. E. Shparlinski, 'Dynamical systems generated by rational functions', *Preprint*, 2002, 1–13.

[439] H. Niederreiter and M. Vielhaber, 'Linear complexity profiles: Hausdorff dimensions for almost perfect profiles and measures for general profiles', *J. Compl.*, **13** (1997), 353–383.

[440] H. Niederreiter and A. Winterhof, 'Incomplete exponential sums over finite fields and their applications to new inversive pseudorandom number generators', *Acta Arith.*, **93** (2000), 387–399.

[441] H. Niederreiter and A. Winterhof, 'On the distribution of compound inversive congruential pseudorandom numbers', *Monatsh. Math.*, **132** (2001), 35–48.

[442] H. Niederreiter and A. Winterhof, 'On a new class of inversive pseudorandom numbers for parallelized simulation methods', *Periodica Mathematica Hungarica*, **42**, (2001), 77–87.

[443] H. Niederreiter and A. Winterhof, 'On the lattice structure of pseudorandom numbers generated over arbitrary finite fields', *Appl. Algebra in Engin., Commun. and Computing*, **12** (2001), 265–272.

[444] H. Niederreiter and A. Winterhof, 'Lattice structure and linear complexity of nonlinear pseudorandom numbers', *Preprint*, 2001, 1–11.

[445] H. Niederreiter and A. Winterhof, 'Incomplete exponential sums and polynomial interpolation of the discrete logarithm', *Finite Fields and Their Appl.*, **8** (2002), 184–192.

[446] R. R. Nielsen, 'A class of Sudan-decodable codes', *IEEE Trans. Inform. Theory*, **46** (2000), 1564–1572.

[447] N. Nisan and M. Szegedy, 'On the degree of Boolean functions as real polynomials', *Comp. Compl.*, **4** (1994), 301–313.

[448] N. Nisan and A. Wigderson, 'On rank vs. communication complexity', *Combinatorica*, **15** (1995), 557–565.

[449] N. Nisan and A. Wigderson, 'Lower bounds on arithmetic circuits via partial derivatives', *Comp. Compl.*, **6** (1997), 217–234.

[450] K. Nyberg, 'Differentially uniform mappings for cryptography', *Lect. Notes in Comp. Sci.*, Springer-Verlag, Berlin, **765** (1994), 55–64.

[451] A. M. Odlyzko, 'Discrete logarithms: The past and the future', *Designs, Codes and Cryptography*, Springer-Verlag, Berlin, **19** (2000), 129–145.

[452] F. Pappalardi, 'On the order of finitely generated subgroups of $\mathbb{Q}^*$ (mod p) and divisors of $p-1$', *J. Number Theory*, **57** (1996), 207–222.

[453] I. Parberry and P. Yuan Yan, 'Improved upper and lower time bounds for parallel random access machines without simultaneous writes', *SIAM J. Comp.*, **20** (1991), 88–99.

[454] I. Parnafes, R. Raz and A. Wigderson, 'Direct product results and the GCD problem, in old and new communication models', *Proc. 29th ACM Symp. on Theory of Comp.*, ACM, 1997, 363–372.

[455] S. Patel and G. S. Sundaram, 'An efficient discrete log pseudo random generator', *Lect. Notes in Comp. Sci.*, Springer-Verlag, Berlin, **1462** (1999), 35–44.

[456] A. Peinado Domínguez, F. Montoya Vitini, J. Muñoz Masqué and A. J. Yuste, 'Maximal periods of x^2+c in $\mathbb{F}_q$', *Lect. Notes in Comp. Sci.*, Springer-Verlag, Berlin, **2227** (2001), 219–228.

[457] R. Peralta, 'Simultaneous security of bits of the discrete log', *Lect. Notes in Comp. Sci.*, Springer-Verlag, Berlin, **219** (1986), 62–72.

[458] R. Peralta, 'On the distribution of quadratic residues and nonresidues modulo a prime number', *Math. Comp.*, **58** (1992), 433–440.

[459] G. I. Perel'muter, 'A bound of the sum along a curve', *Matem. Zametki*, **5** (1969), 373–380 (in Russian).

[460] E. Plaku and I. E. Shparlinski, 'On polynomial representations of Boolean functions related to some number theoretic problems', *Lect. Notes in Comp. Sci.*, Springer-Verlag, Berlin, **2245** (2001), 305–316.

[461] D. Pointcheval, 'New public key cryptosystems based on the dependent-RSA problems', *Lect. Notes in Comp. Sci.*, Springer-Verlag, Berlin, **1592** (1999), 239–254.

[462] C. Pomerance, 'Fast, rigorous factorization and discrete logarithm algorithms', *Discrete algorithms and complexity*, Academic Press, San Diego, CA, 1987, 119–143.

[463] C. Pomerance, 'Factoring', *Proc. Symp. in Appl. Math.*, Amer. Math. Soc., Providence, RI, **42** (1990), 27–47.

[464] C. Pomerance, 'Primality testing: Variations on a theme of Lucas', *Proc. MSRI Conf. Algorithmic Number Theory: Lattices, Number Fields, Curves, and Cryptography, Berkeley 2000*, Cambridge Univ. Press, (to appear).

[465] C. Pomerance, 'Smooth numbers and the quadratic sieve', *Proc. MSRI Conf. Algorithmic Number Theory: Lattices, Number Fields, Curves, and Cryptography, Berkeley 2000*, Cambridge Univ. Press, (to appear).

[466] C. Pomerance and I. E. Shparlinski, 'Smooth orders and cryptographic applications', *Lect. Notes in Comp. Sci.*, Springer-Verlag, Berlin, **2369** (2002), 338–348.

[467] G. Poupard and J. Stern, 'Short proofs of knowledge for factoring', *Lect. Notes in Comp. Sci.*, Springer-Verlag, Berlin, **1751** (2000), 147–166.

[468] K. Prachar, *Primzahlverteilung*, Springer-Verlag, Berlin, 1957.

[469] R. Raz, 'Fourier analysis for probabilistic communication complexity', *Comp. Compl.*, **5** (1995), 205–221.

[470] R. Raz, 'On the complexity of matrix product', *Proc. 34rd ACM Symp. on Theory of Comp.*, ACM, 2002, 144–151.

[471] R. Raz and A. Shpilka, 'Lower bounds for matrix product, in bounded depth circuits with arbitrary gates', *Proc. 33rd ACM Symp. on Theory of Comp.*, ACM, 2001, 409–418.

[472] A. A. Razborov and S. Rudich, 'Natural proofs', *J. Comp. and Syst. Sci.*, **55** (1997), 24–35.

[473] J.-J. Risler, 'Khovansky's theorem and complexity theory', *Rocky Mountain J. Math.*, **14** (1984), 851–853.

[474] J.-J. Risler, 'Additive complexity of real polynomials', *SIAM J. Comp.*, **14** (1985), 178–183.

[475] J. Rivat and A. Sárközy, 'On pseudorandom binary sequences and their applications', *Preprint*, 2001, 1–18.

[476] R. L. Rivest, 'Remarks on a proposed cryptanalytic attack on the M.I.T. public-key cryptosystem', *Cryptologia*, **2** (1978), 62–65.

[477] R. L. Rivest, A. Shamir and D. A. Wagner, 'Time-lock puzzles and timed-release crypto', *MIT Laboratory for Computer Science. Technical Report MIT-LCS-TR-684*, MIT, Massachusets, 1996, 1–9.

[478] R. L. Rivest and R. D. Silverman, 'Are "strong" primes needed for RSA?', *Preprint*, 1999, 1–23.

[479] J. M. Rojas, 'Additive complexity and p-adic roots of polynomials', *Lect. Notes in Comp. Sci.*, Springer-Verlag, Berlin, **2369** (2002), 506–516.

[480] J. M. Rojas, 'Arithmetic multivariate Descartes' rule', *Preprint*, 2002, 1–27.

[481] V. Roychowdhry, K.-Y. Siu and A. Orlitsky, 'Neural models and spectral methods', *Theoretical advances in neural computing and learning*, Kluwer Acad. Publ., Dordrecht, 1994, 3–36.

[482] V. Roychowdhry and F. Vatan, 'Quantum formulas: A lower bound and simulation', *SIAM J. Comput.*, **31** (2002), 460–476.

[483] R. A. Rueppel, *Analysis and design of stream ciphers*, Springer-Verlag, Berlin, 1986.

[484] R. A. Rueppel, 'Stream ciphers', *Contemporary cryptology: The science of information integrity*, IEEE Press, NY, 1992, 65–134.

[485] R. A. Rueppel and J. L. Massey, 'Knapsack as a nonlinear function', *IEEE Intern. Symp. of Inform. Theory*, IEEE Press, NY, 1985, 46.

[486] A. Russell and I. E. Shparlinski, 'Classical and quantum algorithms for Legendre symbol problem with polynomials', *Preprint*, 2002, 1–15.

[487] A.-R. Sadeghi and M. Steiner, 'Assumptions related to discrete logarithms: Why subtleties make a real difference', *Lect. Notes in Comp. Sci.*, Springer-Verlag, Berlin, **2045** (2001), 243–260.

[488] S. Sakata, 'On fast interpolation method for Guruswami-Sudan list decoding of one-point algebraic-geometry codes', *Lect. Notes in Comp. Sci.*, Springer-Verlag, Berlin, **2227** (2001), 36–45.

[489] P. Sarkar and S. Maitra, 'Constructions of nonlinear Boolean functions with important cryptographic properties', *Lect. Notes in Comp. Sci.*, Springer-Verlag, Berlin, **1807** (2000), 485–506.

[490] P. Sarkar and S. Maitra, 'Nonlinearity bounds and constructions of resilient Boolean functions', *Lect. Notes in Comp. Sci.*, Springer-Verlag, Berlin, **1880** (2000), 515–532.

[491] P. Sarkar and S. Maitra, 'Cross-correlation analysis of cryptographically useful Boolean functions and S-boxes', *Theory Comp. Syst.*, **35** (2002), 39–57.

[492] J. E. Savage, *The complexity of computing*, Wiley-Interscience Publ., 1987.

[493] J. E. Savage, *Models of computation*, Addison-Wesley, Reading, MA, 1998.

[494] O. Schirokauer, D. Weber and T. Denny, 'Discrete logarithms: The effectiveness of the index calculus method', *Lect. Notes in Comp. Sci.*, Springer-Verlag, Berlin, **1122** (1996), 337–362.

[495] C. P. Schnorr, 'A hierarchy of polynomial time basis reduction algorithms', *Theor. Comp. Sci.*, **53** (1987), 201–224.

[496] C. P. Schnorr, 'Security of almost all discrete log bits', *Electronic Colloq. on Comp. Compl.*, Univ. of Trier, **TR98-033** (1998), 1–13.

[497] C. P. Schnorr, 'Small generic hardcore subsets for the discrete logarithm: Short secret DL-keys', *Inform. Proc. Letters*, **79** (2001), 93–98.

[498] C. P. Schnorr, 'Security of DL-encryption and signatures against generic attacks – A survey', *Proc. the Conf. on Public Key Cryptography and Computational Number Theory, Warsaw 2000*, Walter de Gruyter, 2001, 257–282.

[499] C. P. Schnorr and M. Jakobsson, 'Security of discrete log cryptosystems in the random oracle + generic model', *Preprint*, 1999, 1–15.

[500] C. P. Schnorr and M. Jakobsson, 'Security of signed ElGamal encryption', *Lect. Notes in Comp. Sci.*, Springer-Verlag, Berlin, **1976** (2000), 73–89.

[501] C. P. Schnorr and S. Vaudenay, 'Black box cryptanalysis of hash networks based on multipermutations', *Lect. Notes in Comp. Sci.*, Springer-Verlag, Berlin, **950** (1995), 47–57.

[502] J. Seberry, X.-M. Zhang and Y. Zheng, 'Nonlinearity and propagation characteristics of balanced Boolean functions', *Inform. and Comp.*, **119** (1995), 1–13.

[503] R. Shaltiel, 'Towards proving strong direct product theorems', *Proc. 16th IEEE Conf. on Comp. Compl.*, IEEE, 2001, 107–117.

[504] M. A. Shokrollahi and H. Wasserman, 'List decoding of algebraic-geometric codes', *IEEE Trans. Inform. Theory*, **45** (1999), 432–437.

[505] V. Shoup, 'Lower bounds for discrete logarithms and related problems', *Lect. Notes in Comp. Sci.*, Springer-Verlag, Berlin, **1233** (1997), 256–266.

[506] I. E. Shparlinski, 'On prime divisors of recurrence sequences', *Izvestija Vysshih Uchebnyh Zavedenii, Ser. matem.*, (1980), no. 1, 100–103 (in Russian).

[507] I. E. Shparlinski, 'On exponential sums with sparse polynomials', *J. Number Theory*, **60** (1996), 233–244.

[508] I. E. Shparlinski, *Finite fields: Theory and computation*, Kluwer Acad. Publ., Dordrecht, 1999.

[509] I. E. Shparlinski, 'Communication complexity and Fourier coefficients of the Diffie–Hellman key', *Lect. Notes in Comp. Sci.*, Springer-Verlag, Berlin, **1776** (2000), 259–268.

[510] I. E. Shparlinski, 'Computing Jacobi symbols modulo sparse integers and polynomials and some applications', *J. Algorithms*, **36** (2000), 241–252.

[511] I. E. Shparlinski, 'On the linear complexity of the power generator', *Designs, Codes and Cryptography*, **23** (2001), 5–10.

[512] I. E. Shparlinski, 'On the Naor–Reingold pseudo-random function from elliptic curves', *Appl. Algebra in Engin., Commun. and Computing*, **11** (2000), 27–34.

[513] I. E. Shparlinski, 'On the uniformity of distribution of the Naor–Reingold pseudo-random function', *Finite Fields and Their Appl.*, **7** (2001), 318–326.

[514] I. E. Shparlinski, 'On the uniformity of distribution of the RSA pairs', *Math. Comp.*, **70** (2001), 801–808.

[515] I. E. Shparlinski, 'On some properties of the shrinking generator', *Designs, Codes and Cryptography*, **23** (2001), 147–156.

[516] I. E. Shparlinski, 'On the distribution of the Diffie–Hellman pairs', *Finite Fields and Their Appl.*, **8** (2002), 131–141.

[517] I. E. Shparlinski, 'Linear complexity of the Naor–Reingold pseudo-random function', *Inform. Proc. Letters*, **95** (2000), 95–99.

[518] I. E. Shparlinski, 'Security of polynomial transformations of the Diffie–Hellman key', *Cryptology ePrint Archive, Report 2000/023*, 2000, 1–9.

[519] I. E. Shparlinski, 'On singularity of generalised Vandermonde matrices over finite fields', *Preprint*, 2000, 1–7.

[520] I. E. Shparlinski, 'Sparse polynomial approximation in finite fields', *Proc. 33rd ACM Symp. on Theory of Comput.*, ACM, 2001, 209–215.

[521] I. E. Shparlinski, 'On the generalised hidden number problem and bit security of XTR', *Lect. Notes in Comp. Sci.*, Springer-Verlag, Berlin, **2227** (2001), 268–277.

[522] I. E. Shparlinski, 'On the uniformity of distribution of the ElGamal signature', *Appl. Algebra in Engin., Commun. and Computing*, **13** (2002), 9–16.

[523] I. E. Shparlinski, 'Security of most significant bits of g^{x^2}', *Inform. Proc. Letters*, **83** (2002), 109–113.

[524] I. E. Shparlinski, 'Playing "Hide-and-Seek" in finite fields: Hidden number problem and its applications', *Proc. 7th Spanish Meeting on Cryptology and Information Security, Vol.1*, Univ. of Oviedo, 2002, 49–72.

[525] I. E. Shparlinski, 'Exponential sums and lattice reduction: Applications to cryptography', *Finite Fields with Applications to Coding Theory, Cryptography and Related Areas*, Springer-Verlag, Berlin, 2002, 286–298.

[526] I. E. Shparlinski, 'Bounds of Gauss sums in finite fields', *Proc. Amer. Math. Soc.*, (to appear).

[527] I. E. Shparlinski, 'Exponential function analogue of Kloosterman sums', *Rocky Mountain J. Math.*, (to appear).

[528] I. E. Shparlinski and J. H. Silverman, 'On the linear complexity of the Naor–Reingold pseudo-random function from elliptic curves', *Designs, Codes and Cryptography*, **24** (2001), 279–289.

[529] I. E. Shparlinski and R. Steinfeld, 'Chinese remaindering for algebraic numbers in a hidden field', *Lect. Notes in Comp. Sci.*, Springer-Verlag, Berlin, **2369** (2002), 349–356.

[530] I. E. Shparlinski and R. Steinfeld, 'Noisy Chinese remaindering in the Lee norm', *Preprint*, 2001, 1–11.

[531] I. E. Shparlinski and S. A. Stepanov, 'On the construction of a primitive normal basis in a finite field', *Matem. Sbornik*, **180** (1990), 1067–1072 (in Russian).

[532] I. E. Shparlinski and S. A. Stepanov, 'On construction of primitive elements and primitive normal bases in a finite field', *Proc. Colloq. on Comp. Number Theory, Hungary 1989*, Walter de Gruyter, New York, 1991, 1–24.

[533] I. E. Shparlinski and S. A. Stepanov, 'Estimates of an incomplete sum of multiplicative characters of polynomials', *Diskretnaja Matem.*, **2** (1990), no. 3, 115–119 (in Russian).

[534] A. Shpilka and A. Wigderson, 'Depth-3 arithmetic circuits over fields of characteristic zero', *Comp. Compl.*, **10** (2001), 1–27.

[535] J. H. Silverman, 'Invertibility in truncated polynomial rings', *NTRU Cryptosystem Tech. Report 9*, 1998, 1–8.

[536] P. J. Smith and C. T. Skinner, 'A public-key cryptosystem and a digital signature system based on the Lucas function analogue to discrete logarithms', *Lect. Notes in Comp. Sci.*, Springer-Verlag, Berlin, **917** (1995), 357–364.

[537] J. Sorenson, 'Polylog depth circuits for integer factoring and discrete logarithms', *Inform. and Comp.*, **110** (1994), 1–18.

[538] P. Štanikă, 'Nonlinearity, local and global avalanche characteristics of balanced Boolean functions', *Discr. Math.*, **248** (2002), 181–193.

[539] S. B. Stečkin, 'An estimate of a complete rational exponential sum', *Proc. Math. Inst. Acad. Sci. USSR*, Moscow, **143** (1977), 188–207 (in Russian).

[540] R. Steinfeld and Y. Zheng, 'An advantage of low-exponent RSA with primes sharing least significant bits', *Lect. Notes in Comp. Sci.*, Springer-Verlag, Berlin, **2020** (2001), 52–62.

[541] D. R. Stinson, *Cryptography: Theory and practice*, CRC Press, Boca Raton, FL, 2002.

[542] C. Sturtivant and G. S. Frandsen, 'The computational efficacy of finite-field arithmetic', *Theor. Comp. Sci.*, **112** (1993), 291–309.

[543] M. Sudan, 'Decoding of Reed Solomon codes beyond the error-correction bound', *J. Complexity*, **13** (1997), 180–193.

[544] M. Sudan, 'Ideal error-correcting codes: Unifying algebraic and number-theoretic algorithms', *Lect. Notes in Comp. Sci.*, Springer-Verlag, Berlin, **2227** (2001), 36–45.

[545] P. Szüsz, 'On a problem in the theory of uniform distribution', *Comptes Rendus Premier Congrès Hongrois*, Budapest, 1952, 461–472 (in Hungarian).

[546] Y. Tarannikov, 'On resilient Boolean functions with maximal possible nonlinearity', *Lect. Notes in Comp. Sci.*, Springer-Verlag, Berlin, **1977** (2000), 19–30.

[547] Y. Tarannikov, 'New constructions of resilient Boolean functions with maximal nonlinearity', *Lect. Notes in Comp. Sci.*, Springer-Verlag, Berlin, **2355** (2002), 66–77.

[548] Y. Tarannikov, P. Korolev, and A. Botev, 'Autocorrelation coefficients and correlation immunity of Boolean functions', *Lect. Notes in Comp. Sci.*, Springer-Verlag, Berlin, **2248** (2001), 460–479.

[549] G. Turan and F. Vatan, 'On the computation of Boolean functions by analog circuits of bounded fan-in', *J. Comp. and Syst. Sci.*, **54** (1997), 199–212.

[550] E. R. Verheul, 'Certificates of recoverability with scalable recovery agent security', *Lect. Notes in Comp. Sci.*, Springer-Verlag, Berlin, **1751** (2000), 258–275.

[551] E. R. Verheul, 'Evidence that XTR is more secure than supersingular elliptic curve cryptosystems', *Lect. Notes in Comp. Sci.*, Springer-Verlag, Berlin, **2045** (2001), 195–210.

[552] E. R. Verheul and H. C. A. van Tilborg, 'Cryptanalysis of 'less short' RSA secret exponents', *Appl. Algebra in Engin., Commun. and Computing*, **8** (1997), 425–435.

[553] I. M. Vinogradov, *Elements of number theory*, Dover Publ., New York, 1954.

[554] C. P. Waldvogel and J. L. Massey, 'The probability distribution of the Diffie–Hellman key', *Lect. Notes in Comp. Sci.*, Springer-Verlag, Berlin, **718** (1992), 492–504.

[555] H. Wasserman, 'Reconstructing randomly sampled multivariate polynomials from highly noisy data', *Proc. 9th ACM-SIAM Symp. on Discr. Algorithms*, ACM, 1998, 59–67.

[556] I. Wegener, *The complexity of Boolean functions*, Wiley-Teubner Series in Comp. Sci., Stuttgart, 1987.

[557] A. Weil, *Basic number theory*, Springer-Verlag, New York, 1974.

[558] K. Werther, 'The complexity of sparse polynomial interpolation over finite fields', *Appl. Algebra in Engin., Commun. and Comp.*, **5**(1994), 91–103.

[559] M. J. Wiener, 'Cryptanalysis of short RSA secret exponents', *IEEE Trans. Inform. Theory*, **36** (1990), 553–558.

[560] A. Winterhof, 'A note on the interpolation of the Diffie-Hellman mapping', *Bull. Aust. Math. Soc.*, **64** (2001), 475–477.

[561] A. Winterhof, 'Polynomial interpolation of the discrete logarithm', *Designs, Codes and Cryptography*, **25** (2002), 63–72.

[562] A. Woods, 'Subset sum "cubes" and the complexity of prime testing', *Preprint*, 2001, 1–17.

[563] A. Youssef and G. Gong, 'Hyper-bent functions', *Lect. Notes in Comp. Sci.*, Springer-Verlag, Berlin, **2045** (2001), 406–419.

[564] A. Youssef and G. Gong, 'Boolean functions with large distance to all bijective monomials: N odd case', *Lect. Notes in Comp. Sci.*, Springer-Verlag, Berlin, **2259** (2001), 49–59.

[565] W. P. Zhang, 'On a problem of Brizolis', *Pure Appl. Math.*, **11** (1995), suppl., 1–3 (in Chinese).

[566] X.-M. Zhang and Y. Zheng, 'Cryptographically resilient functions', *IEEE Trans. Inform. Theory*, **43** (1997), 1740–1747.

[567] Y. Zheng, and X. M. Zhang, 'Improved upper bound on the nonlinearity of high order correlation immune functions', *Lect. Notes in Comp. Sci.*, Springer-Verlag, Berlin, **2012** (2001), 264–274.

[568] Y. Zheng, and X. M. Zhang, 'New results on correlation immune functions', *Lect. Notes in Comp. Sci.*, Springer-Verlag, Berlin, **2015** (2001), 49–63.

[569] Y. Zheng, and X. M. Zhang, 'Connections among nonlinearity, avalanche and correlation immunity', *Theor. Comp. Sci.*, (to appear).

[570] Y. Zheng, X. M. Zhang and H. Imai, 'Restriction, terms and nonlinearity of Boolean functions', *Theor. Comp. Sci.*, **226** (1999), 207–223.

[571] R. Zippel, *Effective polynomial computation*, Kluwer Acad. Publ., Dordrecht, 1993.

Index

Additive complexity, 6, 10, 18, 19, 32, 143, 144, 348, 349
Artin's conjecture, 80, 81, 279
Average sensitivity, 21, 22, 105, 133, 135, 319, 360

Bit security, 8, 13, 62, 189, 191, 200, 212, 217
Blum integer, 250, 261, 268, 269
Blum–Blum–Shub generator, 9, 12, 249, 250, 261, 264, 265, 269, 270
Boolean cube, 4, 20, 338
Boolean function, 3–5, 7, 9, 21, 22, 24, 25, 103–106, 129, 130, 132, 133, 135, 137, 138, 149, 151–154, 179, 181–183, 303–307, 309, 311, 313, 315, 316, 319–322, 333, 337–339, 344, 345, 347, 350, 351, 360, 364, 365
Brizolis problem, 232, 357, 358
Brun–Titchmarsh theorem, 67, 68
Burgess bound, 41, 145, 244

Carmichael function, 18, 68, 71, 81, 198, 200, 250, 265, 267
Characteristic polynomial, 19, 29, 114, 297
Chinese Remainder Theorem, 72, 77, 102, 269, 356, 362
Circuit class
- $\mathrm{AC}_p(d)$, 9, 23, 320
- $\mathrm{BC}(d)$, 9, 22, 23, 132, 133, 320
- $\mathrm{DFAC}_p(d)$, 23, 321
- $\mathrm{PAC}_q(d)$, 9, 23, 171
- $\mathrm{UBC}(d, S)$, 9, 24, 103, 137, 182, 304, 306, 307, 322

Class $\mathcal{P}^r_{k,s}$, 25, 105, 138
Combinatorial complexity, 9, 23, 105
Combinatorial discrepancy, 22, 105, 183, 350, 351
Communication complexity, 8, 9, 13, 22, 105, 182, 304, 351
Computational tree, 6, 9, 144, 308
CREW PRAM, 4, 7, 21, 103, 135, 181, 308, 338, 356
Critical complexity, 21

Decision tree, 5, 9, 24, 25, 104, 140, 305, 306
Differential uniformity, 350
Diffie–Hellman indistinguishability, 11, 192, 361
Diffie–Hellman key, 6, 8, 12, 13, 159, 173, 182, 187–189, 191, 193, 218, 351, 352
Diffie–Hellman protocol, viii, 6–9, 159, 171, 181, 193, 194, 218, 235, 352–354, 361, 365
Diffie–Hellman triples, 11, 12, 50
Digital Signature Algorithm, 12, 13, 63, 201, 202, 206, 207, 210, 235, 241, 242, 351, 352
Discrepancy, 26, 61, 85, 207, 231, 234, 235, 256, 263, 264, 271, 284, 287, 288, 292, 295, 310–312, 357, 360
Discrete logarithm, 2–6, 8, 10, 11, 14, 39, 109, 112–114, 123, 125,

126, 128, 129, 137, 143–145, 147, 159, 160, 191, 232, 343–346, 349, 353, 359, 361, 365
Disjunctive normal form, 5, 141, 345

ElGamal cryptosystem, 189, 215, 362
ElGamal signature, 12, 206, 207, 210, 352
Elliptic curve, 12, 98, 102, 194, 198, 206, 241, 271, 358, 363, 365
Entropy function, 20, 55, 131, 137, 153, 319, 323
Euler function, 18, 267
Exponential pairs, 235
Extended Riemann Hypothesis, 42, 80, 81, 242, 244

Fermat numbers, 55, 309
Fixed point, 232, 298, 357
Formula size, 10, 135, 153, 156, 306
Fourier coefficients, 5, 10, 21, 104, 135, 140, 149, 152, 183, 315, 319, 347, 351, 364

Gauss sums, 41, 47, 359
Greatest common divisor, 303, 304

Hamming weight, 21, 103, 121, 123, 319
Hash functions, 201–203, 207, 210, 299, 325
Hidden number problem, 84, 98, 99, 102, 202, 206, 350

Index, 3
Integer factorisation, 5, 8, 14, 308

Koksma–Szüsz inequality, 61

Lattice, 83, 84, 86–89
Lattice basis reduction, 13, 84, 201, 205, 206, 269, 353, 362
Lattice structure, 19, 290, 291, 333, 360
Linear congruential generator, 13, 269, 284
Linear complexity, 1, 2, 4, 19, 20, 27, 125, 126, 267–269, 275, 287, 288, 291, 292, 294, 336, 357, 360
Linear recurrence sequence, 2, 3, 19, 28, 29, 114, 117, 118, 120, 121, 123, 125, 126, 145, 295–297, 335, 336
LUC protocol, 8, 40, 62, 89, 94, 176, 217, 219, 220
Lucas sequence, 198, 220

$M_f(r)$-invariant, 6, 10, 19, 144, 304
Message concealing, 9, 232–234, 299
Modular inversion, 11, 321, 333, 334, 356
Multiplicative order, 26, 44, 46–49, 53, 54, 56, 62–64, 71, 80, 81, 160, 173, 187–193, 201, 203, 214, 217–220, 235, 257, 258, 271, 273–275, 280, 281, 292, 294, 332, 355

Naor–Reingold function, 9, 12, 271, 275, 358
Non-linear complexity, 4, 20, 128, 170, 271, 288, 291, 293, 358, 360
Non-linearity, 135
Norm, 17, 40, 365
NTRU cryptosystem, 13, 223–225, 363

Parseval identity, 103, 138, 140, 304, 320, 323
Perfect power, 308
Permanent, 10, 330, 331
Permutation polynomial, 9, 333
Pohlig–Hellman algorithm, 81, 159, 191
Power generator, 9, 12, 30, 71, 197, 249, 267, 269, 270, 283, 357

Powering, 333, 334, 357
Primality testing, 5, 14, 81, 303, 307, 308
Primitive root, 3, 6, 9, 30, 31, 38, 49, 50, 53, 55, 122, 159, 173, 187, 188, 193, 207, 255, 284, 294, 333, 334, 337, 338, 354, 357
Pseudorandom numbers, 1, 9, 13, 81, 170, 206, 249, 256, 270, 279, 283, 284, 287, 288, 290–293, 295, 297, 299, 325, 333, 357–360, 365

Quantum computation, 10, 156, 359

RAM, 6, 144, 303, 308
Repeated squaring, 2, 19, 199, 220
Rijndael, 7, 364
RSA cryptosystem, 8, 9, 88, 197, 198, 211, 212, 232, 352–354, 359–362
RSA generator, 9, 12, 249
RSA pairs, 12, 56, 231, 235

Sensitivity, 10, 21, 22, 103, 307, 309, 321, 322, 337
Shamir message passing scheme, 8, 189, 212, 213
Smooth integer, 18, 42, 68, 70, 81, 191, 356, 362
Sparse polynomial, 30, 94, 174, 219, 234, 290, 325, 326, 328, 348, 349, 357, 363, 364
Sparsity, 10, 18, 19, 30, 109, 152, 153, 160
Statistical distance, 9, 11, 188, 241
Strong prime, 81, 197
Strong primes, 81
Subset sum, 9, 13, 232, 295, 297, 299, 360

Threshold representation, 149, 151, 152, 347
Trace, 17, 176, 194, 217, 325, 328, 365

Weight, 141
Weil bound, 11, 13, 37–39, 97, 155, 359, 363, 364

XTR protocol, 8, 40, 62, 89, 94, 176, 217–219, 221

Zech's logarithm, 9, 333, 334, 336

MIX
Papier aus verantwortungsvollen Quellen
Paper from responsible sources
FSC® C105338

If you have any concerns about our products,
you can contact us on
ProductSafety@springernature.com

In case Publisher is established outside the EU,
the EU authorized representative is:
Springer Nature Customer Service Center GmbH
Europaplatz 3, 69115 Heidelberg, Germany

Printed by Libri Plureos GmbH
in Hamburg, Germany